AF600469

Animals in the World

SUNY series in Ancient Greek Philosophy

Anthony Preus, editor

Animals in the World

Five Essays on Aristotle's Biology

Pierre Pellegrin

Translated by Anthony Preus

Published by State University of New York Press, Albany

Printed in the United States of America

For information, contact State University of New York Press, Albany, NY
www.sunypress.edu

Library of Congress Cataloging-in-Publication Data

Names: Pellegrin, Pierre, author. | Preus, Anthony, translator.
Title: Animals in the world : five essays on Aristotle's biology / Pierre Pellegrin, Anthony Preus.
Description: Albany : State University of New York Press, [2023] | Series: SUNY series in ancient Greek philosophy | Includes bibliographical references and index.
Identifiers: LCCN 2022021019 | ISBN 9781438491479 (hardcover : alk. paper) | ISBN 9781438491486 (ebook)
Subjects: LCSH: Aristotle. Historia animalium. | Zoology—Pre-Linnean works. | Zoology—Philosophy.
Classification: LCC QL41 .P44 2023 | DDC 590—dc23/eng/20220616
LC record available at https://lccn.loc.gov/2022021019

10 9 8 7 6 5 4 3 2 1

Contents

Introduction 1

Chapter 1
Is There an Aristotelian Biology? 7
Aristotle and Nineteenth-Century Biology 9
Aristotle and Cuvier 12
Some Remarks on Aristotle's Biological Corpus 24
The Relationships between the *History of Animals* and the *Parts of Animals* 27
An Impossible Chronology 32

Chapter 2
The New Horizon of Teleology 51
The Historical Background 55
Aristotle's Solution and Its Consequences for His Teleology 61
"Many Things Happen because It Is Necessary" 71
Hypothetical Necessity 74
The Two Natures 94
Nature's Excellence 105

Chapter 3
A Philosophy of Life? 111
The Nutritive Soul 118
Sexual Generation and the Female Material 124
Spontaneous Generation 143
A General Theory of Homoiomeries? 164

Chapter 4
Diversity 179
What the Word "Animal" Names 183
Continuity and Diversity, Perfection and Harmony 200
The Revenge of the Special on the General 226

Chapter 5
Animal Nature and Human Nature 237
Animal Pleasure, Human Pleasure 262
Sheep and Men 283

Conclusion 291

Bibliography
Editions of Aristotle 305
Works and Articles 306

Index 313

References to Aristotle's Works 319

Introduction

Perhaps Aristotle conceived his philosophical project as "encyclopedic"—that depends on how we understand the word "encyclopedic"—but it surely would be going too far to attribute to him a "systematic" vision.[1] Nevertheless, it is still true that the Aristotelian corpus looks like a more or less articulated set of *domains*, something that the Platonic corpus does not provide at all. Thus, studies of Aristotle's writings, for the last several centuries, have been of several kinds: there are those who take account of the whole, or of large parts, of this corpus, those who are interested in particular sections, and those who study the relationships between two or more domains; finally, there are those who deal with a question or idea—for example, the question of teleology or chance, appealing to several sections of the Aristotelian corpus.[2] The most noteworthy of these last are perhaps those that include an analysis of vast sections of Aristotle's biological texts. Such studies have recently taken on a new form, thanks to the "biological turn," for reasons that will be provided at the beginning of the first chapter, when we will also define that "turn."[3] Aristotelian studies have developed so much these last twenty or thirty years that it has become more and more difficult for one person to write a work on the whole of Aristotle, unless it be a work of vulgarization, an exercise as dangerous as it is necessary. Thus, it is a time for special studies. But those too have been hit by their own surge, such that it has become difficult to survey certain whole branches of Aristotelianism. Some, but not

1. See Crubellier and Journeau, "Le système de sciences aristotélicien."

2. Johnson, *Aristotle on Teleology*; Dudley, *Aristotle's Concept of Chance.*

3. A remarkable example of this sort of work is provided by the recent book by David Lefebvre, *Dynamis: Sens et genèse de la notion aristotélicienne de puissance.*

all. Thus, the politics and the biology of Aristotle, two domains in which I have worked primarily during my career, seem to me today to be in different situations. As for Aristotle's politics, I decided that it was still possible (for how long is another question) to write a work of synthesis, and that is what I tried to do in *Endangered Excellence*.[4]

That work took the place of another project, that of publishing a collection of articles, more or less revised, of some of those that I had dedicated to Aristotelian politics. Since I also had, as an editor had suggested to me, the project of bringing together articles on Aristotle's biology, I wondered whether I could do for the biology what I had done for the politics. But very quickly the project seemed to me impossible: the extraordinary increase of publications on the subject, specialized analyses more and more profound and subtle that interpreters have provided, made it, in my opinion, impossible, at least for me, to take up again the project that Anthony Preus had successfully carried out in 1975, that of publishing a synthetic work on Aristotle's biology.[5] But I had no intention of publishing a specialized work, so I chose an intermediate solution: deal with the questions that seemed to me important for the understanding of Aristotle's texts dedicated to animals, these questions being chosen, on a purely subjective basis, as those which have particularly interested me in the course of my studies these last forty years. There are five such questions, with a chapter dedicated to each.

This little book turns on two problems—the first being that of knowing whether one may attribute to Aristotle "a biology," the second of judging to what point the idea of perfection applies, for him, to the world of living things. But the work itself is constructed in overlapping layers: the first and third chapters ask whether Aristotle could be considered as the creator of a true biological thought, while the second chapter, which asks about the form given by Aristotle to teleology, finds a natural continuation in the last two chapters, which turn on the idea of perfection, chapter 4 examining the relationships between perfection and diversity and chapter 5 the modeling function of a particular living thing, the human being. If we must isolate the most general and fecund result that this study tries to establish, we may say that it tries to show how much, on fundamental points, Aristotle differentiated himself from what we may

4. Pellegrin, *L'Excellence menacée*, revised English edition, *Endangered Excellence*.

5. Preus, *Science and Philosophy in Aristotle's Biological Works*.

call the unanimity of ancient thought. With, however, two limitations. In the first place, when one works on an author, it is inevitable that one has the ultimately understandable tendency to find that author to be utterly original. Secondly, that originality recalls that which I discerned in Aristotle's political thought in my *Endangered Excellence*, but with an important difference—and it is hard to say whether that difference increases or diminishes the contrast between politics and biology. In biology, Aristotle's originality has a solid foundation in a zoology that had no predecessor and no successor until the nineteenth century, and we will have to say a few words about that extraordinary historical phenomenon, while in the area of political thought, Aristotle has had numerous colleagues.

This book is addressed primarily to two sorts of readers. On sometimes difficult subjects I have tried to be accessible, if not "to the greatest number"—that would be a pious hope—at least to readers who are not part of the circle of specialists—a circle that is becoming less narrow, but still limited. I hope that everyone will find food for thought. Beginners to Aristotelianism will be able to gain a clearer understanding of certain concepts, such as spontaneous generation, hypothetical and other necessities, that Aristotelians habitually use, even if the differences among them are many. I'm not particularly interested in engaging in specialist disputes (even if I need to do that sometimes), but rather I want to *situate* these questions. Thus, I have figured that the Aristotelian doctrine of "hypothetical" or "conditional" necessity, to which the best commentators have applied a great deal of thought, ultimately does not have the theoretical importance that some have thought; Aristotle mainly uses it in his polemics against Presocratic mechanists. But these five chapters are also addressed to specialists in Aristotle, especially specialists in Aristotle's biology. What I propose to them, above all, is a rereading of texts that they know well, but on which I think I sometimes can bring new light. From a certain point of view, this book is above all a collection of texts, put in perspective and commented upon; some will surely find my quotations too long.

Following the French tradition of history of philosophy, this book has primarily the goal, according to Wilhelm Dilthey's famous distinction, of *comprehending* the topics that it touches on, that is, to grasp the internal logic of the questions raised by these texts, and only secondarily to *explain* them by referring them to a larger structure, whether that would be the society in which these ideas arose, or the history in which they occur. Thus, I have hardly yielded to the desire to open the question, so popular among some of our American colleagues, of knowing "what that tells us today." At the same time, there is a very important point for the historian of science, which I too have tried to be, that of trying to elucidate the relationship of Aristotle's biology to biological sciences in their later forms. I have tried to find a path between the naïve continuism of bad historians of science and the absolute otherness between Aristotle and his distant successors, a position derived from a badly applied Bachelardism. I will explain all that in my first chapter.

I obviously recognize the profound influence that my own education and my own intellectual tastes have exerted on both the content and form of this study. My readers will easily discern my excessive taste for naturalists and physicians of the eighteenth and nineteenth centuries, especially Cuvier. For historians of philosophy educated in the French tradition, asking whether Aristotle and Cuvier could be included in the same category, that of "biologist," does not have a whole lot of meaning; the question itself is suspect in that it seems to posit grand transhistorical ideas, like those of "biology," and "biologist." To convey an understanding of the sense in which I have asked myself the question of knowing whether Aristotle could be placed alongside Cuvier in a portrait gallery of biologists, I would want to locate that question in relationship to two others.

The first question, already introduced, asks what Aristotle's biology teaches us today; this question takes two forms. There is a naïve form, critiqued in my first chapter, what is still valid in Aristotle's zoological treatises, how could they help today's biologists. This question, thus posed, with new vitality in a world in which colloquia on Aristotle's biology are financed by pharmaceutical companies, may be answered thus: Aristotle's treatises offer nothing to today's biologists. But there is a more interesting form of this question, which feeds into the larger question, often asked, of the usefulness of the history of philosophy. In the case that concerns us, that of Aristotelian biology, this question can in turn take two forms, or rather bear on two points. First on that which I have elsewhere called "Aristotelian thought."[6] There are, for each one of us, themes, texts, or ideas that we find particularly meaningful, to the point that we may be almost obsessed with them. That's how it is for me with Coleridge's assertion that every man, and we need to add "every woman," is either a Platonist or an Aristotelian. Aristotelian thought, as I have tried to show, is characterized

6. Pellegrin, "De la tradition aristotélicienne."

by several traits, of which antireductionism is one. Knowledge is, to be sure, carved up into officially different sciences that can and must cooperate, although that does not put all the sciences under the control of a single dominant science. Another trait is confidence in empirical evidence, but also an antiempiricism leading to a rejection of facile explanations by way of negentropy—Aristotle obstinately eschews explanations of the more organized by the less organized, as we will see in detail, especially in the second chapter. The requirements of this "thought" have brought it about, over the course of centuries, that thinkers both philosophical and scientific can be included in an Aristotelian tradition that is still very much alive. But the rebalancing, a truly massive task, that the reintegration of Aristotle's zoological treatises, actually close to a third of the corpus taken to be authentic, into our interpretive reading of Aristotelianism, forces us to redefine "Aristotelian thought" from the ground up.

Next, that which has been called the "biological turn" of Aristotelian studies forces us to reconsider the relationships between what, *in modern terminology*, we call "philosophy" and "science." This point will be clarified in the first chapter. But if, to give a rough summary of things, one figures that Aristotle's zoology *also* belongs to the history of science, which is not the case, for example, for his physics, it becomes possible, if not to make Aristotle a precursor to Cuvier, at least Cuvier a successor to Aristotle.

The second question that arises about Aristotle and Cuvier possibly belonging to the same history concerns Aristotle's influence on Cuvier. Cuvier explicitly and strongly attached his project to that of Aristotle. But we will see, if I succeed in making myself understood to my readers, that this attachment needs to be taken with two qualifications. First, Cuvier is universally a partisan of continuist history of science (we will see more exactly what that means in the first chapter), that is, he thinks that he has simply *continued* the work of Aristotle, adding whatever Aristotle had not seen, amending whatever he got wrong. But Cuvier went even further in finding a homology between the Stagirite and himself, as we will see when we speak of the general "laws" that govern the animal kingdom. The second qualification is much more important for us, and I believe that no one or almost no one has noticed it. It's that, in fact, it is not "our" biologist Aristotle that Cuvier took to be his predecessor. For us, in fact, Aristotle's zoology is above all the sublime theoretical construction presented in the *Parts* and *Generation of Animals*, two treatises devoted to the study of causes, especially teleological. I will try to derive as much as I can from the significant fact that Cuvier's Aristotle is above all the

Aristotle of the *History of Animals*. That will allow us to begin a theoretical reevaluation of that undervalued treatise.

But my essay remains basically a contribution to the history of Aristotelian philosophy, and it tries above all to add to the benefits of the "biological turn" in the history of philosophy. If it is also a work of the history of the sciences, it is a philosophical history of the sciences, in the manner of Georges Canguilhem, who was one of my teachers; one may discern his shadow behind many of the following pages. There is an academic practice of thanking those to whom one owes something in the achievement of a work. I won't do it, because there really are too many to whom I owe a debt of gratitude.

Chapter 1

Is There an Aristotelian Biology?

The biological turn taken by Aristotelian studies has been one of the notable facts of the history of ancient philosophy in the second half of the twentieth century. The idea of a "biological turn" does not concern Aristotle himself, but those who read him today, since the Stagirite never took any "turns." But when we apply it to our reading of the Aristotelian corpus, we call the part of Aristotle's work that treats of living things as "biological." The question that we will examine in this chapter can be formulated thus: Is it absolutely anachronistic to say that Aristotle was a *biologist*?

The biological turn that has deeply influenced our reading of Aristotle has tended to move in two opposite directions. First, it's a matter of finding in the biological texts, which represent nearly a third of the genuine Aristotelian corpus, concepts and methods developed in the methodological chapters of the corpus. If Aristotle's biology tries to demonstrate propositions, for example, to what extent do those demonstrations conform to the model outlined in the *Posterior Analytics*? This is also the case for the establishment of first principles. But the biological turn has also served, in return, to use the biological texts for the deepening and refining of our understanding of these concepts and methods. How, in fact, can we understand Aristotelian teleology without the *Parts of Animals*, or the doctrine of potentiality and actuality without the *Generation of Animals*? Thus, the biological turn has especially benefited the history of philosophy, rather than the history of science.

Before the biological turn, the study of Aristotle's zoological corpus was left, with rare exceptions, to historians of biology who generally treated it badly. In the first place, because they were bad historians of science. Nearly all these historians have in fact adopted a position both empiricist and continuist, wonderfully ignoring the revolution in history of science brought about by Gaston Bachelard and his students. They have figured that investigators, over time, ask themselves the same questions about a reality that remains self-identical, so those historians have tried to distinguish in Aristotle whatever was "already" scientific, that is true, from that which was "still" pre-scientific, that is, false.

Abandoned to these mediocre historians, Aristotle's zoological corpus largely escaped historians of philosophy. It is truly astonishing that many of the greatest interpreters of Aristotle, right up to the 1960s, had a very sketchy understanding, or no awareness at all, of the works on animals and life phenomena, which represent, as we said, more than a quarter of the existing corpus, and doubtless a third of the certainly authentic works. This ignorance rests on an implicit split of Aristotle's works into two parts, which may be called "philosophical" and "scientific"; thus, according to the scholars in question, they did not have the same status, and were subject to different treatment, because the history of philosophy and the history of science are not the same thing.

But if the historians we have mentioned were wrong to ignore Bachelard, he is far from having a positive effect on the study of Aristotelian zoology. For the Aristotelian zoological corpus, interpreters, especially French interpreters who were all to some degree influenced by Bachelard, considered it as a perfect example of pre-scientific speculation. Among the specialists on classical antiquity, we may mention Robert Joly[1] and Simon Byl,[2] as well as myself[3] at the beginning of my studies on Aristotelian biology. And the influence of Bachelard is doubtless one of the reasons why there hasn't been a "biological turn" on French soil. Everything happens as if the historians of philosophy, and probably also the Bachelardians, thought that science and philosophy did not need to answer to the same demands, and that the study of Aristotle's biology doubtless did not need to be undertaken by historians of Aristotelian philosophy, because Aristotelian philosophy ought not to be mixed with, and soiled by, "pre-scientific"

1. Joly, "La Biologie d'Aristote."

2. Byl, *Recherches sur les grands traités biologiques d'Aristote.*

3. Pellegrin, *La Classification des animaux chez Aristote*, translation: *Aristotle's Classification of Animals.*

thought, as primitive as that to be found in the zoological treatises. But obviously for Aristotle there is no theoretical gap between the zoological treatises and his texts thought to be philosophical. That is something that the biological turn has also demonstrated.

Not that it is necessary to expel Aristotle from the history of biology: obviously he occupies an important place, but it is also perfectly unique. In the first place, it is necessary to recognize that Aristotle, and he alone for more than twenty-two centuries, has been a real *biologist*. Aiming to see whether such a thesis is historically tenable, we will begin, in what we may call a "documentary" manner, by making a parallel between Aristotelian biology and modern biology, as it was born, or reborn, at the beginning of the nineteenth century. After that, we can return to Bachelard.

Aristotle and Nineteenth-Century Biology

There is an excellent short article by Allan Gotthelf, to which we will return, that establishes that there is an "isomorphism" between Aristotle and Darwin.[4] Both of them show us animals adapted to their environment and their life conditions, since for Aristotle too nature "seeks the adapted" (τὸ πρόσφορον, *History of Animals* 9.12, 615a25). But their opinions about the origin of that adaptation are radically opposed, since, according to Darwin, natural selection should increase the adaptive characteristics of each animal to assure the survival of its species, while for Aristotle, as we will see again in detail when speaking of his teleology, each species has had forever and ever the adaptive characteristics that assure its survival. All the same, Gotthelf is right: Aristotle and Darwin both analyze the situation of a species in terms of the total of advantageous and disadvantageous traits; this establishes an isomorphism between them that brings them together. This isomorphism brings them together to a higher degree than the opposition between fixity and evolution, heretofore thought primary, opens a gap between them. Raptors have talons because they are necessary for their survival, or at least because without them they would find it more difficult to survive, but they do present a disadvantage by making their walking more difficult. But it is better for a raptor to walk with difficulty and to hunt easily than the opposite. A raptor, like every other living thing, can be said to be well adapted when the whole of its positive traits is sufficient

4. Gotthelf, "Darwin on Aristotle."

to counterbalance the negative elements, both internal to the animal and due to its environment. There are in Darwin constraints that Aristotle did not envisage, mainly concerning changes in the environment and in relationships with other living things: climatic disasters, appearance of a competing species, disappearance or decrease of a useful animal or vegetable species, and so on, which makes a species until then well adapted to its environment find new difficulties for survival, sometimes impossible to surmount. Or rather, Aristotle does clearly recognize some of these changes, like ecological disasters, but he includes them in a cyclical eternal return that places changes onto a background of permanence.

Aristotle has often been roundly criticized for his finalism, mainly because critics have understood neither its nature nor its subtlety, and that's something that I will try to remedy in the next chapter. But something that modern and contemporary biology also teaches us is that finalism is not an obstacle to a scientific approach to the living world, as scholars like Ernst Mayr and epistemologists like Georges Canguilhem have shown us. We will come back to that when we speak of Aristotle's conception of life. One may go even further by showing that Darwin is much closer to the fixist Cuvier than to the evolutionist Lamarck, because Cuvier and Darwin were *biologists*, which Lamarck was not—he remained a *naturalist*. We will see a little later what that means. Cuvier too was fixist—thus his contortions in trying to explain fossils—and no one disputes the scientific character of his work. What Gotthelf's article also shows is that Aristotle's teleology also is not an obstacle to a scientific approach, contrary to what has been repeated for at least two centuries. Because we must note here that, as Gotthelf says, when Darwin declares that such and such a character is present in an animal *because* it is advantageous, he introduces, whether he knows it nor not, final causes into nature. Perhaps it will be necessary, in order to understand Darwin's position, to resort to the concept of *teleonomy*, which seems to have been introduced in 1958 by Colin S. Pittendrigh, the "father" of the biological clock, because for Darwinians, evolution does not have intentions—does not *want* anything—but everything happens as if an intention was involved in the choice of certain characteristics and the elimination of others. That remark will turn out to be valuable when we examine Aristotelian teleology.

In fact, the grasp of the degree of adaptation of a living thing in terms of a relationship between advantages and disadvantages is both the expression of a finalist position and the mark of a true *biological thought*, deployed in the extremely theoretical treatise, *Parts of Animals*. When we have clarified what I think is the real nature of Aristotelian finalism, we will discover that the *Parts of Animals* constructs a real *biology*, and that thus it is not absurd to find in it an isomorphism with Darwinian biology. In any case, it was in a letter that he sent to William Ogle, February 22, 1882, thanking him for sending his translation of the *Parts of Animals*, that Darwin wrote the sentence, which has become very famous and often cited, "Linnaeus and Cuvier have been my two gods, though in very different ways, but they were mere schoolboys to old Aristotle." Nevertheless, even if one takes this declaration as something more than a polite remark on Darwin's part, who by the way knew no Greek, and even if one thinks that he wound up reading Ogle's translation (although he admitted in a letter of 1879 that he had read nothing by Aristotle), all things difficult to establish firmly, the connections between Aristotle and Darwin have remained historically very tenuous.

With Georges Cuvier, the other great biologist of the nineteenth century, on the contrary, Aristotle has more than an isomorphism; one may speak of a real *homology*. All the more in that this relationship was recognized and underlined by Cuvier himself, and by historians of the biology of his epoch. Cuvier has been called "the Aristotle of the nineteenth century," while no one would think of making Darwin into an Aristotelian. However, as we will see, it is not, one may say, the same Aristotle to whom one may relate Darwin and Cuvier. That difference can be read especially in the fact that the isomorphism between Aristotle and Darwin refers to the *Parts of Animals*, while it is the Aristotle of the *History of Animals* that Cuvier recognizes as his precursor:

> The *History of Animals* is not a zoology as such, that is, a set of descriptions of various animals, rather it is a sort of general anatomy, in which the author deals with the generalities of organization presented by various animals, where he presented their differences and similarities, resting on the comparative examination of their organs, and where he posits the bases of large classifications with the greatest exactitude.[5]

5. Cuvier, *Histoire des sciences naturelles depuis leur origine jusqu'à nos jours*, 147I.

Nevertheless, it remains that Cuvier, like everybody of his time and later, up until the biological turn, above all examined the contents of the *History of Animals*, noting whatever it contained, "true" or "false," properly insisting on the importance of correct observations carried out by the Stagirite.

All that presents us with two immediate tasks: first, to clarify, at least in summary, the character of the homology between Aristotle and Cuvier, then to find in Aristotle's zoological corpus the two sources of a *biological thought*, one which relates him to Darwin, the other to Cuvier.

Aristotle and Cuvier

Thus we may turn to Cuvier, first to see why he deserves to be called a biologist. To do that, we will appeal to the very well-known analysis of Michel Foucault in *Les Mots et les choses* (*The Order of Things*, 1966). It is a very controversial analysis, but for the most part for wrong reasons, by historians unable to liberate themselves from their continuism. Better than others, Foucault describes and analyzes the *paradigm shift* that occurred between the taxonomy and natural history of the seventeenth and eighteenth centuries, and the birth of *biology*, properly so-called, with Cuvier. That too is a "biological turn," brought about by Cuvier. Risking, for the moment, to be bad historians of science, we can see that if we take them out of their respective contexts, Cuvier's ideas have a startling resemblance with those of Aristotle.

Taxonomists, skilled at what has been called "natural history," tried to grasp the differences between living things by taking note of the visible characteristics that distinguish one living thing from another. In that sort of approach, each trait can be just as significant as any other. Taxonomists, to be sure, also recognize the role of function that Aristotle attributed to what he calls the *parts* of animals, but as a parameter that should be placed on the same level as structure:

> In Classical analyses, the organ was defined by both its structure and its function. . . . The two modes of decipherment coincided exactly, but they were nevertheless independent of one another—the first expressing the *utilizable*, the second, the *identifiable*. It is this arrangement that Cuvier overthrows: doing away with the postulates of both their coincidence and their independence, he gives function prominence over the

> organ—and to a large extent—and subjects the arrangement of the organ to the sovereignty of function.[6]

These two aspects, structure and function, are also present in Aristotle. One may even say that these two projects, taxonomist and biological, coexist in him, since we find, in his works on animals, procedures for classifying by considering visible structure. He says several times that the parts of animals can be distinguished by their form, size, and presence or absence. Birds may have longer or shorter wings, insects may have two or four wings, the lung may be long or short, damp or dry, located on the side or below, the same for the uterus. There is an important point there: if this method of differentiation had been Aristotle's last word on the question of similarities and differences between animals, Aristotle would have been a taxonomist. Anyway, we can hardly imagine how a zoologist like Aristotle or Cuvier could avoid an at least approximate classification of the animals he studies. As I tried to show in the earlier book I cited previously, Aristotle surely had the means to put forward a "successful" taxonomy of animals, but that was not his intention. As soon as he resorts to a criterion other than that of structural resemblances, notably that of analogy (the lung is for some living things what gills are for others), Aristotle brings in a relationship to function of such a kind that the subversion that Foucault attributes to Cuvier had already been carried out by Aristotle, in whom biology largely dominates natural history. In Aristotle too, in fact, function rules, because it is definitory of the organ: the lung of some particular species cannot be defined as "a long spongy organ, suffused with blood, located on the side," but as "spongy organ that cools the blood."

Some have blamed Foucault for having exaggerated Cuvier's role in the appearance of biology, but we will follow the trail, partly because the role of Cuvier was impressive, partly because it doesn't matter for us very much since we are not trying to understand the biologists of the nineteenth century with the help of Aristotle, but Aristotle thanks to them. Biology initiated, or at least practiced, by Cuvier introduced two major changes. First, Cuvier considered living organisms as unities with functions both integrated and hierarchical, and he applied to them two general laws. First, the law of *coexistence*, which "designates the fact that an organ or system of organs cannot be present in a living being unless

6. Foucault, *Les Mots et les choses*, 276; translation: *The Order of Things*, 264; 2002, 287.

another organ or system of organs, of a particular nature and form, is also present."[7] Secondly, the law of the *subordination of characters* distinguishes characters and functions as they are more or less fundamental. From the point of view of classical taxonomy, on the other hand, there is no organ that is more significant or less significant; organs have more or less taxonomic significance, which means that they characterize the living thing more or less well *for the observer*. Thus, Linnaeus classified plants by their reproductive organs because they all have them, and they are easily describable.

That also led Cuvier to introduce what Foucault describes as a change in the relationships between Same and Other:

> What to Classical eyes were merely differences juxtaposed with identities, must now be ordered and conceived on the basis of a functional homogeneity that is their hidden foundation. When the Same and the Other both belong to a single space, there is a *natural history*; something like *biology* becomes possible when this unity of level begins to break up, and when differences stand out against the background of an identity that is a deeper and, as it were, more serious identity than that unity.[8]

As has already been mentioned concerning lungs and gills, we will see Aristotle adopt this attitude in relation to the identity and difference which bring it about that organs formally quite different can be called identical because they have one and the same function.

But this "profound identity" does not cancel out the radical difference that living things have between them, which makes impossible a scale of beings in which all beings are located in a unique series with no gaps: "Living things, because they are alive, can no longer form a tissue of progressive and graduated differences; they must group themselves around nuclei of coherence which are totally distinct from one another, and which are like so many different plans for the maintenance of life."[9] Contrary to people like Étienne Geoffrey Saint-Hilaire, Cuvier thus thought that there is no single plan for the entire animal kingdom. Thus, one understands

7. Foucault, *The Order of Things*, 265, 289.

8. Foucault, *The Order of Things*, 265, 288.

9. Foucault, *The Order of Things*, 272–273, 297.

that this means that the evolutionist Lamarck, who placed all animals on one scale of increasing complexity, needs to be put on the side of natural history, rather than biology, and consequently makes him further from Darwin than is Cuvier.

Finally, one might be able to define, at least in outline, a *biological thought* by this double relationship, at first glance contradictory, between the Same and the Other. On the one hand, that which natural history understands as a profound otherness is in fact based on a radical identity, because the function of different organs can be the same; on the other hand, since each living entity is a system of correlated and hierarchical functions that cannot all be included in the same chart. Of course, animals that are generally homologous, that is, that have the same way of combining and ordering their functions, ought to be ranked in one and the same group. But there are irreducibly different ways of constructing such living systems. Thanks to this new approach to living things, Cuvier posed an in-depth classification of animals. The law of subordination of characters permitted, in fact, a division of the animal kingdom that was very different from that of natural history: by the various form that it takes, the nervous system, which is the most basic system (Cuvier says it "is the basis of every animal"), to which all the other systems are subordinated, serves to define the most general classes of animals, the branches. In vertebrates "the nervous system is located on the digestive tract," while in mollusks it is located below. But that is not a variation of the arrangement in vertebrates that would make it possible to include these two arrangements in the same chart, but two radically different dispositions, two different *plans*, that is, two irreducibly different ways of being a living thing. Cuvier distinguishes four branches, each exhibiting its own proper plan: vertebrates, mollusks, arthropods, and zoophytes. The more superficial systems, like circulation and respiration, provide classes, then systems even more superficial will provide orders, families, genera, and species. "Superficial" means subordinate, since always according to the law of subordination, certain systems and organs have, in the functioning of the organism, a greater effect than others; Aristotle says that they are "prior by nature" to the others. This in-depth taxonomy is very well characterized by the very title of the fundamental work, published by Cuvier in 1817 (in English translation), "The Animal Kingdom Arranged after Its Organization; Forming a Natural History of Animals, and an Introduction to Comparative Anatomy."

Comparative anatomy, of which Cuvier was, if not the founder, at least the renewer, and in any case the most illustrious practitioner in modern times, is precisely the most effective means of recognizing that the animal world, and even the living world, including vegetable, is irreducibly apportioned into several ways of living, and of recognizing that, in spite of the fact that classes of animals do not belong to the same space, one can weave between them ties of similarity. But historians know that by doing that, Cuvier returned, in a most conscious fashion, to an Aristotelian project.

A few more words about the era of Cuvier. There is a last characteristic that Foucault associates with the birth of biology in the nineteenth century, that is that the new relationship that is put in place in relation to living things presupposes in both philosophy and epistemology vitalist positions. A fundamental difference is put in place between the living and nonliving that is foreign to natural history, for which life is not a taxonomic characteristic. It's even the case that for the mechanistic Cartesians, the living thing and the machine belong to the same category of existent. For vitalists, as Foucault sees them, the living thing is more completely a being than the nonliving. In his inimitable style, Foucault writes: "Life becomes a fundamental force, and one that is opposed to being in the same way as movement to immobility, as time to space, as secret wish to visible expression. Life is the root of all existence, and the nonliving, nature in its inert form, is merely spent life; mere being is the nonbeing of life."[10] We may add that, as is often the case when something new appears, the elements of the next shape are already there, like *disjecta membra*, but not organized, before the emergence of that shape. As a good historian, Foucault reasonably gives particular attention to the narrow time frame that saw the end of the taxonomic enterprise of natural history steering itself toward a properly biological approach. People like Jussieu, Vicq d'Azyr, or Lamarck, had already adopted viewpoints that would be taken up by Cuvier. These naturalists, by means of the concept of *organization*, distinguished and subordinated some of the characteristics of living things to others, attributing functions to them. The principle of life is thus not entirely in the visible order. But biology was yet to come.

Foucault cites this remarkable passage from Lamarck's "System of Animals without Vertebrae":

> Consideration of the articulations of the bodies and limbs of the crustaceans has led all naturalists to regard them as true

10. Foucault, *The Order of Things*, 278, 303.

> insects. . . . But since it is recognized that organic structure is of all considerations the most essential as a guide in a methodical and natural distribution of animals, as well as in determining the true relations between them, it follows that the crustaceans, which breathe solely by means of gills in the same way as mollusks, and like them have a muscular heart, ought to be placed immediately after them, before the arachnids and insects.[11]

A remark that, even if it is clearly aimed at putting together a *scala naturae*, reminds us of the procedure of classification in depth of Cuvier, so to that extent, Lamarck already has, one may say, a foot in biology . . .

But it is remarkable that this structure that combines laws of correlation and subordination of characteristics with an original accounting of identity and diversity, a structure that I claim, following Foucault, characterizes *biological thought*, finds a homologous form in Aristotle, and for a very long time, only in him. We will have the occasion to show this precisely for each of the aspects of this structure, but we can already provide a quick summary. Aristotle too, thanks to his doctrine of the different kinds of unities, will search for a functional unity more profound than the diversity of appearances. This is demonstrated, among many other passages, by the project proposed by Aristotle at the beginning of the *History of Animals*:

> There are some animals whose parts are neither identical in form nor differing in the way of excess or defect, but they are the same only in the way of analogy, as, for instance, bone is only analogous to fish-bone, nail to hoof, hand to claw,[12] and scale to feather; for what the feather is in a bird, the scale is in a fish. The parts, then, which animals severally possess are diverse from, or identical with, one another in the fashion thus described. (*History of Animals* 1.1, 486b17, Thompson translation)

Thus, the many works that begin comparative anatomy with Aristotle are right, and even more right are those who do not find resurgence of this

11. Paris 1801, 143, as quoted in Foucault, *The Order of Things*, 229, 248.

12. Χηλήν: talon or claw.

kind of research until the beginning of the nineteenth century, principally with Cuvier. We will see from other places that, contrary to what a cursory reading of two passages often invoked seems to suggest, there is in Aristotle neither a *scala naturae* nor a unique plan for animals. Not only is it the case that the great division of all animals into those with blood and those without blood is not thinkable except analogically, but Aristotle also divides those without blood into four groups that also cannot be considered together except in terms of comparative anatomy: mollusks, shellfish, crustaceans, and insects. Although Aristotle does not use the word, it is really a matter of "plans." One may cite numerous passages in the Aristotelian zoological treatises that remind us of Cuvier's circumscribed classifications. Thus, in the *Parts of Animals*, Aristotle mentions that mollusks and insects have an organization "contrary" (cf. ἐναντίως, II.8, 654a10) to that of crustaceans, which have organizations "opposed" to each other (cf. ἀντικειμένως, 654a11). A little farther along, insects are called "contrary" (ἐναντίως, 654a26) to blooded animals. Thus, this very interesting remark: insects do not have, like other animals, a hard part and a soft part (other animals combine these two parts in different ways: some have the hard part inside the soft part, others on the outside), but they are made of an entirely hard material, nevertheless keeping a mean between the hardness and softness of others: "the whole body is hard, the hardness, however, being of such a character to be more flesh-like than bone, and more earthy and bone-like than flesh" (654a28, Thompson). In any case this passage appeals to analogy in order to break through the blurring of the boundaries between these different families.

As for organic correlation, that is expressed in Aristotle in the multiple correlations that frame the Aristotelian text, notably the "moriological"[13] part of the *History of Animals*, but also in the *Parts of Animals*. But this law appears in a much more relaxed form in Aristotle: all viviparous quadrupeds have hair, and all the oviparous quadrupeds have scales, but "not all vivipara have hair," "all animals with fingers have nails, except the elephant," "no eunuch becomes bald" (*History of Animals* 9.50, 632a4), "no animals have both projecting teeth and horns, and none of those that have sharp teeth have either projecting teeth or horns." We'll say more about those correlations later. But it is the law of the subordination of characters that seems to present the closest formulation in both Aristotle

13. I proposed the term "moriology" in *Aristotle's Classification of Animals* to designate the study of the parts (*moria*) of animals.

and Cuvier. Though Aristotle does not explicitly propose a *law* (we must remember that the term "law" [*nomos*] has in Aristotle only a political usage),[14] he distinguishes more or less basic characteristics and functions: digestion, which, in animals with blood, produces blood from food, is more fundamental than the circulation of blood, which in turn is more fundamental than the function of cooling brought about by respiration. We will see that these distinctions rest on a doctrine that does not belong to biology, that which distinguishes essential attributes, proper, per se, and accidental, which one finds, among other places, in the *Topics*, for example. That poses a serious problem for the theoretical unity of Aristotelianism, but does not compromise the biological efficacy of the "law."

As for the vitalist positions that Aristotle takes, and which will be reflected in those of nineteenth-century biologists described by Foucault, those correspond to a profound structure of Aristotelian thought. Because a very general explanatory scheme, which runs through the entire work of Aristotle, and to which we will frequently return, refers the less to the more, and not the more to the less. Thus, when it is a matter of understanding what is ethical virtue, one must refer to its developed form, that of a virtuous free adult Greek citizen, and not to its deficient forms, embryonic or truncated, like the virtues of a woman, child, barbarian, slave, or vicious person. Thus, we must agree with Marwan Rashed when he thinks, in his edition of the treatise *On Generation and Corruption*,[15] that the "chemistry" of that treatise is potentially biological. How, for example, can we understand the analysis of growth found there without referring to the most developed form, namely the growth of the living thing by assimilation of food? Thus, increase is a simplified growth, rather than growth a complex increase. It is a far from absurd idea that Aristotle's teleology is, among other things, a way of expressing the idea that a living being is more completely being than a nonliving being. That is why Aristotle cannot conceive of a perfect being (god, or the universe) other than as living. Although Aristotle believes that physiochemical laws are never violated by living beings, nevertheless he considers life as a *supplementum entis*: "The soul is better than the body, the animate better than the inanimate because of the soul, being than nonbeing, living better than nonliving" (*Generation of Animals* 2.1, 731b28). Aristotle too could, like Foucault, have written, "the nonliving, nature inert, is nothing more

14. One exception at *De Caelo* I, 1, 268a14.

15. Aristotle, *De la génération et la corruption.*

than spent life." We will have occasion to say again that does not mean that Aristotle is committed to that which Georges Canguilhem calls a "vitalism of exception" like that of people like Xavier Bichat (1771–1802), who would liberate the living world from the laws of inert matter.

A particular form of this schema, absolutely crucial for the understanding of Aristotle's thought taken together, can be seen in the thesis according to which "generation is for the sake of substance, and not substance for the sake of generation" (*Parts of Animals* 1.1, 640a18). Immediately after that declaration, Aristotle gives it meaning by criticizing Empedocles, who held that animals have a particular characteristic, to wit that their vertebral column is divided into vertebrae, "because it is bent (in the womb) so that it gets broken" (640a21). In fact, it is necessary that there be a developed parent, logically and chronologically prior to its offspring, that provides the characteristic "separated into vertebrae" to the embryo by transporting it, one way or another, in its seed, for according to the famous formula, here repeated, "human being engenders human being" (640a25). All that comes to affirming that which the moderns have called the principle of entropy, of which the Parmenidean prohibition of getting being from nonbeing is the ontological version.

Let us return to the perspective of the beginning of this chapter, that of the history of the sciences. To make Aristotle a biologist in the sense in which Foucault takes Cuvier as the founder of biology is not only to overturn Foucault's historical approach; it also casts doubt on the Bachelardian structure according to which we French have done history of science since the 1950s. Let us first recall some historical facts that people today have perhaps begun to forget, but that will clarify what was rapidly summarized earlier. When, in 1938, he published *The Formation of the Scientific Mind*, Gaston Bachelard broke radically, and a bit provocatively, with the continuist practice of the history of the sciences. According to the latter approach, science develops gradually by the accumulation of discoveries, but also by the elimination, no less continual, of errors. A paragon of continuist history of the sciences was the Australian historian of science Alistair Crombie (1915–1996), a man of exceptional scientific, and other, culture. His leading work is his *Augustine to Galileo: The History of Science A.D. 400–1650* (1952). A year later he published *Robert Grosseteste and the Origins of Experimental Science, 1100–1700*, whose very title is, one may say, an entire program.

An approach like that is based on a naïve empiricism, according to which the natural world, at last globally, remains the same over time, and scientists, relying on the discoveries of their predecessors but also learning their errors, observe and describe this world in an increasingly detailed and exact way. From a perspective like that, Aristotle was obviously a precursor of modern biology, and more than others to the extent that he was able, more than anyone else before modern times, to observe correctly, to ask pertinent questions, and to infer accurate conclusions, and despite his many errors in which modern biology, as they say, "does not recognize itself." Even the word "precursor" ought to be refined. For when one says that Democritus was the precursor of atomic theory, one cannot fail to notice that between Democritus and Niels Bohr there is a radical paradigm shift, while that is not the case, at least not to the same degree, between Aristotle and Darwin, and especially between Aristotle and Cuvier. When, then, Georges Canguilhem proposed his famous thesis that there are no precursors, that surely applies to Democritus, but a good deal less surely to Aristotle.

Everyone, at least on the continental side of the English Channel and the European side of the Atlantic, knows that according to Bachelard, the sciences were born as a consequence of an "epistemological break," that this happened suddenly, and that a science does not have parents in the sense that it is not *engendered* by that which precedes it, because generation is the production of a new organism by organisms of the same species. But there is nothing, or very little, in common between Robert Grosseteste and Galileo. It is primarily the gap between science and pre-science that Bachelard studies in his *The Formation of the Scientific Mind*. Thus, Galilean physics was constructed not out of Aristotle's physics, but against it. Doubtless Bachelard's analysis has Auguste Comte in its background. In a very interesting article, Laurent Fedi[16] argues that Aristotle, although a typical example of "metaphysical" thought in the Comtean sense of that word, tends in the direction of a "positive" approach, still in Comtean terms. We will see that Fedi is more correct than he thinks. Certainly, there have been, and continue to be, continuists among the French and discontinuists among English-language writers. Thus, Thomas Kuhn's *The Structure of Scientific Revolutions* (1962) was, to the great astonishment of the French, a thunderclap in the nearly calm sky of English-language history of science. Today, no one continues to doubt that scientific knowledge develops by the construction of models and the movement from one

16. Fedi, "Le prince des philosophes."

paradigm to another, and not, for the most part, by observation of new phenomena through the accumulation of observations.

Finding a homology between the biologies of Cuvier and Aristotle, a homology in whose name I intend to argue that Aristotle was a real *biologist*, obviously does not mean that Aristotle's science was at the same level as that of Cuvier. Those who, like Robert Joly and Simon Byl, offer a Bachelardian reading of Aristotle's biology propose two main points for refusing him any scientific status. First (and no one disputes this point), Aristotle's zoology is full of mistakes and areas of ignorance: Aristotle does not know what a muscle is, nor that blood circulates; he attributes digestion to a kind of cooking, believes that glands are filters, and so on. This discontinuity from modern science reaches its culmination perhaps concerning the reproduction of living things, especially of animals. The *Generation of Animals* is perhaps the treatise in which the second phase of the program of the biological turn has turned out the best, because cardinal notions like those of potentiality and actuality would have remained partially empty without the study of that actualization par excellence that is animal conception. This treatise is equally precious methodologically, for example in its long passage on the reproduction of bees. Nevertheless, it remains that the *Generation of Animals* runs into a formidable obstacle, that this treatise is extremely far from "reality" and mixes ideological considerations (especially sexist), and thus must be classified as pre-science. One may say that in the *Generation of Animals*, nothing, so to speak, is "true." Of course, as people like Sophia Connell[17] have shown, the female, according to Aristotle, is far from being pure passivity, as we will see in detail in a later chapter; nevertheless, it is clear that the analogy between male/female and artisan/material has been a formidable epistemological obstacle to the scientific investigation of the reproduction of living beings. And how could such an approach be possible when one does not know about the existence of spermatozoa and ovules? That is why, in his article on Aristotle's biology, Robert Joly (1968) uses the *Generation of Animals* most of all, since his goal is to show that Aristotelian zoology belongs to archaic thought. But if the presence of errors is enough to disbar a speculation from "true" science, that would become an empty set.

The second criticism of Bachelardians seems to be more serious. It consists of saying that in elaborating his biology Aristotle pursues goals that one may call "metaphysical," which are not in any case scientific in the modern sense of the term. These interpreters begin by invoking the

17. Connell, *Aristotle on Female Animals.*

prejudices that Aristotle aims at justifying by appealing to his zoological observations, and usually it's Aristotle's sexism that is cited, claiming that he considers females as generally, or sometimes totally, inferior to males. We need to temper these accusations of sexism brought against Aristotle and his zoology, but we cannot eliminate them entirely. Next, it is Aristotle's finalism that, in the eyes of these interpreters, disqualifies any scientific pretension to his biology. We return to this question in detail in the next chapter. But we may for the moment remark that it is difficult to find in the history of science a scientist who is totally innocent of metaphysics.

As for Foucault, as a good Bachelardian he doubtless imagines, though he does not say it this way, the moment of Cuvier as an epistemological break. Actually, Foucault is closer to Kuhn than to Bachelard since he finds between the natural historians and Cuvier a paradigm shift rather than a movement from nonscience to science. Foucault is all the more inclined to see in the biology of Cuvier a radical break because the term "biology" was introduced in his epoch. He is not, however, the inventor of the word. It was Lamarck who introduced the word "biology" into French, possibly around 1800 in a manuscript whose date is uncertain, but in any case in 1815 in volume 1 of his *Histoire naturelle des animaux sans vertèbres* (*Natural History of Invertebrates*): "All that which is generally in common in plants and animals, like all the faculties that belong to each of these beings without exception, ought to constitute the unique and vast object of a particular science that has not yet been established, and that does not have a name; I will give the name 'biology' to it" (49). It would then be Gottfried Reinhold Treviranus in his *Biologie oder Philosophie des lebenden Natur*, published in six volumes between 1802 and 1822, who would have been the real initiator of the term.

Even if one remains at a rather formal and general level, this parallelism between Aristotle and Cuvier in their understanding of the animal world will lead us to several revisions, none of them truly distressing. As far as the relationships between Aristotle's physics and that of Galileo, we may agree that, on the whole, Bachelard is right: they don't belong to the same history.[18] It is not the same for biology. To try to analyze the

18. Despite what I have called the "pockets of scientificness." In the introduction to my French translation of Aristotle's *Physics*, I mentioned the Aristotelian theory of projectiles as revealing, despite its obviously metaphysical aspects (like "horror of vacuum"), "a theory of false dynamics, rather than a false theory of dynamics." It is also the case that René Thom has drawn my attention to the exceptional scientific fertility of the Aristotelian doctrine of the continuum as it is presented in *Physics* book 6.

reasons for this difference would demand a book, but we can at least give this indication: to the degree that Aristotle's physics does not concern the relations between variables, but rather objects perceived by the senses and as such constituted of a set of qualities, there is no continuity, and even less paternity, between that of Aristotle and that of Galileo. The fact that the object of biology, the living being, is a self-organized being, separating itself from an environment that is not it, even if it has need of this environment to live (the distinction between interior and exterior has, in contrast, no meaning other than pragmatic for an inanimate object), and carrier of values at least vital (the objects in the world dividing for it into "good" and "bad"), establishes a temporal permanence for this biological object, which means that Aristotle as biologist and Cuvier are talking about the *same thing*, which is not the case for Aristotle the physicist and Galileo. The Aristotelian substance (*ousia*), the central objective of Aristotle's physics and metaphysics, can, in modern science, get a biological sense, but not a physiochemical sense.

All that establishes between Aristotle and Cuvier a profound common structure that persists despite the enormous differences between the Aristotelian zoological corpus and the works of Cuvier. One may find all the indications one may like of an "archaic" attitude in Aristotle's biological treatises, one may be surprised by the "fact" that he found fewer sutures on the female skull than the male skull, one may mock his theory as both fantastic and dogmatic of the generation of animals; nevertheless, it remains that the Bachelardian analysis does not apply very well to Aristotle's biology, and possibly to all biology. We may add that, although Cuvier does not avoid mentioning Aristotle's mistakes in observation (and how could he have done otherwise), it is not Aristotle the finalist that he invokes as a predecessor, but Aristotle the author of a comparative anatomy. Not the Aristotle of the *Parts* and *Generation of Animals*, but that of the *History of Animals*. To appreciate that fact, we need to take a detour to consider rapidly the Aristotelian zoological corpus in all its diversity.

Some Remarks on Aristotle's Biological Corpus

Aristotle takes account of living things, particularly animals, in essentially all his works that have come down to us. But there are, in the corpus as we have it, several treatises consecrated to the study of animals: *History of Animals*, *Parts of Animals*, *Generation of Animals*, three large treatises that are strongly united by reciprocal citations and allusions. There are also two shorter treatises, the *Movement of Animals* and the *Progression of Animals*. But, as we will see especially in chapter 3, we may consider the *De Anima* as a work of general biology, and the eight or nine (depending on how they are separated) treatises brought together in the *Parva Naturalia* as a sort of appendix to the *De Anima*, logically but not necessarily chronologically after that treatise. They examine certain functions of living things: perception, memory, sleep, dreams, youth and old age, respiration, life and death. We will also see that Aristotle's biological doctrines are in the background of some writings that are not formally consecrated to that, and we will recall several times that Aristotle also alludes to a treatise on plants; we don't know whether he wrote it, and whether it is lost, or if it has been assimilated into the botanical treatises of Theophrastus. Here it is a matter not of proposing a general presentation of the Aristotelian zoological corpus, but of selecting some questions that interpreters have raised about it, all concerning the general problem posed in this chapter: Can Aristotle be considered a biologist?

As for the first three treatises cited, which bring together the essentials of what Aristotle has to say about animals, the question most often asked is that of the relationship between the *History of Animals* and the *Parts of Animals*. Aristotle asserts that there is a division of labor between them. We could cite several passages; here are two of them: "I have already described in detail the nature and number of the parts of which animals are composed in the *Studies of Animals*, but we must now investigate the causes that in each case have determined this composition, a subject different from that dealt with in the *Studies*" (*Parts of Animals* 2.1, 646a8, first lines of book 2). "In the studies on animals" (ἐν ταῖς ἱστορίαις ταῖς περὶ αὐτῶν, the αὐτῶν referring to τῶν ζῴων on the same line) refers, in the unanimous opinion of scholars, to the *History of Animals*, a treatise here designated by its traditional title, while the examination of "causes" is the objective of the *Parts of Animals*. Elsewhere the two treatises are designated differently: the *History of Animals* is supposed to be concerned with the φαινόμενα (*Parts of Animals* 1.1, 640a14), a word that doubtless is better translated by "facts" than by "phenomena," unless we take this term in its etymological sense, while the examination of "causes" is, in this passage too, assigned to the *Parts of Animals*. That is repeated in the second passage:

> All that, put forward in this way, is for the moment a schematic foretaste of all the subjects and all the properties that we must

> consider in order that we may first get a clear notion of their actual differences and common properties. After that we will discuss these matters with greater accuracy.
>
> Next we will have to go on to a discussion of the causes. For to do this when the investigation of the details[19] is complete is the natural method; for from them the subjects and the premises of our proof become clear.
>
> In the first place, we must look at the constituent parts of animals. For it is in terms of these parts, first of all, that animals in their entirety differ from one another: either in the fact that some have this or that, while they have not that or this; or by their position or arrangement; or by the differences that have been previously mentioned, depending upon form, on excess, on analogy, or on contrariety of properties. (*History of Animals* 1.6, 491a6)

This programmatic passage, often invoked, teaches us at least two things. It repeats the division of work between the two treatises: first, to grasp that which is in common and that which is proper to different animals is the task of the *History of Animals*, then search for their causes, task of the *Parts of Animals*, the causes both of the properties of each part and of the identity and differences that they have among each other. He then asserts that the differences between animals should be grasped by means of the differences of their parts. It must be added that this passage, by using the terminology of the *Posterior Analytics* (it's a matter of "proof" [ἀπόδειξις, 491a13]), "subjects" designates the class whose attributes will be demonstrated, "from them" the principles from which the proof begins, attributes a strongly *scientific* character to zoological research.

There are also passages that seem to include the third of the large zoological treatises, the *Generation of Animals*, in this division of labor. Thus, at the beginning of this treatise, Aristotle considers his zoological research by way of the Aristotelian theory of the four causes, attributing, in a recognizable though not explicit way, the study of the parts of animals via the material cause to the *History of Animals*, the form and final cause to the *Parts of Animals*, and the efficient cause to the *Generation of Animals*: "It remains to speak of those parts which contribute to the generation of

19. "Details" means the properties and parts more than whole animals. But whole animals may also be considered.

animals and of which nothing definite has yet been said, and to explain what is the moving cause" (1.1, 715a11, Platt). The passage from the *Parts of Animals* cited earlier seems to support the same program: one must first grasp the phenomena, then the causes, "and also the generation" (καὶ επὶ γενέσεως, 640a15). But it is more likely that that account fails to understand that Aristotle means that in the study of generation one must also first consider the phenomena and subsequently the causes.[20] Even if, in fact, the *Generation of Animals* describes in detail the *formation* of male and female seeds, and resorts for that to efficient causality, this treatise is above all located in the perspective of the formal and final causes.

The Relationships between the *History of Animals* and the *Parts of Animals*

Concerning that more or less "classic" question, both for interpreters and for Aristotle himself, the programmatic passages like those previously cited seem to support the idea that the *History of Animals* is a collection of facts, a sort of database of givens, thus located outside, or anyway on the margins, of purely scientific research, in the Aristotelian sense of the term, which can only be causal. The *History of Animals* thus looks like an impoverished cousin in the zoological corpus. Besides, its usefulness does not seem obvious, because if the *Parts* and *Generation of Animals* had been lost, not only would we have a much more summary approach to Aristotelian teleology and theory of actuality and potentiality, as we have seen from the useful materials drawn by the readers of Aristotle of the biological turn, but we would have been more or less ignorant of what one may call the biological thought of Aristotle. If the *History of Animals* had been lost, the textual loss would have been immense in that many of the animals that are described and studied in it, notably in the latter books and especially among the fish and birds, would have been simply unknown to us, even their names. But the properly theoretical loss would have been ultimately rather little. To be precise, there is a point that must be recognized: the theoretical cooperation that the programmatic texts propose, when they tell us that the *History of Animals* identifies and describes the parts of animals that the *Parts of Animals* explain is an illu-

20. That is also the interpretation of James Lennox in his commented translation of the *Parts of Animals*, *Aristotle: On the Parts of Animals*.

sion, since there is no animal part explained in the *Parts of Animals* that is not identified and described in that treatise.[21] Furthermore, the study of parts in the *History of Animals* is slightly less long than in the *Parts of Animals*, an imbalance that only increases when we recall that in the *History of Animals* there are parts that are not discussed in the *Parts of Animals*, for example tendons. Thus, the *History of Animals* is generally useless for the *Parts of Animals*.

But that is true only, in part, for the four first books of moriology, and, to a lesser degree, for the three following books that concern the reproduction of living things. In these, in fact, Aristotle describes the genital organs of various species, their manner of intercourse, gestation, periods of fertility, spontaneous generation, and so forth, subjects that are for the most part treated in the *Generation of Animals*, but from a causal perspective. The two last books, which have been called the psychology and ethology of animals, and which have largely contributed to making the *History of Animals* a contribution to zoology unsurpassed for twenty centuries, are in fact descriptive, limited in theoretical application. But it is this collection, composite but incredibly rich, that, more than the more scientific treatises in the Aristotelian sense of the word, roused Cuvier's enthusiasm. That enthusiasm is doubtless enough to urge us to a theoretical rehabilitation of the *History of Animals*.

This composite character of the *History of Animals*, as well as its relationship with the *Parts of Animals*, has contributed to two lively debates among the interpreters, and we must get involved in them. First, a debate about the authenticity of the *History of Animals*. Even in the six first books, whose authenticity no specialist has really contested, the problem does arise. Thus, in book 4, chapter 3, there are things repeated that have already been said; other passages reveal an astounding psittacism: "The urchins are devoid of flesh, and this is a character peculiar to them; and while they are in all cases empty and devoid of any flesh within" (4.5, 530a32, Thompson), "dipterous . . . have their stings in front. The coleoptera are, without exception devoid of stings; the diptera have the sting in front, as the fly, the horsefly, the gadfly, and the gnat" (1.5, 490a17, Thompson). Such facts are precious, notably useful for discussions of questions of chronology, in showing us that the *History of Animals* is closer to a collection composed of successive additions than to a "work" in the usual, and modern, sense of the word. But, as we will see, "collection

21. Lennox has recognized that fact: Lennox, "Aristotle's Biological Development."

of givens" does not mean either a disordered accumulation of observed facts (as would be later collections, for example those of Aelian), nor a totally unstructured presentation.

This is not the place to get involved in the scholarly debates about the authenticity of the text of the *History of Animals*, but rather to try to decide what the various positions taken by editors and commentators mean. As we cannot go through the whole text in detail, we will satisfy ourselves with certain objections that have been made to the authenticity of certain books and passages.

D'Arcy Wentworth Thompson, who translated (in 1910) the *History of Animals* in the great edition of J. A. Smith and W. D. Ross,[22] accepted the position of Aubert and Wimmer, and of Dittmeyer, authors of the two great critical editions of the *History of Animals* of the nineteenth and beginning of the twentieth centuries, and declared the probable inauthenticity of book 7 of the *History of Animals*. Their two major arguments were, on the one hand, that this book includes numerous parallels with books 3 and 4 of the *Generation of Animals*, and on the other hand that it includes assertions that are consonant with certain doctrines of the Hippocratic corpus. But why would it be surprising if this part of the *History of Animals* includes facts presented, and often explained, in the *Generation of Animals*, when we have just seen that the moriological section describes organs that are also described in the *Parts of Animals*, and furthermore are given a causal explanation? As for the fact that the text of the *History of Animals* includes doctrines called "Hippocratic," that agrees with Aristotle's usual procedure, of appealing to the theoretical and technical competencies of various scholars and technicians. Descended from physicians on both sides, Aristotle would have known particularly well the medical literature of his time. Besides, the expression "Hippocratic doctrines" doesn't mean much for anyone who knows the corpus from which they are thought to be drawn. We will see, in a later chapter, what use Aristotle makes of the descriptions of the human vascular system offered by some physicians.

But it's mainly books 8 and 9 of the *History of Animals* that have roused the suspicions of editors. An argument often proposed by the partisans of the inauthenticity of large parts of these two last books is that the sections are not in accord with Aristotle's usual philosophical-scientific "style." That sort of apprehension does not fail to impress the translators of the *History of Animals*, me included. Long acquaintance with Aristotle's

22. Aristotle, *Historia Animalium* (Thompson).

texts assures us that the notion of an "Aristotelian style" is not senseless. But what does such an affirmation rest on if one cannot compare the last two books with other Aristotelian texts that also deal with animal psychology and ethology? An interesting counterexample is furnished by *Politics* 1.11, which deals with agriculture and other methods of acquiring a familial fortune: mining, commerce, and so forth. This chapter is right where it belongs among the subjects raised in this part of *Politics* 1, but its content, its form, are so different from the rest of the book (Aristotle divides the art of acquisition into three species, instead of the usual two, and he uses a vocabulary including an important number of hapax legomena), that the hypothesis that we are dealing with an independent text, possibly Lycean, seems possible. We have nothing like this in the case of the *History of Animals*, since we cannot just compare the last two books with the others. What has made great stretches of the *History of Animals* suspect in the eyes of some is that there does not exist, whether in the Aristotelian corpus, or elsewhere, and for a long time, anything that can be compared with such a work.

The text of the *History of Animals* has undergone a critique of another sort, that it may contain interpolated passages. An example well known to translators and commentators is that of book 7, chapters 21 to 28, which, in the middle of considerations about the influence of the environment on animals, turns to the illnesses of swine, dogs, cows, horses, donkeys, elephants, and bees, to return later to the influence of places on animals. One might diminish the oddity by recalling that chapters 19 and 20 talk about the illnesses of fish, and one might also imagine a scenario of the kind that editors of ancient texts love: a copyist, having accidentally copied the chapters in doubt instead of those we now find, did not want to suppress them, because copying is hard work, paid by the page, and support is expensive. But one cannot base anything on hypotheses like that . . . And as the fact of introducing here veterinary considerations does not break the train of exposition, there is not a lot of damage. However, that does not give us the right to decide that the *History of Animals* is a catalogue of facts in which the order of exposition has no, or very little, importance, as would be the case with compilations of later authors. Nor can we suppose that after a relatively well-structured exposition in its moriological part, the work piles up a jumble of observations, which might inevitably lend some support of probability to the hypothesis that the last books are by another hand than that of Aristotle. In circumstances like that, commentators have too often made use of the locution "later Peripatetics."

Right from its first chapter, the *History of Animals* posits its objective: it's a matter of seeing in what respects animals differ from each other; "Differences are manifested in modes of subsistence, in their actions, in their habits, and in their parts" (1.1, 487a10, Thompson). There follow examples of the first three sorts of differences: terrestrial or aquatic animals, this difference including numerous divisions, methods of local movement, group or solitary life, ways of nutrition, wild or tame animals, differences in sound production, differences in courage, propensity for sexual pleasure, but also differences in character between good temper and ferocity, withdrawn or vain, and so on. Following that, right to the end of book 4, Aristotle turns to noting and describing various parts,[23] various animals, taking as a model and common thread the parts of the human being, because that is best known to us, "just as any group tests coinage against that with which it is most familiar" (1.6, 491a20, Thompson). We will comment, in the last chapter, on this order and its inversion, when it is a matter of describing parts devoted to reproduction, for which Aristotle says that one must begin with the simplest animals.

Books 8 and 9 deal with three kinds of differences announced in book 1: ways of life, activities, characters, but without devoting a section to each one of these differences, and adding quite a few morphological remarks, whence the aspect of a catalogue of these books, in which are named, and sometimes described, at least in certain of their aspects, many fish and birds, some of them known, in ancient literature, only in the *History of Animals*, sometimes difficult or impossible to identify. It therefore is not a matter of a spineless exposition, but it can include some disorder in presentation, to the point of the extrapolation of some passages and the repetition of some facts. Decidedly, the form of the *History of Animals*, and especially the last books, argues strongly in favor of a work that is not a "book" in the modern sense of the word, but a collection that has undergone additions and amendments, and doubtless composed over a long period of time. Which brings us to the second debate, that of chronology.

One can hardly deny that the *History of Animals* is logically prior to the *Parts of Animals* since Aristotle himself said so expressly. But it is

23. We will again have occasion to notice a regular practice in Aristotle to chiasmic exposition: he says that he will deal with subjects A and B, and begins with B. Here, the parts are last on the programmatic list, and the first to be examined. Ignoring this chiasmic arrangement has led to notable errors about certain texts.

still necessary to determine what sort of priority: priority "by nature" or "per se," or priority "for us." One can certainly not conclude that there is a chronological priority, as many commentators have done.

An Impossible Chronology

Chronology is one of the sins of Aristotelians, never really abandoned, though it ought to have been, long ago. The fundamental cause of its uncertainty can be found in the very status of Aristotle's texts; since the great majority of them are not from Aristotle's hand, they cannot be ordered chronologically by stylistic criteria. There remain doctrinal criteria, and the insurmountable difficulties that they raise. In the first place, there is the question of locating in time the whole of the biological corpus. But that cannot be accomplished without having a global image of Aristotle's intellectual development: was he, under the influence of his medical family, first attracted to natural science, before becoming a metaphysician? Was he at first a Platonist before he turned toward more "positive" studies, of which zoology would be part? Did he forcefully reject Platonism at first, before coming back to it, guiltily, after the death of the master? Did he carry on his studies of animals during his whole life, along or with the assistance of disciples and studies, or at first alone, and then in a group? The human mind ought not to pose itself questions it cannot, and cannot ever, answer, but people cannot stop themselves from doing that.

Perhaps the only really documented attempt to assign a temporal location to Aristotle's zoological research remains that of Henry Desmond Pritchard Lee who, in a short article published in 1948,[24] believed that he could affirm that many of the fish cited in the *History of Animals* are found around the island of Lesbos and especially in the channel of Pyrrha that separates Lesbos from the coast of Asia Minor. Lee concluded from that, as had D'Arcy Thompson before him in a "prefatory note" to his translation of the *History of Animals*, mentioned earlier, that Aristotle's studies of fish, found in the *History of Animals*, had been carried out during Aristotle's stay in that region around 345 BCE, which goes against the conclusions of the famous work by Werner Jaeger, whose central thesis was that Aristotle became progressively distanced from Platonism, a thesis that, at a certain time, gained a nearly complete unanimity among the

24. Lee, "Place Names and the Date of Aristotle's Biological Works."

chronological interpreters.[25] Lee was roundly criticized in an article by Friedrich Solmsen.[26] It would have been impossible, Solmsen notes, that between the death of Plato (347) and the moment when he was called to be the tutor of Alexander (343), Aristotle could have carried out such observations on fish, some of them extending over several years, like those concerning their alimentary habits or their reproductive comportment. Doubtless Solmsen was right in thinking that Aristotle could not himself have observed the facts that he reports about fish, but that does not destroy Lee's hypothesis, because even if Aristotle did not make all these observations himself, he could have interviewed fishermen, as he says he did, which would allow Lee's chronological hypothesis to remain intact. Lee responded to Solmsen,[27] but we will leave them to their debate, not without mentioning the following point: Solmsen remarks[28] that Aristotle had close ties to Lesbos, since at least four of his disciples, including his successor as head of the Lyceum, Theophrastus, were originally from that island. Thus, Aristotle could have had access to information about the fish that live there even if he didn't live there any longer. So even if one accepts his analysis of a few passages (in fact, just one seems significant) on which Lee rests his argument, there is not enough there to overturn Jaeger's thesis; I will say more about that later, in detail. We will never know, minus some astounding discovery, when Aristotle began his work in biology and when he stopped, if in fact he stopped during his lifetime. The composite, redundant, sometimes contradictory character of the *History of Animals* argues rather, one may say, for a composition that stretches over a very long time.

But what we are especially interested in is the question of the relative chronology of the zoological treatises. The great majority of interpreters, as the literature shows, does not hesitate to proceed from the logical priority given by Aristotle himself to the *History of Animals* to a chronological priority. Some have tried to give a more scientific basis

25. Jaeger, *Aristotle: Fundamentals of the History of His Development*. This is the second edition of the English translation, which must be used because it was revised by Jaeger. The original work, *Aristoteles, Grundlegung einer Geschichte seiner Entwicklung* appeared in Berlin in 1923, and could not have been known by D'Arcy Thompson.

26. Solmsen, "The Fishes of Lesbos and Their Alleged Significance for the Development of Aristotle."

27. Lee, "The Fishes of Lesbos Again."

28. Solmsen, "The Fishes of Lesbos," 471.

to their position by invoking internal references. According to Hermann Bonitz's count,[29] the *History of Animals* is cited twenty-eight times with a title including the word *historia* and several other times in the form "we have said previously," or "elsewhere," or "elsewhere previously." Eighteen of these citations refer to identifiable passages in books 1 to 6, and it is possible, but not certain, that others refer to books 7 to 9. Typically, it is said that all these citations or allusions are made in the past tense, as if *History of Animals* were an already available text. And it is also true that the *History of Animals* does not refer explicitly to any other zoological treatise. Still, there are passages in the *Parts of Animals* that seem to refer to the *History of Animals* as a work yet to be written. Thus at the end of *Parts of Animals* 2.14, Aristotle says that he allowed himself to be led to the examination of hair on the head by that of eyelashes, "because of the connection between the two," and that he must defer dealing with them until the "proper occasion," which apparently refers to *History of Animals* 2.1, 498b11ff. Reciprocally, the *History of Animals* 2.1, 500b16, alluding to the differences between ways of urinating by male animals, says that "males thus differ from each other, as we have said," which could very likely refer to *Parts of Animals* 4.10, 689a33.

In opposition to the dominant position, David Balme, the best authority on the *History of Animals* of the second half of the twentieth century, fought in favor of a late date for the *History of Animals*, making it out to be the last-composed biological treatise. The first objection one may offer to Balme's position, even before considering his arguments, is that a thesis like that is more or less forced to consider the various zoological treatises, especially the *History of Animals*, as a "work" in the modern sense of the term, or something approaching that, which is far from being more likely than the opposite hypothesis of a collection gathered over a long period of time, as we have seen. This last hypothesis generally disqualifies the question of a relative chronology of the treatises. The second objection that one may make to Balme is aimed at one of his arguments: he figures that the *History of Animals* is later because among the differences that it reveals in comparison with the other treatises, it adopts the "truer" versions, for example, while the *Parts of Animals* claims that the squid has no brain, the *History of Animals* says that it does, which modern biology has recognized as true. This continuist position, according to which science discovers more and more truth, is precisely unsustainable

29. Bonitz, *Index Aristotelicus*, 103a43–55.

in the case of ancient observational science. Balme's last argument is just as weak as the others, but it has interesting consequences. It's a matter of noticing something that seems to be generally true, that the versions offered by the *History of Animals* of facts presented in the other treatises appear more condensed in *HA* than in the other treatises. It's a weak argument, because one may just as well turn it round the other way: why not think, in fact, that the *History of Animals* offers results that the treatises deal with more completely? In fact, the reversibility of Balme's arguments is remarkable. Thus, to recall an example noted elsewhere,[30] in *Parts of Animals* 4.2, 677a1ff., Aristotle reports that at Naxos sheep have a great deal of bile, while at Chalcis of Euboea they do not have any. In the *History of Animals*, he notes the same fact, but also tells us that at Naxos, foreigners who do not know that natural peculiarity are frightened by the quantity of bile when they offer sheep in sacrifice to the gods, thinking that they see in it a bad omen (1.17 496b25ff.). If the *History of Animals* is prior to the *Parts of Animals*, then Aristotle does not bother recalling that anecdote from a work already written, while supporters of Balme would argue that Aristotle learned this fact between the times of composition of the two treatises . . . But it's this argument that leads us toward the least bad solution of the problem of the relationships between the *History of Animals* and the other treatises, notably the *Parts of Animals*.

As for *History of Animals* book 10, Balme, contrary to most editors, insists on claiming that it is authentic, which leads to remarkable contortions: unable to deny the important differences from other books in the treatise as much in the form as in the content that this book exhibits, Balme decides that it is a matter of a youthful composition later added to the set. The *History of Animals* would thus be an assemblage consisting of late texts to which have been added an old book . . . Chronological thought thus begins to resemble magical-religious thought, able to explain everything and to justify everything.

We have seen that the goal assigned to the *History of Animals* was to indicate in what respects animals are "the same and different." If one follows the text of the treatise as we have it, one finds there a frantic search and apparently a rough draft of differences, whether that's in the first four books dedicated to parts of animals, the three next books concerning the genital parts and certain characteristics related to reproduction, but also

30. In the introduction to my French translation of the *History of Animals*: Aristotle, *Histoire des animaux*.

the last two books, dealing with the psychology and ethology of animals. Perhaps it is via this relationship between the same and other that the division of labor between the *History of Animals* and *Parts of Animals* can be best understood. We may see that by studying the parallel versions of passages that the two treatises dedicate to the same "part."

If we take the example of "bone," the passages dedicated to it in both treatises (3.7 in *History of Animals* and 2.9 in *Parts of Animals*) say generally the same thing, with the first difference that, although being of comparable length (60 Bekker lines in one, 61 in the other), the *History of Animals* reports more facts than the *Parts of Animals*, which implies that the later treatise says more about the points that both touch on. The *History of Animals* begins, like the *Parts of Animals*, by saying that the bone system, like that of blood vessels, is continuous, and depends on a starting point, the vertebral column, with the expected difference that the *Parts of Animals* gives reasons for that fact: an isolated bone could not fulfill the principal function of the bone system, that of assuring bending and, in addition, it could wound the flesh where it would be found. In a general way, the *Parts of Animals* explains the function of bones by comparing them to the armature on which sculptors form their clay—they serve to support the body. But then the *History of Animals* wanders off into a sort of free association, since it goes from the vertebral column to the vertebrae, then the skull, to note that some animals have a skull made of a single bone, others of several, and men have three sutures on the skull, women only one, a "fact" already noted previously (1.7, 491b2), and also reported in the *Parts of Animals*, but in the study of the brain, since the explanation lies in the maintenance of thermal balance of the brain. The *History of Animals* notes, finally, that "a human head without sutures has been observed," a fact neither repeated nor explained in the *Parts of Animals*. It's easy to understand why: it would be difficult to give a cause for a fact like that without engaging in a long digression. Continuing the same sort of "drift," the *History of Animals* goes on to the jawbone, repeating in passing that "only the river crocodile moves its upper jaw" (516a24), an error well known to readers of Aristotle, then to the teeth, which are bones that are simultaneously full and hollow, and which are the only bones impossible to carve. For the *Parts of Animals*, the jaw needs to be studied elsewhere, not with the bones, because this treatise considers parts from the viewpoint of their function.

The rest of the two texts shows that the two passages have drawn from the same source, unless one was the source of the other. The same facts are invoked: large animals need larger bones, some bones have marrow and others seem not to have marrow, males and carnivores have harder bones, to such an extent that the bones of lions strike fire when one rubs them against each other, that fish and snakes have spiny bones,[31] but the dolphin has bones and not fish-bones. The *Parts of Animals* adds several points: large animals are notably found in Libya, large snakes have large bones because the flexible bone of small snakes would not be able to support their body, viviparous animals need bones that are more solid than oviparous (one imagines why, but Aristotle doesn't say). As for a fact noted in both texts, that there is no bone in the belly, the *Parts of Animals* provides an explanation (they would impede expansion due to the absorption of food, and pregnancy), and the *History of Animals* does not.

Thus, we see that the *Parts of Animals* has a double program, of which there is no trace in the *History of Animals*: discover the cause of the properties of bones, and show that according to a certain number of parameters (size, mode of life, and reproduction) bones vary in size and solidity, to the point that they become spinous in animals in which that sort of support is sufficient. For the *Parts of Animals*, teeth have, after all, the nature of bone, but they are not bones because they do not have the function of bones. Thus, we see that in the *History of Animals* differences lead to each other, including differences that are not those that Aristotle usually lists for the parts (presence or absence, size, shape, disposition); as for example the possibility of being carved. Besides, bone not being defined by its function, its field extends to include the teeth. In the *Parts of Animals*, on the contrary, differences are referred to the functional identity, of which Foucault would say that that is "more serious" than the others.

One understands from that why the impression of disorder that the *History of Animals* sometimes leaves is baseless. Thus, in *History of Animals* 2.1, Aristotle deals with hair, a topic to which we will soon return. After having described the hair of many quadrupeds, he turns to talk about camels, or particularly female camels, since this species is referred to in the feminine,[32] and has numerous characteristics worth noting: humps, knees, nipples, genital organs, ankle bone, hooves. As the

31. Translator's note: Greek and French both make a verbal distinction between the mammal bone (ὀστέον, os) and fish bone (ἄκανθα, arête). In English, it's all bone.

32. Αἱ δὲ κάμελοι. A fact heretofore left unexplained by commentators; it also applies to the bear, ἡ ἄρκτος. But after all, in French a swallow is *une hirondelle* and a carp is *une carpe*.

History of Animals, at least in the first six books, is not a natural history in the manner of Buffon, dealing with each animal, one by one, and in any case, it is necessary sooner or later to notice the characteristics of this remarkable animal, there is no reason not to take the opportunity of the inspection of quadrupeds to talk about the camel. That would have been more difficult in the *Parts of Animals*: in that treatise, it does happen that an animal is examined, sometimes at length, apart from others, but it is almost always on the occasion of examining one function. Thus, there is a long discussion of the elephant's trunk, but relative to the examination of the role and mechanism of the sense of smell.

The examination of hair also demonstrates the differences between the two treatises. In the *History of Animals*, Aristotle enumerates the different kinds of pelt that different animals possess: hair over the entire body, fur on the back, mane, tufts, among others; the text also mentions next that the antelope has a beard on its larynx, then takes the occasion to specify that the antelope and the *pardion* (an animal impossible to identify) have split horns and hoofs, that the female antelope does not have horns, that the antelopes live in Arachosia (Punjab), like the water buffalo, which leads to a brief discussion about buffaloes, which are to cattle what boars[33] are to pigs, then about their horns, to come back to those of the antelopes that are "pretty much similar to those of the goat" (2.1, 499a9). Next Aristotle comes back to pelt to note that the elephant is the least hairy of quadrupeds. Then comes the camel.

The *Parts of Animals* deals with hair in book 2, chapter 14, with chapter 15 dedicated to eyebrows. The study begins with eyelashes, peculiar to human beings (at least the fact of having lashes on both upper and lower eyelids), all that being referred to a general rule: hair serves to protect, and thus it is reasonable that the animal that stands upright would not have it distributed in the same place as quadrupeds. Because it's a matter of protecting the parts "that have the most value" (τοῖς τιμιωτέροις, 2.14, 658a22), but quadrupeds already have the belly protected from the fact that it is to be found under the animal. For human beings,

33. To be precise, boars did not exist for the ancient Greeks; they spoke of "wild pigs." Because for something to exist, it is not enough that the world offers us empirical examples, it is also necessary that there is a concept that defines that thing. The Romans, on the other hand, distinguished, as do we, the boar (*aper*) from the pig (*porcus*). Thus, they doubtless distinguished, as do we, wild pigs and boars. [Translator's note: The wild pig is *sus scrofa*, the tame pig is *sus scrofa domestica*. Other species of the genus *Sus* would (probably) have been unknown to the Greeks and Romans.]

it is notably necessary to protect the head (they alone have head hair) and the genital parts. And Aristotle notes, in a list shorter than that in the *History of Animals*, various cases: "Some are covered with hair over the whole dorsal surface, as is the case for instance in dogs, or, sometimes, it forms a mane, as in horses and the like, or as in the male lion, where the mane is still more ample" (658a28, Thompson with minor change). It is remarkable that Aristotle simply notes these specific cases and is satisfied with subsuming them under a general law, that hair is a protection. He does not attempt any finalist explanation of the lion's mane, for example. But he could doubtless have found reasons for these phenomena; for this exact animal, nature acted for the best in giving a mane to male animals, as he emphasizes she has done in the case of human beings (658a23). And all the more, since at 658a32, Aristotle calls on a finalist principle to explain the diversity of tufts of hair on tails: animals with a short tail have long hair, and those with a long tail have short hair, "because nature always gives to one part what she has taken away from another" (658a35).

Thus, everything happens as if Aristotle was not looking to adapt to each particular case the explanation he gives for hair. There could be several reasons for that: maybe he doesn't feel the theoretical force; maybe the lion's mane doesn't have a final cause but depends solely on material causes, which puts this phenomenon outside the task of the *Parts of Animals*, which is a teleological treatise; maybe he didn't have time to pursues the question adequately, which could be indicated by the last sentence of chapter 14: "These, however, are matters which by their close connection with eyelashes have led us to digress from our real topic, namely the cause to which these lashes owe their existence. We must therefore defer any further remarks we may have to make on these matters till the proper occasion arises" (2.14, 658b10, Thompson). We may thus conclude that if the program announced at the beginning of the *History of Animals* consisted of determining in what respects animals are "different and the same," it is also, to a great extent, the program of the *Parts of Animals*, the *History of Animals* puts the accent on diversity, while the *Parts of Animals* concentrates on determining the functional unity of that diversity. Thus, from that point of view we meet up with a paradoxical situation: while the *History of Animals* is Aristotle's zoological treatise with which Cuvier felt the most affinity, it's in the *Parts of Animals*, more than in the *History of Animals*, that this properly *biological* procedure, consisting of relating an organic diversity to the unity of function, is deployed.

There is an important aspect of the exposition proposed in the two treatises that can allow us to explicate the relationships a bit more. As we noted earlier, the Aristotelian text is peppered with correlations that are expressed in the form of premises in all the possible forms of predication: A (all blooded animals have a liver), E (no oviparous animal has mammaries), I (some blooded animals have a lung), O (some blooded animals do not have feathers). Some are reciprocal (all blooded animals have a heart, and all those which have a heart are blooded), others are not ("the animals with hair are absolutely all viviparous," *History of Animals* 1.6, 490b27), but "not all viviparous animals have hair" (b25). Most of these correlations are explained in the *Parts of Animals*. Thus, this correlation, is remarkable for the number of factors put in play: "no animal has simultaneously projecting teeth and horns, and none of those that have sharp teeth have projecting teeth or horns" (*History of Animals* 2.1, 501a19), which is explained in *Parts of Animals* 3.1, 661b23, by the fact that nature wants to avoid redundance in the means of defense that she gives each animal that needs them.

Most, but not all. There are, in the first place, correlations that are not formally explained, but for which Aristotle provides the means for an explanation. Thus, we read in *History of Animals* 1.2, 489a3: "Now the residue of food is twofold, those animals that have parts receptive of fluid residue also have parts that receive the dry residue; but those that have the latter do not always have the former. That is why animals with a bladder also have an intestine, while those with an intestine do not all have a bladder."

The expulsion of solid residue is tied to the most fundamental function of a living thing, digestion, while plenty of animals do not produce urine. In fact,

> not all animals have a bladder, but it seems that nature has intended to give it only to those whose lung contains blood, which is reasonable. For the superabundance in their lung of its nature[34] causes them to be the thirstiest of animals, and makes them require a more than ordinary quantity not merely of solid but also of liquid nutriment. This increased consumption necessarily entails the production of an increased quantity of residue; which thus becomes too abundant to be concocted

34. Some manuscripts, followed by William of Moerbeke and Thurot, read "heat" instead of "nature," which would doubtless be a gloss: it is because they are blooded that the lungs have a hot nature, which causes the animal to be thirsty, provoking an intake of liquid, making the bladder "reasonable."

> by the stomach and excreted with its own residual matter. The residual fluid must therefore of necessity have a receptacle of its own. (*Parts of Animals* 3.8, 670b32)

The presence of a bladder is not *subordinated* to that of an intestine, even if the latter performs a more fundamental function; it is rather subordinated to the presence of a bloody lung.

There are also correlations that are explained, or at least explicable, but which involve exceptions that perhaps are not. Thus, it seems explicable from an Aristotelian point of view that males should be *bolder* than females. Since the male is hotter than the female, one might imagine that its blood would also be hotter, and nature has sufficient foresight to attribute to the male the role of protection. But why, in bears and panthers, would the females be more courageous than the males (*History of Animals* 9.1, 608a33)? Aristotle is careful not to mention a fact of this kind in the *Parts of Animals*, devoted as it is to teleological explanation. As I noted in the introduction to my French translation of the *History of Animals*, these exceptions are fairly numerous.

There are, in addition, correlations that remain unexplained. Thus, the *History of Animals* asserts that animals with many teeth live a long time (2.3, 501b22), though Aristotle never explains this. Similarly, he mentions in both the *History of Animals* and in the *Generation of Animals* that what we call ruminants have cotyledons in the uterus, the *Generation of Animals* extending this property to animals that have several vessels, and not just one, terminating in the uterus (2.7, 745b30). These cotyledons, the *Generation of Animals* explains, are reservoirs of blood, just as the breasts are reservoirs of milk, nourishing the embryo, such that they diminish proportionately as the embryo grows. The *History of Animals* passage is worth quoting at length:

> There is also a difference that distinguishes uteruses from each other. In fact, the females of horned animals that do not have teeth in both jaws are furnished with cotyledons in the uterus when they are pregnant, and that is also the case among animals with teeth in both jaws, the hare, the mouse, and the bat; while all other animals that have teeth in both jaws, and are viviparous and furnished with feet, have the uterus quite smooth, and in their case the attachment of the embryo is to the uterus itself and not to any cotyledon. (3.1, 511a27)

The fact of having a smooth uterus is not rationally related to any other characteristic, or, in any case, Aristotle's text does not relate any other characteristic, since it concerns both ruminants and animals that are not ruminants. Thus, it is a pure "fact." Because, besides correlations that remain unexplained, the *History of Animals* presents many *facts* similarly unexplained; it is unknown whether they will be explained eventually, or remain unexplained forever. Why is the viper the only viviparous serpent? Why is it only the crocodile that moves its upper jaw? Why do all serpents have "an excessive taste for wine" (8.4, 594a10)? We will have more to say about these kinds of facts when we discuss the diversity of animals, in a later chapter.

Finally, we should note that, although these correlations are found especially in the moriological books of the *History of Animals*, there are also some in book 8, about the habits of animals. Thus, in 8.2, 591a17: "All fishes, except the mullet, eat their own species; the conger is especially ravenous in this respect," or at 8.3, 593b25: "Birds of prey feed on any animal or bird that they may catch, but they never touch one of their own genus, whereas fishes often devour members their own species." This long comparison between the *History of Animals* and the *Parts of Animals* will allow us, still thanks to a confrontation between Aristotle and Cuvier, to establish on a larger foundation the thesis that there is a true Aristotelian biology. According to what we have seen, particularly when we relied on Michel Foucault, the *Parts of Animals*, but also on the question of generation, the *Generation of Animals*, seem to be closer to Cuvier's biological project than the *History of Animals* in that they detect more neatly the unity of function beneath the diversity of organs, formulating, and ultimately rather neatly, the two great laws that lie at the foundation of Cuvier's biology, particularly his comparative anatomy, the law of organic correlation, and the law of the subordination of characters.

It's still Cuvier who, from a different angle, often neglected, of his theoretical construction, will allow us to give considerable scientific weight to Aristotle's *History of Animals*, if everything I have said is correct. Cuvier often insists on the necessity of observation in the natural sciences, precisely as he sees in the practice of observation a distinctive trait of Aristotle's philosophy. In the "Preliminary Discourse" of his research on the fossil bones of quadrupeds, published in 1812,[35] we read:

35. This preliminary discourse was republished in 1824, with some modifications, under the title, *Discourse on the Revolutions of the Surface of the Globe, and on the Changes That They Have Produced on the Animal Kingdom.*

> One has a general idea of the necessity of a more complex digestive system in species where the dental arrangement is less perfect; thus, one might say that certain ought to more likely be ruminant animals, if they lack one or another sort of teeth; one might deduce from that a certain sort of esophagus, and corresponding forms of the neck vertebrae, and so forth. But I doubt whether one would have imagined, had it not been learned from observation, that ruminants all have cloven hooves.[36]

Obviously, I have chosen this example because it includes a well-known Aristotelian reference to the *Parts of Animals*, as we will see in a later chapter. Five lines farther along, Cuvier gives us, whether he knows it or not, one of the best possible introductions to the theoretical project of the *History of Animals* as he simultaneously provides the best analysis of its relation to the *Parts of Animals*:

> However, since these relationships are constant, there must be a sufficient cause; but as we don't know what it is, observation must make up for the lack of theory; observation establishes empirical laws that become almost as certain as rational laws, when they rests on observations sufficiently repeated, to such an extent that today if someone sees only the footprint of a cloven hoof, one may conclude from that that the animal that left that print was a ruminant, and that conclusion is absolutely as certain as any other in physical or moral philosophy.[37]

And, in fact, among the facts cited and the correlations put forward in the *History of Animals* one discerns a large collection of observations that have resulted in the situation described by Cuvier: it is possible to represent the animal kingdom as a set of organisms obeying the law of organic correlation. Some of these correlations are explicable by what Cuvier calls "theory," which corresponds to Aristotle's final cause, and some are not, or aren't yet. One may doubtless consider that it is this explanation that allows applying to these organisms the law of subordination of characters. In fact, it is because they have a defective dentition that ruminants have need of several stomachs, and not because they have several stomachs that

36. Cuvier, *Recherches sur les ossements fossiles de quadrupèdes*, 101.

37. Cuvier, *Recherches sur les ossements fossiles de quadrupèdes*, 102.

they have an incomplete dentition, which will allow us to construct a scientific syllogism in which the middle term is the cause of the conclusion.[38]

In the introduction to his edition of the *History of Animals* in the Budé series, Pierre Louis offers an opinion that is simultaneously false, contradictory, and pregnant, when he says that the *History of Animals* is one of those "writings in which there seem to be no concern for explaining the phenomena and searching for their causes, but are only preoccupied with classifying them. Such are the collections of facts, gathering remarks and observations, intended to furnish the material for didactic treatises."[39] False, and we have seen why,[40] and contradictory, because classifying givens is an extremely theoretical procedure. Furthermore, if it's the *Parts of Animals* that Louis calls a "didactic treatise," one wonders what he means by "didactic." But Louis's remark can lead us toward a fundamental characteristic of the *History of Animals* that harmonizes with the opinions of Cuvier that we have just now mentioned. With its correlations and its unexplained facts, it makes observation *surpass* explanation. That's a matter of a *scientific* trait, one that prevents Aristotle's biology from being "metaphysical," in the Comtean sense of the word, as we will see in the next chapter. The *Parts of Animals*, in contrast, always has an explanation, most often finalist, available to explain observed facts, and (a fundamental point) this treatise passes over in silence whatever escapes explanation. Thus, the *Parts of Animals* explains that the differences that exist between their parts, notably the blood, bring it about that females are less prudent and less courageous than males (2.2, 648a12), simply not mentioning the cases of bears and panthers. It was doubtless this characteristic of the *History of Animals* that caught Cuvier's attention, and which is perhaps one of the reasons for his special admiration for this treatise, bringing Aristotle closer to the biologist Cuvier.

In fact, a true biology, that is, according to Georges Canguilhem, a science of living things that is not reduced to an application to living

38. As in the famous example at *Posterior Analytics* 1.13, 78a25ff.: planets do not glitter because they are close, not because they do not glitter, they are close.

39. Aristotle, *Histoire des animaux* (Tricot), xi–xii [trans. A.P.].

40. In the introduction to my translation of the *History of Animals*, I offered the example of eyelashes to show that this treatise gives causal explanations that do not appear elsewhere: if the *Generation of Animals* does indeed explain why hair grows, it's actually in the *History of Animals* that Aristotle explains why eyelashes form where they do. Aristotle, *Histoire des animaux*, 49 (Pellegrin).

things of physiochemical laws, cannot be an entirely deductive science. To conclude that the footprint of a cloven hoof belongs to a ruminant it is necessary to rely on what Cuvier calls an "empirical law," one that is established by observation. It's the same, in Aristotle, for the empirical law that animals with many teeth live long lives. That's an important aspect of Cuvier's biology, one that Foucault has missed, and it has an Aristotelian duplicate.

Hence the division of labor posited by Aristotle between the *History of Animals* and the *Parts of Animals* takes on a new dimension. If it's a matter of finding in the *History of Animals* facts and correlations that will be explained in the *Parts* and *Generation of Animals*, the *History of Animals* is largely useless, since the other treatises also present the facts and correlations in question. It's the excess, what I have described as the *surpassing* of the *History of Animals* in relation to the other treatises that gives it perhaps its main epistemological interest. We could consider many examples to show that; we will limit ourselves to two. If among serpents of comparable size living in an identical environment the viper is the only one to be viviparous, it is difficult to imagine a final cause for that fact. It is, on the other hand, impossible to imagine that Aristotle would attribute that fact to chance, since it is of a regularity that makes it a natural characteristic, and it is necessary that this characteristic would have, as Cuvier says, a "sufficient cause." It remains that this cause, in Aristotelian language, must be sought among material and efficient causes. We know that the extreme chromatic diversity among birds has a crucial role in their reproduction in that the colors that they exhibit simultaneously indicate their presence to birds of the same species, and their sexual availability, and this significant system does not only concern the birds. Ethology was not sufficiently advanced in his day so that Aristotle could be conscious of that; consequently, he could not attribute a final cause to this diversity of colors, something that modern ornithologists do easily. Aristotle thus counts the color of birds among the brute facts without final cause, and that's why he talks about it in the *History of Animals* and nowhere else.

We will see, in our fourth chapter, that the flourishing diversity that is the proper object of the *History of Animals* does not have (either) a final cause, something that separates Aristotle from providentialist teleologists for whom every fact, every characteristic, is what it is for the sake of some good. Jacques-Henri Bernardin de Saint-Pierre, who pushed that sort of finalism to ridiculous extremes, thus asserted that carnivores attack their prey necessary for their survival at night, while they are asleep, to make their death gentler. We shall return to this example. It will be an important point of my present task to try to show that animal diversity is not a consequence of an intention of nature.

The Aristotelian zoological corpus, in which the *History of Animals* thus finds an eminent *scientific* place, proposes a model of zoology that would not be subverted except by the theory of evolution and, to a certain degree, by modern genetics. Starting from a homologous model, twenty-three centuries later, Cuvier constructed the monumental picture of the animal world that we know. That permanence over the centuries, even over millennia, did not escape the notice of nineteenth-century Aristotelians. Thus Jules Barthélemy-Saint-Hilaire (1805–1895), translator of the whole of the Aristotelian corpus, published in 1883 the three volumes of his translation of the *History of Animals*, a remarkable and still useful work. He introduced the work with a preface of 190 pages, followed by a 33-page dissertation "On the Authenticity and Composition of the *History of Animals*." Barthélemy-Saint-Hilaire was out of step with the biology of his time: he did not know Mendel and his laws, but in that he was not alone, or Darwin, whose theories he ridiculed. For him, zoology ended at Cuvier, for whom he avows a limitless admiration, almost as great as he expresses for Aristotle. Like Cuvier, he places a particular importance on the *History of Animals*, and he defines its place in the history of zoology. For him, it was obvious that Aristotle was a *precursor* of Cuvier, and "at over two thousand years distance, these two geniuses understand each other, and the second pursues and extends the work of the first" (CX).

Barthélemy-Saint-Hilaire was definitely not (paleo)-Bachelardian: "Thus, whether for his style, or his method, or for the order that zoology needs to use in its descriptions, or for the scale of beings and for the unity of composition, that is, in general and specific questions, we can believe that Aristotle belongs to our time; he was the first to discover and to discuss the problems which still separate the scholars of our century" (CLI). But he was not continuist either, since for him, "before Aristotle, Greek philosophy, despite its marvelous activity and extremely ingenious curiosity, could not found anything scientific in zoology; after Aristotle, the human spirit was too weak to follow him; it was only in this last century that the science given birth by him could be reborn and developed" (CX). Barthélemy-Saint-Hilaire is not the only one, including among his contemporaries, to be surprised at the lack of posterity of Aristotelian zoology, while mathematics and astronomy progressed remarkably. Obviously, the question arises of the availability of Aristotle's zoological works in antiquity, and there have been erudite studies of the question, and very ingenious answers. The citations and allusions of later authors, Aristophanes of Byzantium, Pliny the Elder, Athenaeus of Naucratis, Plutarch, Aelian, are sufficient to show us that texts on animals, attributed to Aristotle, were circulating. Which texts these authors were looking at is a question not without interest for us; some of their statements leave us stunned, like this one of Pliny: "After having interrogated all these people (hunters, herdsmen, etc.), Aristotle wrote his famous works on animals, about fifty of them, which I have summarized, adding whatever he left out" (*Natural History* 8, 17, 44).

Sticking to the *History of Animals*, one of the most adventurous interpreters, Ingemar Düring, thought that he could establish that this title could refer to at least three texts. First, to "our" *History of Animals*, in nine or ten books. But, and Düring is not the only one to think this, this edition was preceded (or succeeded, the argument of Düring is not solid on this point) by another in six books, the one that Athenaeus cites under the title "Parts of Animals," these six books dealing, in fact, with the parts of different animals, with uncertainty concerning book 7.[41] Düring then imagined the following scenario: the edition used by Athenaeus would have been pre-Andronican, and it would be Andronicus of Rhodes who would have added the psycho-ethological books,[42] our book 7 still remaining "up in the air." But Athenaeus also cites passages that can be referred to book 9, under the title "The Character of Animals" and "The Character and Way of Life of Animals." On the other hand, calling our *History of Animals* the "Parts of Animals" is perhaps an example of practice, usual in antiquity, of designating a work by its first words, which are, for the *History of Animals*, "among the parts that comprise animals." Similarly, John Keaney cites a

41. It must be noted that all the manuscripts reject book 7 to the end of the work, and it was Theodore of Gaza, in his Latin translation of 1458 (published in 1476), who introduced the order generally accepted ever since, except by Balme.

42. Thanks to Strabo, we are aware of a story, bizarre but doubtless partially true, of the odyssey of Aristotle's manuscripts hidden in a basement, then finally arriving in Rome after Sulla's military conquest in 86 BCE, when they were edited following serious revisions by the last Scholarch of the Lyceum, Andronicus of Rhodes.
For the Andronicus edition one must now consult the article by Myrto Hatzimichali. A notable result of this research is that this edition was far from authoritative in the time of Athenaeus. Hatzimichali, "Andronicos of Rhodes and the Construction of the Aristotelian Corpus."

papyrus from the beginning of the third century CE locating a passage from *History of Animals* book 9 "in the parts of animals."[43] Finally, we find in Athenaeus a large group of citations referred to Aristotle concerning animals, but we cannot locate them in the zoological works as we have them. Thus, the hypothesis of Düring and others that there existed a third collection of zoological data, possibly compiled very early. As for saying, as some have done, that it was done by Clearchus of Soles, a student of Aristotle, or by Aristophanes of Byzantium, born in the middle of the third century BCE, whose *Epitome of the History of Animals*[44] has come down to us, is more than uncertain. The existence of such a collection is all but improbable, and it is even likely that several works of that kind existed, given how great was the authority of Aristotle. According to David Balme, Aristophanes of Byzantium, Pliny the Elder, Plutarch, and Aelian would have found their citations in this kind of collection, and Balme thinks that Athenaeus got all his citations in such a collection, without knowing that his source borrowed from different parts of the *History of Animals*, for Athenaeus was far from being an expert in zoology.[45] That shows us at least two things. First, that the *History of Animals* was not a collection of facts as were the works of Pliny or Aelian,[46] second, that Pliny, Aelian, and the other later "zoologists" were simply not interested in the theoretical or methodological considerations to be found in Aristotle's zoological treatises to which they doubtless had access.

That so to speak no one, except Theophrastus and his botanical treatises, obviously following Aristotle, attempted to compose zoological

43. Keaney, "Two Notes on the Tradition of Aristotle's Writings."

44. A very precious work for the study of ancient zoology; the text is available in the supplements of the famous collection of *Commentaria in Aristotelem Graeca* of the Berlin Academy. This work, in addition to the Aristotelian materials, using for them a clearly Aristotelian vocabulary, gathers in an order that looks more like free association than anything systematic, facts and anecdotes about many species of animal. Aristophanes's text has come down to us incomplete; it consisted of four books, and we have fragments only of the first two (and much more of the second than of the first), compiled by order of the Emperor Constantine VII Porphyrogenites, in the tenth century.

45. Aristotle, *Aristotle: History of Animals*, 5 (Balme).

46. I have tried to show that in the introduction to my translation of the *History of Animals*, in recalling notably how far Cuvier was from attributing such a status to the *History of Animals*, which he considered, notably, the first treatise of comparative anatomy. Cuvier, *Histoire des sciences naturelles*, 1, 159.

treatises comparable to those of the Stagirite, one may understand from the very fact of the existence of these treatises. Aristotle's method in the natural sciences certainly remained a set of theoretical tools to which one might resort, as was for example the case for some medical writers. A physician as important and innovative as Herophilus of Chalcedon used a method inspired by Aristotle,[47] and it was certainly from Aristotle that Galen borrowed the finalism of *De Usu Partium*. But as for properly biological research in the Aristotelian line—nothing. Possibly we may say a bit more about that later, in the conclusion of this work.

Jules Barthélemy-Saint-Hilaire thus was right in thinking that Aristotle's zoology remained without successor, like a torch waiting more than two thousand years to be taken up and rekindled by Cuvier. Except that, however, he thought that that was the final state of biology. For him, Aristotle clearly outlined the contours of a biology that would attain its culmination with Cuvier. It's a matter of an approach to living things that one may call structural and functional. By way of a criticism, cursory and in fact ridiculous, of what he calls "transformism," a doctrine that he attributes to "M. Darwin," Barthélemy-Saint-Hilaire rediscovers, by way of Cuvier, the fundamental positions of Aristotle's zoology. For Barthélemy-Saint-Hilaire, transformism is based primarily on the question of the origin of beings, a question which belongs, he says, to cosmogony, while "whatever idea one constructs about the origin of things, zoology need not say anything about these impenetrable obscurities, lost in the night of bygone centuries" (141). According to Barthélemy-Saint-Hilaire, transformism "above all rests on" "embryogeny and paleontology," and these "lead us back to the puerile theories of Empedocles, roundly defeated by Aristotle" (140). To provide a foundation for their theory, the transformists were forced to attribute to species a faculty of variation that they do not have. That, even if Barthélemy-Saint-Hilaire is not totally aware of it, is a reprise of the Aristotelian critique of the Presocratics in the name of the primacy of substance over genesis, a critique that we will find again when we discuss teleology.

That is something that Cuvier expresses in several ways, notably like this: "Every organic entity forms a set, a closed and unique system, in which the parts mutually correspond with each other, and work together in a single action by a reciprocal reaction. No part can change unless the

47. See Pellegrin, "Ancient Medicine and Its Contribution to the Philosophical Tradition," 670.

others also change."[48] We know that Cuvier, faced with the evidence of fossils, preferred to propose the absurd theory of successive creations of different species, rather than to have some species to come from previous species. The development, as much of paleontology as of genetics, showed that Cuvier was wrong by transcending him, but not abolishing him. We will see, in the next chapter, that this refusal to look for the initial origin of living things, as well as that of conceiving the animal world as adaptable and transformable, is a fundamentally Aristotelian position. To make "embryogeny" the essential chapter of zoology is, on the contrary, a position that was that of the Presocratic philosophers, criticized by Aristotle for their "mechanism."

48. Cuvier, *Recherches sur les ossements fossiles de quadrupèdes*, 97. Idem.

Chapter 2

The New Horizon of Teleology

Until we define these terms more precisely, and identify the realities to which they refer, let us call "mechanistic" an explanation that accounts for a process and states that result from it by the necessary play of the properties that belong to that process or to those resultant states, which we will call physiochemical laws. On the other hand, a "finalist" or "teleological" explanation accounts for phenomena by indicating their end, finding that they were intended. In Aristotelian etiology, mechanistic explanations present material and efficient causes. Aristotle also talks about a "necessary" causality, because the effect follows necessarily from the material properties of the entity under consideration; it's a matter of an absolute necessity, one that may be called "Democritean,"[1] because Democritus is the premiere representative of this sort of explanation. In the Aristotelian etiological system, finalist explanations rely on formal, final, and also efficient causes. The efficient causes can, in fact, operate in two ways: either it affects the bodies to which it is applied by working on them by virtue of properties possessed by their matter or the matter around them (a hot organ heats the region around it), or it is directed toward an end and depends strictly on formal and final causality (in fertilization as Aristotle understands it the form of the male parent directing, in the midst of the female materials, the movements that give form to the embryo). On the other hand, the end that a finalist explanation identifies is conceived as a good, either real or apparent. Nobody denies that Aristotle was a finalist, but everything or almost everything in his

1. Cf. Cooper, "Hypothetical Necessity and Natural Teleology," 259.

finalism is debated by interpreters. Even before we consider the subtleties of Aristotelian teleology, we can consider three questions that are rather preliminary, and that will not be answered until later.

First, when Aristotle writes that "nature is a cause, and a cause in this sense: with a view to something, that is clear" (ὅτι μὲν οὖν αἰτία ἡ φύσις, καὶ οὕτως ὡς ἕνεκάτου, φανερόν, *Physics* 2.8, 199b32), does that extend to all realities that he considers as natural? So, we will have to take part in a famous hermeneutic quarrel, and decide whether rain falls in view of making the grain grow, or is it a simple automatic result of material and efficient causes that make water evaporate and then condenses it in the atmosphere, and it falls back down in the form of rain. Perhaps Aristotelian teleology goes beyond the limits of what Aristotle calls "nature" (as we will soon see), but there is in Aristotle a *natural* teleology, as the passage we have just cited, among others, affirms. But, since Aristotle defines natural beings as those "that possess in themselves a principle of motion and rest" (*Physics* 2.1, 192b14), and this definition applies most completely to living beings (besides, "animals and plants" are the primary examples of natural beings, cf. 192b9), finality is about the living world, and notably the animal world, in a manner that is both more obvious and more complete than for nonliving objects. In any case, natural phenomena like the eclipse are brought about by a necessary cause, and in view of nothing (cf. *Metaphysics* Eta 4, 1044b12). We will also see that not all the characters of all animals will be found to have a final cause. But the question of knowing the difference between natural beings and living beings will be dealt with in the next chapter.

In many passages, including several that we will cite later, Aristotle never stops asserting that final causes are not the only ones to be active in nature, and thus that anyone who studies nature must also take account of material causality. But that is a position taken by all the teleologists in antiquity, since none claimed to explain every natural event as being directly brought about by the simple will of a god or some other creative reality. Mechanism, on the other hand, claimed to occupy the explanatory terrain alone, and to totally do without teleology, considering it an illusion. Some, like Wolfgang Wieland,[2] have tried to attribute to Aristotle a teleology of this kind, which would have issued from *als ob* judgments, as Kant says: everything happens *as if* some intelligent being had arranged the eye in

2. Wieland, *Die aristotelische Physik*. The principal thesis of this very rich book is that Aristotelian concepts all derive from an elaboration of speech patterns.

such a way that it sees well enough for the one who has it. That's how Aristotle reports the position of mechanists (he cites Empedocles) in a famous passage from the *Physics* (2.8), examined later.

Next, it is difficult to be completely clear about the place of human beings in Aristotle's finalized nature; an entire chapter will be mainly dedicated to this problem. There is at least one passage that clearly seems to attribute to the Stagirite an anthropocentric conception of teleology. Among other commentators, David Sedley, in an often-cited article, proposed a reading of this kind of passage from the *Politics*:[3]

> Even at the moment of childbirth, some animals generate at the same time sufficient nutriment to last until the offspring can supply itself—for example all the animals that produce larvae or lay eggs.[4] And those that bear live young have nutriment within themselves for their offspring for a time, the substance called milk. Hence it is equally clear that we should also suppose that, after birth, plants exist for the sake of animals, and other animals for the sake of men—domesticated animals for both usefulness and food, and most if not all wild animals for food and other assistance, as a source of clothing and other utilities. If, then, nature makes nothing incomplete or pointless, it is necessary that nature has made them all[5] for the sake of men. (*Politics* 1.8, 1256b10–22)

A difficult passage, because by aligning the relationship of human beings to animals (and to plants) on the fact that nature furnishes the means of nutrition to the offspring of various animals, Aristotle would seem to say that one may find the same sort of teleological relationship in both cases.

---

3. Sedley, "Is Aristotle's Teleology Anthropocentric?"

4. "The difference between egg and larva is this: an egg is that from a part of which the young comes into being, the rest being nutriment for it; but the whole of a larva is developed into the whole of the young animal" (*Generation of Animals* 2.1, 732a29, Platt, with minor changes). The larva is "like an imperfect egg" (3.9, 758b20) and *larvipara* are less perfect than *ovipara*.

5. "Them all" (1256b22) can refer to three things: "plants and animals," "animal," "wild animals." The strict logic of the passage militates in favor of the second option, but one may think that the fact that plants are "for the animals" does not prevent them from also being "for humans."

We will see that an anthropocentrism as naïve as that does not provide the foundation of Aristotelian finalism.

Finally, there is a historic question of first importance. Aristotle opposes the mechanism of his predecessors, sometimes in a nuanced manner, but sometimes in a very summary way, as at the beginning of the same *Physics* 2.8: "For all writers ascribe things to this cause [necessity], arguing that since the hot and the cold and the like are of such and such a kind, therefore certain things *necessarily* are and come to be—and if they mention any other cause (one friendship and strife, another mind), they barely touch on it, and then say goodbye" (198b12). But then what to do with Plato, for example? In a well-known passage of the *Parts of Animals*, Aristotle notes that "in the time of Socrates . . . inquiry into nature ended, and philosophers turned toward useful virtue, that is, politics" (1.1, 642a29). That suggests that Aristotle, unless this passage was written before the *Timaeus*, which is far from impossible, considered the Platonic attempt to think about nature as at best a philosophical nonevent. And in fact, one gets the impression, and absent explicit texts this remains an impression, that it is an heir to the Presocratics that Aristotle represents himself, thinks himself, if not exclusively then at least principally, as a *physicist* in the ancient sense of the word. And, in fact, although he interests himself in all areas of knowledge, Aristotle's theoretical studies, including those in the group of treatises brought together under the title *Metaphysics*, principally concern physics. It seems, in any case, that Aristotle did not regard Plato's response to the mechanism of his predecessors as adequate. And, in a more general way, by describing *all* other philosophers as strict mechanists whose appeals to other sorts of cause were nothing but mirages, Aristotle seems to ignore supremely the existential crisis that Parmenides provoked in the inquiry into nature. For Aristotle, Parmenides was, in the last analysis, a Presocratic like the others.

Aristotle presents himself as simultaneously the heir and the reestablisher of natural philosophy. On the one side, in fact, he is anything but a philosopher of the blank slate, and situating his position in relation to those of his predecessors is not simply, for him, a matter of performing an academic ritual—anyway, that ritual did not yet exist. There is a well-known point that is decisive in the study of the relationships between Aristotle and the philosophers who preceded him, but seems to be of more interest to those who study Presocratic thought than to students of Aristotle; that is that our documentation on the Presocratics is almost entirely derived from Aristotle himself and his ancient commentators. If we had lost the text of Aristotle, we would have known almost nothing about the pre-Platonic philosophers. In fact, the reformulation of the problematics of the Presocratics and Plato into Aristotelian terms is a great help to the interpreter of Aristotle, but an important obstacle in our study of the Presocratics. But, from another angle, we will see that the Aristotelian refoundation constituted one of the most impressive ruptures in the history of philosophy.

The Historical Background

The mechanist approach, for explaining both the composition of the universe and the structure of beings and events, notably generations, in the universe, is one of the principal achievements of Presocratic thought. One normally attributes the beginning of the kind of rational thought that we call philosophy to Thales, when he maintained that the universe is composed of water, and that physical and chemical actions on water explain the different forms of things and the events in the world. These operations include boiling, freezing, condensation, expansion, compression, and several others, all processes about which we still do not have a very clear idea. The world of Thales, to the extent that we can form a not-too-inexact image, seems quite empty of intention. In contrast, mythic thought, to which philosophical thought is usually opposed, was doubtless not shy about giving finalist explanations, if only by attributing the origin of events to human and/or divine intentions.

It's this kind of philosophy, beginning with Thales, that seems to have called itself, and Plato and Aristotle call it, "inquiry into nature" (τὸ ζητεῖν τὰ περὶ φύσεως, *Parts of Animals* 1.1, 642a29), more often called *historia peri physeōs*, this expression having apparently served as the title of several Presocratic works. Because this inquiry was *physical* (since it was about nature, *physis*), Aristotle called those who carried it out *physikoi* or *physiologoi*. It does not matter much here whether these two terms are totally equivalent or not. The most gifted description, and the best known, remains that given by Socrates in the *Phaedo*:

> When I was a young man, Cebes, I was wonderfully keen on that wisdom which they call natural science, for I thought it splendid to know the causes of everything, why it comes to be, why it perishes and why it exists. I often changed my mind as I

> investigated questions such as these: Are living things nurtured when heat and cold produce a kind of putrefaction, as some say? Do we think with our blood, or air, or fire? (*Phaedo* 96a)

We know what comes next: it's difficult to measure the exact degree of irony in Socrates's next words, but he says that at first he was delighted (ἄσμενος, 97c6) to think that Anaxagoras could, thanks to his "Mind" putting everything in order (διακοσμῶν, 97c2) in the world, "assign to it the causes of things accessible to Mind," and goes on to declare that he was disappointed that he found in Anaxagoras just another inquirer into nature, a *physiologos*, just like the others. What Socrates expected to find in this revelation of the causes would have been an explanation of things by showing, for example about the Earth, "that it was best (τὸ ἄμεινον), and that it is better that it be as it is" (97e). This brings in one of the characterizations of finalism, that of explaining things by showing that they are "good" and/or "beautiful." And, for Platonists, for example, things are good because a god, or an agent directed by a god (the demiurge of the *Timaeus*) wanted them that way.

This is not the place to return to the difficulties, particularly historical difficulties, that Plato's text raises, for example that it seems to leave him wide open to the criticism addressed earlier to Aristotle, that of ignoring the Parmenidean crisis; indeed, he pays careful attention to it in other dialogues. It's appropriate to notice, right away, that starting with Plato finalism critiques mechanism, and in a way that is systematized and developed by Aristotle. This is a matter of a finalism that is in a sense "adult," one that could not be developed except in response to Presocratic mechanism. But this critique was preceded by another one, that of the naïve finalism of myth criticized by the birth, and the triumph of mechanism. Among the remaining bits of evidence of philosophical mechanism aimed at criticizing the finalism that preceded it, one is particularly interesting here, because it is reported by Aristotle in one of his passages that critique, in a somewhat sarcastic way, the philosophical mechanism in question. It's worth quoting the passage at some length, because we will need to come back to it:

> What prevents nature working, not for the sake of something, nor because it is better so, but just as the rain falls from the sky, not in order to make the grain grow, but of necessity? (In fact, what is drawn up must cool, and what has been cooled

> must become water and descend, and coincidentally the grain grows.) . . . Why then should it not be the same with the parts in nature, for example, that our teeth should come up by necessity—the front teeth sharp, fitted for tearing, the molars broad and useful for grinding down the food—since they did not arise for this end, but it was merely a coincident result. Whenever then all the parts came about just what they would have been if they had come to be for an end, such things survived, being organized spontaneously in a fitting way; whereas those which grew otherwise perished and continue to perish, as Empedocles says of his "man-faced ox-progeny."[6] (*Physics* 2.8, 198b16)

This passage owes its fame to several factors. It has impressed modern readers because it seems to attribute to some Presocratics, including Empedocles, a pre-Darwinian theory in which, among a group of modifications happening to animals by chance, only the viable and effective modifications survive. It seems that Darwin attributed to Aristotle himself this prefiguration of his own positions. In fact, a hypothesis like this has nothing Darwinian in it, notably because it does not think that the deviations intervene on a precise character in the course of *reproduction*, in the case of animals within species (longer wings or more eggs), but seems to attribute a stochastic development to reproduction itself. The passage is also one of the most cited from the Aristotelian corpus because the example of rain, of which an important aspect has been omitted here, has served as the basis of a fundamental dispute, already noticed, concerning Aristotelian finalism: Is rain for something or not?

And in fact, the great theoretical durability of Presocratic mechanism is reflected in this passage. It seems to me that there is little doubt that the wording of the phrase "it is thus there that everything happens as if things occurred for the sake of something" (ὅπου μὲν οὖν ἅπαντα συνέβη ὥσπερ κἂν εἰ ἕνεκά του ἐγίγνετο, 198b29) reports, if not the letter, at least the spirit, of statements of mechanists like Empedocles. Here's how I think one must reconstruct his argument: in beings that have a function (because no one would deny that teeth have a function, some to cut,

6. B61 DK. This can be understood in two ways: (i) oxen with human faces have been destroyed and continue to be destroyed; (ii) oxen with human faces are not put together as if they had been constructed for something.

others to grind), those who are not philosophers suppose that there is an intention at the origin of the arrangement of teeth, while in fact the play of mechanical causes alone is sufficient to explain this formation, so long as one grants to nature the possibility of retaining only viable combinations. As we see, in fact, that the incisors are made *for* cutting, in living beings that we see provided with incisors, we infer from that, that they were made *for* that, while in fact it is by chance that some teeth can cut and others grind. Finalist explanation is thus, for people like Empedocles, an illusion that affects those who are incapable of correctly grasping the causes of the formation and structure of things. The passage does not say to whom this "common" thought attributes this intention. Doubtless in mythic thought, which philosophy was trying to destroy, it was a matter of divine intervention.

But Aristotle's critique of this mechanist position of people like Empedocles, in the name of finalism, does not consist of a pure and simple return to the previous situation. Whatever the insufficiencies of this mechanist explanation, the mechanism attributed to Empedocles represents a decisive theoretical progress in comparison with the finalism that preceded it, of which one may say, as does Spinoza in the appendix of the first book of the *Ethic*, that it is an "asylum of ignorance." That's the finalism of the lines on the melon of Bernardin de Saint-Pierre, that nature has drawn so that the fruit may be divided for the family without making anyone jealous. It is more rational to understand that a natural selection has operated among beings that have been produced by the pure play of natural entities and forces, rather than to appeal to an intervention of intentions of a god or a nature. Thus, we have here, on Aristotle's part, a critique of a critique, a negation of a negation.

The "mechanistic moment" is thus a definite step forward of thought, one that will be critiqued and surpassed, but never abolished. We notice, among other things, the fact that none of the philosophers who have been partisans of finalism and sometimes violent critics of mechanism ever try to *substitute* teleology for mechanism, not even Plato. We mentioned that before; now let's take a closer look. A little after the passage in the *Phaedo* cited previously, Socrates explains that the cause due to which he can be found sitting in prison does not reside in his bones and tendons that place him where he is, but in his decision to obey the laws of the city, according to his concept of justice. And Socrates distinguishes between real causes, and necessary conditions, because without his bones and tendons he wouldn't be there; "but to call those things 'causes' is too absurd" (99a). Despite Plato's extreme demotion of mechanistic causality, he does not eliminate that sort of explanation, even if Socrates's intention to stay in prison in order to obey his concept of justice is not only the fundamental cause, but in fact the only thing to deserve the name of cause, the movements of his bones and tendons being "that without which the cause could not be a cause" (*Phaedo* 99b4).[7] Aristotle goes much further than Plato in taking into account this "mechanistic moment," and that is why it is only with him, or with him above all, that one may speak of a negation of the negation.

The first modification that Aristotle brings to this Platonic schema consists of completely rehabilitating material causality. We will see that Aristotle too introduces a hierarchy between types of causes, but the material cause is really a cause alongside the others. To the question of whether there are cases in which material causes are the only ones to act, Plato, like Aristotle, answers in the affirmative, but in conditions completely different, as we will see. Aristotle grants an important place to "mechanistic" explanations, that is, to material and efficient causes, in his study of living things, and that is even what is asserted, two lines before the passage cited earlier, right at the beginning of the chapter in the *Physics* we were discussing: "We must then explain first why nature belongs to the class of causes that act for the sake of something; and then the necessary and its place in nature" (2.8, 198b10).[8] "The necessary" being a way of designating material and efficient causalities. At the same time, in the first book of the *Parts of Animals*, when he declares about natural beings, particularly animals, that "there are then these two causes, the 'for the sake of something' and the 'by necessity'" (1.1, 642a1), Aristotle shows clearly that an explanation is not complete if it does not present material causality, which is a real causality. We may summarize that by saying that finalism needs mechanism, but the latter does not need the former, because, in reducing finalist explanations to the level of anthropocentric illusions, mechanism claims to occupy the entire explanatory field. This historical circumstance is reflected in Aristotle's position that there are sometimes mechanical causes that function without being combined with final causes, but no final cause can function alone. Nevertheless, it remains that when it functions with final causes, necessity takes, or can take, a particular form, as we will see.

7. Cf. Johnson, *Aristotle on Teleology*, 127–129.

8. Λεκτέον δὴ πρῶτον μὲν διότι ἡ φύσις τῶν ἕνεκά του αἰτίον, ἔπειτα περὶ τοῦ ἀναγκαίου, πῶς ἔχει ἐν τοῖς φυσικοῖς.

The Democritean approach does not suffer much damage from Socrates's objections in the *Phaedo*, because it is not impossible to interconnect, as Kant was to do, for example, the domains of physical necessity and human liberty: the possibility of choosing freely whatever seems good and right to us does not prevent the construction of a mechanistic physics. Even if one rises above, in one way or another, the crisis begun by Parmenides, in which we saw a crisis in the foundations of Presocratic philosophy, it remains that making the *order* of things depend on the play of physical forces acting at random is tough, hence the introduction of organizing principles, "for example, hate, love, mind, or spontaneous action" (*Parts of Animals* 1.1, 640b7), as Aristotle says, alluding to Empedocles, Anaxagoras, and the atomists. Two domains are particularly crucial for thinking this setting things in order, that of the cosmos as a whole, and that of the reproduction of living things.

The organizing principles proposed by the post-Parmenidean philosophers to explain the appearance of the cosmos, a word carrying in itself the signification of order and harmony, and thus of beauty, are ultimately of the same nature. They had to presuppose one or several forces (the whirl, attraction/repulsion, weight, etc.) acting on the material of the universe and introducing order to it, this process occurring once and for all, or repeating itself at more or less regular intervals. The rupture introduced by Platonic finalism is that it makes the construction of the universe depend on a rational will of a superhuman being. We can see that the structure of all these theories is fundamentally identical: it's a *narrative* structure, that is, it recounts a series of events, with a characteristic just as much in common, of resting on the distinction between two different natural states, of which one is the result of the other. It's a matter of the great distinction between one order, which is in fact a pre-cosmic disorder, absolute or relative, and the cosmic order itself. We can see clearly, and this is essential, what these *cosmogonies* have in common with their mythic parents, since a myth is an explanatory story that accounts for the actual state of things by relating it to a previous state depending on other laws than those that rule the present world: in myths too there is a preliminary phase, the Golden Age, Garden of Eden, or something else, which is followed, often as the consequence of a catastrophic event in the etymological sense of the word, for example eating the forbidden fruit, by a new order.

It's the same for the question of the reproduction of living things, with which all mechanists in the course of history have collided. Were the first living things constituted the way Empedocles says, according to Aristotle? And how did they pass on their structure to their descendants? The atomists, for example, did not need a demiurgic entity to constitute their complex beings, but they were forced to weave their atoms with affinities and repulsions, if only due to their geometrical shapes, which could explain, notably, the *regular* reproduction of living things within their species. Obviously, Aristotle was going to ask these questions.

Aristotle's Solution and Its Consequences for His Teleology

The tradition, whose first version goes back to Plato, has given the Parmenidean critique a fundamentally ontological orientation: it is impossible to get being from nothingness, or the contrary, which means that one may assert nothing but the presence of an unchangeable being, ungenerated and eternal, of which Parmenides says that it is "like a well-rounded sphere" (fragment B8.43). But what interests us here is that which may be called the cosmological version of this doctrine. Because it is what concerns the generation of the cosmos and of that which it contains that the Parmenidean critique put an end to what Jonathan Barnes, in his work on Presocratic philosophy, called "the age of innocence."[9] Philosophical thought was really ejected from the "Eden" (Barnes's word) in which it had naïvely lived. If we stick to the least contestable case, that of the Milesians, that is, Thales, Anaximenes, and Anaximander, the fatal criticism of Parmenides and his disciples (we know only of Zeno and Melissus) amounts to blaming them, to speak in modern terms, of ignoring the second principle of thermodynamics, that of entropy, by deriving the organized cosmos such as we see it simply from the properties of the first matter of which it is composed. But a house does not build itself out of the bricks and cement left simply to their physiochemical properties.

The responses to this existential challenge consisted, in various forms, of introducing into the world ontological poles of stability, for example, atoms that are as it were little Parmenidean beings, and to posit alongside the first matter of the universe adventitious realities capable of introducing information into this material, like the love and hate of Empedocles, or the Mind of Anaxagoras. Thus, according to these philosophers, one would be

9. Aristotle, *Posterior Analytics*, 155.

able to get around the entropic tendency of the universe, which makes it go toward simpler states, making it more probable and more stable. Plato, as we will see once again, is a member of this family of philosophers, telling a story about a negentropic ordering of the universe, even if he is perhaps the only one to give it a teleological spin.

Aristotle's solution to these problems is extremely original, putting him into opposition with all other ancient philosophers, and possibly even all other later thinkers. Not only does Aristotle not imagine the emergence of the cosmos from nothingness, any more than other Greek philosophers, but he rejected any distinction between a pre-cosmic phase and a cosmic phase of the universe, and by that fact, ended the cosmogonical period of ancient thought, and proposed a *cosmology* pure and simple. Friedrich Solmsen, in a marvelous book, knew how to decode this Aristotelian approach and to restore its originality.[10] After Aristotle, thinkers returned to their old habit of asking how the universe was constituted. From a modern point of view, Aristotle's stance can be understood in two ways, positive and negative. One may think, and one likes to think, that on this fundamental question, Aristotle, who is otherwise presented as the paragon of "metaphysical" thought in the Comtean sense of the word, demonstrated a really scientific spirit in refusing to engage with questions that were outside the range of his investigation. Because, pace Karl Marx, the human spirit often poses for itself, and has posed for it, questions to which it cannot respond. But we may also remark that, without demanding that Aristotle formulate the second principle of thermodynamics, he could have had some prescience of it, as was the case for most ancient philosophers.[11]

Aristotle, who usually agrees with all or most people, on this fundamental point has a clear consciousness of his own originality: "Everyone says that the world was generated, but some say it was generated to be everlasting, others say that it is destructible like any other natural formation. Others again, like Empedocles of Acragas and Heraclitus of Ephesus, believe that it alternates, sometimes being as it is now and sometimes different and in a process of destruction, and that this continues forever" (*De Caelo* 1.10, 279b12). After a short argument that Simplicius, in his

10. Solmsen, *Aristotle's System of the Physical World.*

11. On this matter, see the passage from Lucretius cited at the end of the last chapter of this book.

commentary on the *De Caelo* (301, 31) calls "inductive," Aristotle offers a battery of reasons that may be called "logical," right to the end of chapter 12, the end of book 1. These passages, difficult and fascinating, are, one may say, essentially Parmenidean. The first of these arguments, for example, the second in chapter 10, establishes that whatever exists eternally cannot change, because if there were a cause able to bring about this change, it would have done so already. Nothing can be simultaneously eternal and corruptible, or generated and incorruptible. By the same token, change from disorder to order is unthinkable.

Not only is the world a constant mass (which for Aristotle means something precise because it is finished), but *globally* it is what it has always been and will be forever. In rejecting the idea that the elements constituting the universe would have a different destiny in the pre-cosmic and cosmic phases, Aristotle rids Greek thought of the last remnants of mythology. Within this world, there are many changes, including the birth of new substances, but there is never an appearance of really new entities. The process of the transmutation of the elements into each other by exchange of their elementary qualities (when dry and hot fire becomes dry and cold air, for example) is cyclical and does not bring about any entity of a sort that never existed before. In the Aristotelian universe, therefore, nothing is created, nothing is lost. Thus Aristotle avoids one of the stumbling blocks on which many ideas have been broken, that of negentropy: the best way to think of the very difficult question of movement from less organized to more organized (stones, mortar, and wood to house) is to posit that the more organized already existed, in one form or another, in the plans that the architect has in his mind or in the blueprint, for example. "And if that's not how it is [i.e., if one does not accept Aristotle's explanation], the world would have proceeded out of night, and 'all things together,' or nonbeing" (*Metaphysics* Lambda 7, 1072a19), Aristotle writes, after the previous chapter dedicated to demonstrating, perhaps with a touch of irony, the difficulties that overwhelmed his predecessors.

The difficult, even unsolvable, questions about the origin of the world, or its order, are thus disqualified. Aristotle's radical originality has been, like everything original, difficult to understand, and more than one interpreter, both in antiquity and down to our day, has tried to soften its cutting edge, for example by giving the Unmoved Mover, Aristotle's god, an efficient causality. Its pure and eternal identity certainly represents a model for the cosmos, and it is indeed that which Aristotle says, when he shows that the first heaven tries to imitate, as best it can, the unchangeability of the First Mover by means of an everlasting circular movement. Without the First Unmoved Mover there would not be any movement in the world, and thus no nature, but the First Mover does not introduce changes into the course of things, and there is obviously no period prior to, or posterior to, its intervention, since there is no intervention. We will have to return to the relationship between the First Mover and finality. Aristotle is thus revealed to be the philosopher who has taken the Parmenidean critique most seriously, since he did not try to evade it by conciliating eternity and immutability of the All, and the birth of the world. That has decisive consequences for finality, and notably finality in the animal world.

There is a finality when certain states of affairs or certain processes are explained by reference to an end. Thus, according to an example in the *Physics*, one goes somewhere *in order to* see someone, or, in the myth in the *Protagoras*, Hermes gives human beings a sense of shame and of justice *so that* they can live in society, and that is *for* them being able to survive when confronted by other animals. In these two examples, there is an intention, and a subject who has this intention. If thoughts like those of Democritus, or before him, that of the Eleatics, do not seem to have any room for finality, it is ultimately difficult to decide how it is for people like Empedocles or Anaxagoras. Doubtless Aristotle is right to group them among the people who do not have a teleological perspective, even though they are able to offer a little room for teleology, being forced by the facts.[12] But the notion of a goal or end seems to presuppose, if not always that of a subject that achieves or at least pursues this goal, at least the idea that things take place in an after, preceded by a before. If the goal is reached, or ought to be reached, it's that it has not always been thus, and there was a time when it was not yet thus. Sometimes the finality is present as temporally coextensive with the concerned reality: nature, or the demiurge, has given, since their creation, fins to fish so they can move more easily, providing many advantages. But there has to be a time when fish were created, and thus there would have been a world without fish. That's what might be called the normal frame of all teleology.

A finality of this kind is certainly that of other philosophers, notably Plato, but it is impossible for Aristotle due to his cosmological position. The most important consequence of his position for the living world is

12. Cf. *Parts of Animals* 1.1, 642a18, on Empedocles. Aristotle says the same for Democritus at 642a27.

that living species, animal and vegetable, are everlasting. One of the most famous passages of the Aristotelian corpus, at the beginning of book 2 of the *Generation of Animals*, says that living things, unable to be numerically everlasting (Aristotle doesn't explain why here, but one easily understands why material beings that undergo so many transformations during their lives could not exist forever), must find another way to be everlasting.

> Since it is impossible that such a class of things [as animals] should be of an everlasting nature, therefore that which comes into being is everlasting in the only way possible. Now it is impossible for it to be everlasting as an individual—for the substance of things that are is in the particular; and if it were such it would be everlasting—but it is possible for it as a species. This is why there is always a class of human beings and animals and plants. (2.1, 731b31)

This passage is difficult to interpret in detail, especially the phrase between — and —. A. L. Peck, in a note to his translation,[13] proposes considering that Aristotle uses the term "eternal" (ἀίδιον) in two senses, first an "absolute" sense, which applies to that which is incorruptible and divine, second in a derived sense described by the redundant expression, "in the way it is possible for it, in this way" (καθ' ὃν ἐνδέχεται τρόπον, κατὰ τοῦτον, 731b32), which applies to that which can and cannot be. We are certainly dealing with a "diminished" eternity, since it does not concern the *substance* (*ousia*) of the being under consideration, but its *genos*, but genera and all universals are *ousiai* only in a derived sense, as is shown in *Metaphysics* Zeta. What is important for us in this reading, which is not absurd, is that this specific and not numerical eternity (everlastingness) gives species, though eternal (everlasting), the possibility of including individual variations.

We will return more than once to the permanence of species, but we may note immediately that if species are permanent, that means that they did not appear nor are they transformed at a given moment. That is not only impossible to imagine in the habitual frame of teleology, but hard to imagine at all. Thus, Aristotle is forced, in order to account for the finalized character of certain organs and characteristics of living

13. Aristotle, *Generation of Animals* (Peck).

beings, to use a highly ambiguous vocabulary. Aristotle's zoological texts are scattered with statements like "nature always achieves the best," "does nothing in vain," "gives on one side what she takes on the other"; we will return to these phrases in order to develop a more complete understanding, but we can say already that they ought not to be taken as indicating that nature achieves a plan: nature did not *choose* the best among several possible choices in the usual sense of the verb "choose." In an example to which we will return, Aristotle explains that nature assures the survival of little fish by making them prolific, because they suffer important losses to predators. But that is not the result of an observation, much less a deliberation,[14] nature, seeing the situation, deciding to take measures to prevent the disappearance of the little fish: little fish have forever found in fecundity the means to survive. When, thus, we read in the *Progression of Animals*: "The reason [that birds flex their limbs as quadrupeds do] is that nature's workmanship (δημιουργεῖ) is never purposeless, as we have said previously, but everything for the best possible in the circumstances" (12, 711a19), such a passage may be confusing even in its vocabulary, since it seems to allude to nature as a demiurge, but what is a demiurge that never has an *occasion* to act? The permanence of the world, and thus of living species, is certainly a crucial aspect of the difference separating Aristotelian finalism from the mechanism of Presocratic philosophers, and also from Platonic finalism.

For mechanistic thinkers, to give an account of what such and such an animal is, is to show how it is constituted, for example, given the mutual affinities among such and such sorts of atoms, how such a species could be produced and could reproduce itself. If the odds that an organism so complex as that of a "superior" animal would be constituted by the play of chance alone are very slim, one may always appeal to a version of the "principle of plenitude," according to which everything that is not impossible will be actualized in infinite time,[15] so a given event may reoccur

14. Art per se doesn't deliberate either: "Art does not deliberate either, and if the art of ship building was in the wood, it would work the same way as nature" (*Physics* 2.8, 199b28). It's the artist who deliberates.

15. It was Arthur Lovejoy, in his famous work, *The Great Chain of Being: A Study of the History of an Idea*, published in 1936, who first used the term "principle of plenitude." He applied this principle to two different situations, attributing it to Plato, but not to Aristotle: "not only the thesis that the universe is a *plenum formarum* in which the range of conceivable diversity of *kinds* of living things is exhaustively exemplified, but also . . . that no genuine potentiality of being can remain unfulfilled" (52). Cf.

an infinite number of times; the species "dog" would thus be actualized an infinite number of times. Still, the problem of reproduction needs to be dealt with . . . Certainly, Aristotle seems to be closer to Plato than to Democritus, since he shares with him a finalist approach to nature. But ultimately, the gap between Aristotle and his teacher is hardly less wide than between him and Democritus. The difference between them is that for Plato things can arrange themselves, or be arranged, for the better or for the best. One may show quickly, while keeping in mind that the account in the Timaeus has a status that is difficult to nail down, that commentators are not always in agreement, and that Aristotle shows a certain levity in taking the account completely literally. It is certain that the world of the *Timaeus* was not constituted at random from its constituent elements, and that the demiurge made it by looking at the eternal and the good (29a). Nevertheless, it remains that the world cannot be eternal (everlasting), because it is material, and was generated (28b), and that goes also for the beings that it contains:

> The things we see were in a condition of disorderliness when the god introduced as much proportionality into them and in as many ways—making each thing proportional to itself and to other things—as was possible for making them be commensurable and proportionate. For at the time, they had no proportionality at all, except by chance (ὅσον μὴ τύχῃ), nor did any of them qualify at all for the names we now use to name them, names like fire, water, etc. (*Timaeus* 69b, Zeyl)

This remarkable passage shows us at least three things. First, that the difference between the pre-cosmic disorder and cosmic order, and all the difficulties that a distinction of that kind brings with it are both at least as marked in Plato as in previous philosophers, and that the passage from disorder to order is done by a teleology of the intentional kind, on which Aristotle turns his back. Next, for Plato, it is not excluded that this setting in order is done in some cases by chance, in the Democritean manner, but cases of that sort can only be marginal. Finally, one may deduce from this passage that there are, for Plato, two kinds of generation, primordial generation, which puts the animal world (since we're interested in animals)

---

the critique of this position in the fundamental work of Jaakko Hintikka, *Time and Necessity*. I save the appellation "principle of plenitude" for the second thesis, using "completeness" for the first.

and its various parts in their places, then the reproduction of living things within their respective species. Aristotle disagrees on all three points. He resolves the formidable question of the appearance of the cosmic order by positing that it was always there; we will see that he absorbs the Democritean moment completely within his etiology; since species were never *introduced* into the world, they have only one kind of generation, specific generation, whether sexual, spontaneous, or some other kind.

To say it again, and before a good many more times in the following pages, the principal difference that separates Aristotle from other philosophers on the question of the birth of natural beings and especially living beings, may be expressed by the notion of *efficiency*. In a justly famous article, Michael Frede argues that it was with the Stoics that the modern notion of causation was put in place, that which, following the Galilean scientific revolution, is reduced to what Aristotle calls the efficient cause.[16] But we see that all of the philosophers prior to Aristotle, including Plato, deploy an efficient (i.e., moving) causality: Plato's distinction between necessary conditions and divine causes does not change the fact that both need to find the means to *effectuate* something. Once again, Aristotle, although he too resorts to efficient causality, alone stands up against all the others. We can see that concerning animals.

A species of animals is, for Aristotle as for all naturalists until the twentieth century, the group of animals sharing distinctive traits that are transmitted to them genetically. Aristotle characterizes the genetic patrimony as *form*, which, in animals reproducing themselves sexually, is transmitted to the female material by the male sperm by the means of a system of movements, all accompanied by a hot "innate breath" whose role is still debated among commentators. In animals that do not reproduce sexually, there are analogues of these different factors (male, female, etc.) to be examined later. In any case, an unresolved question among interpreters of Aristotelianism is knowing the relationships between the general form of an animal species and the individual form of each of its members. Let's stick to the empirical observation that two dogs are closer to each other in form than either is to a cat, and that two Molossian hounds are closer to each other in form than either of them is to a poodle.

16. Michael Frede: "Quite generally our use of causal terms seems to be strongly coloured by the notion that in causation there is something which in some sense does something or other so as to produce or bring about an effect." Frede, "The Original Notion of Cause," 125.

But, it's one of the cardinal doctrines of Aristotelian physics that form is the unmoved principle of movement, and Aristotle goes so far as to say that, in view of its immobility, it does not belong to nature, that is, that it doesn't belong to physics. Natural beings are, in fact, by nature the beings that have in themselves a source of movement and rest:

> In respect of coming to be, it is mostly in this way that causes are investigated—"What comes after what? What was the primary agent or patient?" and so on at each step of the series. But the principles that naturally cause motion are two, of which one does not belong to nature, as it has no principle of motion in itself. Of this kind is whatever causes movement, not being itself moved, such as that which is completely immobile (i.e., the first being of all), and the essence of a thing, the form; for this is the end and that for the sake of which. (*Physics* 2.7, 198a33)

For natural beings, the form plays the role of unmovable mover, for it also causes motion. It's worth the trouble to say a little more about the question of the role of the immobile in natural beings, that is, those that are essentially mobile.

Natural beings are, in fact, subject to multiple movements, notably to movements that characterize them as natural beings, namely those which have their origin in these beings themselves. Among these movements, the most fundamental is that of coming to be (genesis), which brings it about that in the uterus of a female, for example, the embryo progressively acquires its characteristics. For the physiologists, one must determine the *order of events* that leads to the final state that one wants to explain:

> Another question that must not be passed over without consideration is, whether the proper subject of our exposition is that with which the earlier writers concerned themselves, namely, the way each thing is naturally generated, or rather the way it *is* (πῶς ἕκαστον γίγνεσθαι πέφυκε μᾶλλον ἢ πῶς ἔστιν). For there is no small difference between these two views. . . . For even in house building, these things come about because the form of the house is such and such, rather than its being the case that the house is such and such because it comes about thus. For the generation is for the sake of the substance (οὐσία) and not the substance for the sake of the generation. That is

> why Empedocles did not speak correctly when he said that many of the characters belonging to animals were merely the results of incidental occurrences during their development; for instance, that the backbone is as it is because it happened to be broken owing to the turning of the fetus in the womb. In so saying he overlooked the fact that the seed that has been constituted[17] needs to be present with a power of this kind. Secondly, he neglected another fact, namely that the parent animal preexists, not only logically, but temporally. For a human being is generated from a human being; and thus it is because the parent is such and such that the generation of the child is thus and so. (*Parts of Animals* 1.1, 640a10)

Not that Aristotle finds the course of intrauterine events leading to the appearance of vertebrate animals insignificant, nor does he refuse to this course of events all explanatory character, any more than he is indifferent to the manner in which the construction of a house occurs, notably the order in which various processes happen. Efficient causality is for him full-fledged causality. What he rejects is thinking that the sequence of movements that leads to the fracture of the dorsal spine is set in motion simply by the properties of the matter constituting this dorsal spine, and of the surrounding material.

We will return in more detail to the generation of living things and to the role that the properties of the female matter play. What interests us, at this stage of the analysis, is Aristotle's affirmation of the preeminence of substance in generation. This preeminence is a necessary consequence of the eternalist character of his cosmology. There is also, in Aristotle, a narrative logic in the description of the generation of living things. For example, the *Generation of Animals* furnishes a kind of story about the formation of the embryo during that generation, showing how various organs are constituted successively, the most important first, with the heart or its analogue being formed first. But such sequences cannot be the only principle of the movements that mark the appearance of a new living thing; it is also necessary that they be directed by a form that is at the same time their goal, for example, the human form. That's what

17. Keeping, at 640a23, the reading of the manuscripts, συστὰν rather than συνιστὰν, "constituent," adopted by editors since Platt. The seed has been constituted in a manner that it can carry hereditary characters.

the passage from *Physics* book 2, cited earlier, says. But that form does not exist separately from concrete living humans, and that's what the famous Aristotelian formula means: "human being begets human being." A generator is necessary, of the same species with few exceptions, for an offspring to exist. But that immobile form is in a way the trace, at the level of each living thing, of the eternity and thus the global immobility of the universe. Just as that global immobility guarantees that one may think the universe and its movements without falling under Parmenides's critique, in the same way it is of this immobile form that one may think the coming-to-be of an animal. The manner in which each animal actualizes this form is a problem that one may, for the moment, set to one side.

It is thus decisive for the understanding of Aristotle's natural teleology to consider it as a teleology of structures and functions prior to being a teleology of process, and to understand that it is a teleology of process because it is a teleology of structure and function.[18] Thus there are in nature, since forever and for forever, animal species that are so many forms transmitted via reproduction. That's why Aristotle can designate the species with the word *eidos*, one of the usual words with which he designates the form, but he more often designates the species with *genos*, a term that indicates a genetic community. *Eidos* can also refer to the fact that a group is taken to be a subdivision of a larger group, that called *genos*. Since it is not generated, this form (species) cannot be constructed by an efficient cause.

"Many Things Happen because It Is Necessary"

Aristotle's rehabilitation of necessary causality demonstrates that he completely accepts the legacy of the physiologists. We have seen that Plato himself recognizes the activity of other causes than the final cause, since the world itself was established by a cooperation of intelligent and necessary causalities, but he sometimes hesitates to attribute to necessity a true causal status. But here again there is a remarkable difference between the teacher and the student, even if both affirm that there are two kinds of

18. "Aristotle is not concerned primarily with the process of a certain end coming into sight and with its gradual materializing, but with the way in which certain materials are the components of more highly organized structures. It is a matter of structures and functions." Kullmann, "Different Concepts of the Final Cause in Aristotle," 169.

cause. Plato calls them Mind and Necessity. But the second kind of cause produces "only haphazard and disorderly effects every time" (46e, Zeyl). A little later in the dialogue (68e), Plato explains that the god has used these mechanical causes in order to construct a good and beautiful universe. In a remarkable expression, Timaeus, who has just recalled the distinction between two sorts of causality, divine and necessary, the second including only supportive causes, concludes: "We must search for the divine cause in all things if we are to gain a life of happiness. . . . We must search for the necessary cause for the sake of the divine" (τὸ δὲ ἀναγκαῖον ἐκείνων χάριν, 69a). Farther along, he will deny this necessary cause the right to be called "cause."

In opposition to Platonic positions, we must therefore see the importance, and the limits, of Aristotle's rehabilitation of "necessary" causality. The first chapter of *Parts of Animals* book 1 is hardly concerned with the Platonists (who will be critiqued, but only on the question of division, in the two following chapters), but develops, more than any other text, the Aristotelian critique of Presocratic mechanism in the name of finalism. In a passage cited many times, we read: "It is plain then that there are two modes of causation, and that both of these must, so far as possible, be taken into account, or that at any rate an attempt must be made to include them both; and that those who fail in this tell us in reality nothing about nature" (642a13).[19] First, we must note that the sentence is badly constructed, although that does not really get in the way of our understanding. The term δῆλον ("it is plain") at 642a15 ought to govern the three words that follow it and introduce the two *hoti* (ὅτι) clauses, the one before it, the other after.[20] So we can paraphrase the passage thus: "It is clear that there are two sorts of causes; if one cannot apprehend both that one must try to clarify, it is clear that those who do not apprehend both do not do natural philosophy." That which one must try to make "clear" is not, in fact, clear, but it would be "that those who do not apprehend both sorts of causes do not do natural philosophy," that is, "that those who do not say why, in some cases, or cannot give both types of causes, are not doing natural philosophy."

19. Ὅτι μὲν οὖν δύο τρόποι τῆς αἰτίας, καὶ δεῖ λέγοντας τυγχάνειν μάλιστα μὲν ἀμφοῖν, εἰ δὲ μη, δῆλόν γε πειρᾶσθαι ποιεῖν, και ὅτι πάντες οἱ τοῦτο μὴ λέγοντες οὐδὲν ὡς εἰπεῖν περὶ φύσεως λέγουσιν.

20. That's why it is impossible to put a comma after this δῆλόν, as does Ogle, followed by Peck.

A few lines previously (642a1), Aristotle writes, "There are then two causes, namely, necessity and the final end. For many things are produced, simply as the results of necessity." Elsewhere, in the *Physics* for example, in appealing to a more developed version of his etiology (that which identifies four causes), Aristotle writes: "Since the causes are four, it is the business of the student of nature to know about them all, and if he refers to all of them, he will assign the 'why' in the way proper to his science—the matter, the form, the mover, and that for the sake of which" (2.7, 198a22). But it is a matter of the same doctrine, the binary version gathering several causes of the four into a single one, necessary causality including in general the material and efficient (moving) causes, "that for the sake of which" the final and formal causes. But one has to say "generally," because sometimes, as we have seen, the efficient cause falls on the side of "that for the sake of which" and the formal cause combines with the material and efficient causes.[21] In other words, the natural philosopher absolutely ought to know the "by necessity" of the objects that he studies.

Necessity alongside finality, that does not mean that Aristotle does not introduce a hierarchy between the two. Thus, in an often-cited passage from the *Generation of Animals*, we find both the coexistence of these causal approaches and their ranking: "Why animals are generated and why there are both female and male, to the extent that that results from necessity, that is, from the proximate mover and a certain sort of matter, our exposition needs to try to explain those step by step, but to the degree that it is for the better and from the cause for the sake of something, this principle is higher" (2.1, 731b20). But as soon as Aristotle has reintroduced necessity into its rightful place by making it a full participant in the explanation of nature, he seems to return necessity back under the control of finality by introducing the notion of "hypothetical necessity" (or "conditional necessity"), which is applicable to "beings that undergo generation" and notably, therefore, to animals. This concept of hypothetical necessity has aroused debates, and publications, among interpreters.

21. Cf. my article, "De l'explication causale dans la biologie d'Aristote," where I have tried to show that for certain animal parts and vital processes it would be the material or moving cause that blends with the formal cause: although the lung is defined by its final cause (cooling the organism), semen is defined by its moving cause (its manner of production from food) and not by its final cause (producing an offspring); the same for sleep, which is defined by its material cause (the capture of the primary sensory organ, i.e., the heart or its analogue).

Hypothetical Necessity

Aristotle presents this idea thus:

> Is it the case that "by necessity" exists in natural things hypothetically, or also[22] absolutely? Because, in fact, [the physiologists] are of the opinion that necessity is to be found in the generation of things in the following way: it's as if one were to think that the wall has been produced necessarily, because heavy things are naturally carried downward and light bodies upward, so that the stone foundation will be on the bottom, earth above because of its lightness, and the pieces of wood on top, because they are the lightest. But even though the wall would not come to be without these elements, nevertheless it is not because of them, except as a material cause; it is for the sake of sheltering and preserving certain things. Similarly in all other things that involve that for the sake of which: the product cannot come to be without things that have a necessary nature, but it is not due to these (except as its material cause); it comes to be for an end. For instance, why is a saw such as it is? To be in a certain way, and for the sake of something. This end, however, cannot be realized unless the saw is made of iron. It is, therefore, necessary for it to be of iron, if we are to have a saw and its function of sawing. It is thus that necessity intervenes hypothetically, but not as an end. Necessity is in the matter, while that for the sake of which is in the definition. (*Physics* 2.9, 199b34)

Two clarifications: To say that the "by necessity" intervenes hypothetically but not as an end means that it is hypothetically necessary that the saw be made of iron, but the saw does not exist for the sake of being made of iron. As for the "definition" (λόγος), it's another description of the form or the substance, and we will later on translate the crucial formula ἡ κατὰ

22. This καί has been interpreted in various ways. It can mean "also," and Aristotle means to say that beings that undergo generation come about by both absolute necessity and hypothetical necessity, but Cooper, "Hypothetical Necessity," 265, claims that this καί means "in fact," relying on Denniston, *The Greek Particles*, 317, and that Aristotle posits a disjunction: in natural things, the necessity is either hypothetical, or absolute.

τὸν λόγον φύσις (*Parts of Animals* 3.2, 663b23) by "Nature according to reason," not without justifying the capital N in Nature.

Besides this chapter 9 and last of *Physics* 2, three passages from the first chapter of *Parts of Animals* 1 deal with hypothetical necessity. We must quote them:

> Anyway, that which is by necessity is not met with in the same way in every natural being, although almost all try to refer their reasons to this, without having distinguished in how many senses "necessary" is said. But necessity in the absolute sense belongs to eternal [everlasting] beings, but hypothetical necessity also[23] belongs to all beings that undergo generation, as it belongs to artifacts, for example a house and any other object of this sort. It is thus[24] necessary that such a matter be present if there is going to be a house or some other goal. (1.1, 639b21)

> For many things are produced because it is a necessity. But one may doubtless ask which necessity one is talking about when one says, "by necessity," because neither of the two kinds defined in the philosophical treatises is suitable. In fact, a third sort exists among beings that undergo generation. We say, in fact, that food is something necessary according to these two sorts, but because it is not possible to exist without it. That's a necessity that is in a way "hypothetical," because it is the same as, since it belongs to an axe to cut, it is necessary that it be hard, and if it must be hard, it needs to be made of brass or iron, so too, since the body is a tool (in fact each of its parts is for something, as is also the whole body), it is thus necessary that it be such as it is and constituted of elements which are such that are needed for this tool. (1.1, 642a2)

23. This καί has to be understood in the same way as the one in the preceding note (cf. Lennox, *Aristotle's Philosophy of Biology*, 127): it doubtless means "also," and Aristotle means to say that beings that undergo generation demonstrate simultaneously absolute and hypothetical necessity, but Balme (1992, 84) claims that this καί means "in fact," still relying on Denniston, *The Greek Particles*, 317.

24. Emending the δέ of the manuscripts to δή, 639b26, as does Thurot. Gaza has *enim*.

> It is necessary to demonstrate things in this way: for example, that on the one hand, respiration is for this, and on the other hand, it is necessarily going to be made from these things. Necessity means, at the same time, that if this, for the sake of this, needs to exist, it is necessary that such and such things be the case, and at the same time that things are such because they are such by nature. It is necessary, in fact, that heat goes out and comes back in when it meets an obstacle, and that air be brought in. That is necessary. As the interior heat makes an obstacle, the introduction of exterior air occurs when there is a cooling. (642a32)[25]

This means that the Aristotelian philosopher of nature will not be content with explaining the saw by saying that it is a tool designed to cut, but he must also explain what it is by saying what it is made out of (material cause). A saw cannot be made out of wool. It's thus the idea, that is, the form, that indicates the end (sawing), which determines in the last analysis the type of material of which the saw is made. There are two questions that have interested the commentators, and to which they have given divergent answers. The first question bears on the fact of knowing if hypothetical necessity, which Aristotle doubtless applies to natural beings and notably to animals, is the only kind of necessity that concerns these natural entities; the second question consists of appreciating the degree of autonomy that necessity retains in this new configuration. Hasn't Aristotle ultimately adopted a Platonist position by making the necessary cause a simple appendix to the final cause? In any case, the idea of "hypothetical necessity" seems definitely destined to make necessity, which is in a way the heritage of Presocratic philosophy, function together with finalism. Philosophers like Democritus or Empedocles think, on the contrary, that the only necessity at work in nature in general and in the animal world in particular is that which one may call "material necessity," that is, the necessity that regulates the movement of matter by virtue of the properties

25. Most manuscripts have "the entrance and the exit." This description is only for the sake of a methodological exposition, and does not correspond to Aristotle's position on the question. Respiration is hypothetically necessary for the cooling of the organism, but here Aristotle presents the absolute necessity that guides the process (he leaves to one side the third sense of necessity, namely that which is produced by movement that is forced and contrary to nature).

of that matter: composite materials constituted mainly of earth go down, when those that are made of fire go up, and so forth, barring special circumstances, themselves thinkable in terms of material necessity. Let us undertake, therefore, in our turn, to reexamine (rapidly) this famous notion of hypothetical necessity.

It is necessary to take seriously both terms in the expression "hypothetical necessity." First, hypothetical necessity is *necessity*, which means that it is applied to something that cannot be otherwise than it is, the most "basic" definition of necessity. From this fact, it is imprecise to say that it is hypothetically necessary that a saw be made of iron, or that the survival of an organism obligatorily occurs by eating meat, because the saw can be made of bronze, and one may be vegetarian. What is necessary is that the saw be rigid and the food nourishing. But iron and meat do the job, because iron cannot be anything else than rigid, nor meat anything else but digestible, *and that is by virtue of a material necessity*. Doubtless one may say that an iron saw does a better job than a bronze saw, and that, for an individual in given circumstances, a carnivore regimen would be better than a vegetarian regimen. Let's take an anatomical example: a blooded animal cannot live if its large vessel (its vena cava in modern terms) is not located in the front of the body; but the liver plays, at the same time as another function (namely the concoction of food), the role of an anchor point for the large vessel. That's a hypothetical necessity: if the end, that the vessel be anchored, is given, it is necessary that the liver exist in the location where it is, as it is. But here again, that which can, in all rigor, be called "necessary," is that there be an organ that can anchor the blood vessels, and not the specific means that nature has found to realize this end. Another organ could have fulfilled this function, possibly less well, but possibly better. We will have the occasion to rediscover, in the animal world, this distinction between that which suffices in fulfilling the function, and that which fulfills it better.

But this necessity is "hypothetical" (ἐξ ὑποθέσεως). A hypothesis is a proposition that is posited to serve as the basis for an argument, one that has been demonstrated or accepted without demonstration. Among the forty-two uses of the expression ἐξ ὑποθέσεως that we find in the Aristotelian corpus, a majority refer to a deduction of a conclusion from premises. Here's an example from the *Posterior Analytics*: "If it is not possible to know the first premises, it is also not possible to know absolutely and in a proper sense that which follows from them, but one may know it hypothetically (ἐξ ὑποθέσεως) by assuming that the premises are true" (1.3, 72b13). There is a small difficulty here, that notably happens due to the fact that hypothetical necessity is sometimes called "conditional necessity." Because if that which is necessary, that the saw be made of iron for example, is indeed a *condition* for achieving the end, sawing, it is indeed the goal, sawing, that is posited as *hypothesis* to make it necessary that the saw be made of iron. In the same way, from the fact that the organism of a particular blooded animal could not function without its large vessel having a fixed point in the middle of its course, the presence of a liver is indeed a condition for the survival of the animal, but it is the end, which would be intermediate (the fixing of the vessel), or general (the survival of the animal), that is the *hypothesis* that makes necessary the presence of the liver. We will find there one of the main differences between Aristotle and the Presocratics whom he criticizes. For them, in fact, it's the necessary properties that determine the end. Thus, according to an argument that Aristotle attributes to Empedocles, if it is found that sharp teeth are formed simply by the play of the properties of the material, they can serve as front teeth for an animal. For Aristotle, on the contrary, the end determines the means.

The subject has been sufficiently exhausted that we should be able to avoid taking it up from the start again. Two interpreters, David Balme and John Cooper,[26] can serve as a basis for our discussion; they engaged in a dispute that ended in a partial retraction by Balme that can serve as a starting point. In his translation with commentary of *Parts of Animals* book 1 (1972), Balme understands lines 639b23–25 as signifying "the absolutely necessary is present in what is eternal, but it is the *hypothetically necessary* that is present in everything that comes to be, as it is in the artefacts,"[27] meaning by that, that only hypothetical necessity is at work in the case of animals, for example. Balme, in his retraction, claims that only the passage at 642a6 suggests that it is only hypothetical necessity that would be at work in perishable natural beings, but that is not certain: the text says, "but, in fact, a third sort exists among beings subject to generation" (ἔστι δ᾽ἔν γε τοῖς ἔχοθσι γένεσιν ἡ τρίτη), which does not necessarily mean that it exists alone. Cooper, on the other hand, argues that absolute necessity, which he calls "Democritean," is indeed at work in perishable natural beings, but in a specific way. In the formation of a living being, Democritean necessity indeed plays a role, because the

26. Aristotle, *Aristotle's De Partibus Animalium I*, and Balme, "Teleology and Necessity"; Cooper, "Hypothetical Necessity."

27. Aristotle, *Aristotle's De Partibus Animalium I*, 4.

matter of which that being is constituted necessarily obeys the laws that govern it (as we have said, earthy composites tend downward, except in special cases, themselves necessary, while certain other composites rather go upward, etc.), but this necessity is exercised exclusively hypothetically, that is, in the *hypothesis* that the living being is the goal to be achieved. According to Cooper, for Aristotle, material composites of an animal indeed act according to the "laws" of matter and according to these "laws" only (and that's why I called attention earlier to the fact that iron is suitable for being the matter for a saw "by virtue of a material necessity"), but the process that set these material components in motion to result in a particular organism only obeys a hypothetical necessity. Cooper is certainly right about that.

Cooper takes an important step when he says that the initiating movement of material processes (in the formation of an embryo for example) always obeys a hypothetical necessity: if these are indeed processes directed by a material necessity that, for example, forms a membrane deployed around the fetus, it is because the materials are *first* ordered according to a hypothetical necessity to form this fetus.[28] "Aristotle," writes Cooper, "'subsumes' Democritean necessity under hypothetical necessity, in the explanation of living things. But that does not mean that he *reduces* it to hypothetical necessity."[29]

If then there are certain organs or life processes that are present or are produced without having a goal, a point strongly affirmed by Aristotle, one ought to consider them as due solely to the interaction of their material components, but they occur in an organism that has itself been formed necessarily, but in that case, according to hypothetical necessity. It is necessary, in fact, that the end, the organism, be given (i.e., be *posited* as a hypothesis) so that the material components make that which is required at the moment needed. Thus, we have here a reaffirmation of the primacy of structure over generation: "For since health, or a person, is of such and such a character, it is necessary for this or that to exist or be produced; it is not the case that, since this or that exists or has been produced, that of necessity exists or will exist" (*Parts of Animals* 1.1, 640a4). A case particularly interesting, because it involves an example that interpreters habitually cite to remind us that in Aristotle not everything is purposive in a living being, is that of bile:

28. Cooper, "Hypothetical Necessity," 267.

29. Cooper, "Hypothetical Necessity," 264.

> But even though the bile that is found in the rest of the body seems to be a certain residue or a certain dissolution, similarly it seems that the bile in the region of the liver is a residue, and is not for the sake of something, just like the excretions in the stomach and the intestines. Of course nature sometimes makes use even of residues; nevertheless it is not appropriate to look always for a purpose in every case; some things being as they are, bring about other things that happen necessarily. (*Parts of Animals* 4.2, 677a11)

Not only is bile not for the sake of some good for the animal that secretes it, but it is even harmful, and Aristotle claims that the less bile that an animal secretes, the better are its chances for a long life (*Parts of Animals* 4.2, 677a35; *Posterior Analytics* 2.17, 99b5). But the nature and the functioning of the liver bring it about that it produces bile *necessarily* in most animals. In fact, it is not necessary that every liver produce bile. Thus, at *Parts of Animals* 4.2, 676b25, Aristotle notes that "there are animals that do not have bile, such as horse, mule, ass, deer, and goat." One may justly say that the liver is hypothetically necessary if one is to have the large blood vessels fixed to the trunk, and that the food be concocted, so that the rule discerned by Cooper is indeed respected: although the bile does not have a final cause, it is produced as a result of a sequence that begins with a hypothetical necessity, because the liver, which produces bile, indeed is the consequence of a hypothetical necessity, as we have just noted. Thus, we find a phenomenon that is harmful to the animal that is incidentally brought about by a hypothetical necessity, that is, a necessity aiming at an end, in other words a good.

But it seems that there are many important exceptions to this schema; the principal one is noticed by Cooper himself. A passage in the *Generation of Animals* needs to be read carefully; it is particularly useful for us:

> It is possible that each thing exists for the sake of something, and is thus brought into being by the final cause and by the others, whichever are present in the definition of each thing, or which either are for the sake of some end or are ends for which something exists. As for generated things that are not like that, we must look for the cause in the movement and generation, considering that these things differ in the course of their very constitution. In fact, it is by necessity that an animal

> will have an eye, because one assumes that it is an animal of this kind (τοιόνδε γὰρ ζῷον ὑπόκειται ὄν), but it will have an eye of a particular kind of necessity in another sense, not the sense mentioned just above, because it is its nature to act or be acted on in this or that way. (5.1, 778b10)

The first sentence provides a list of cases in which the final cause intervenes in the explanation of vital realities. A characteristic, especially an organ, can belong to an animal as a consequence of the essence of that animal. It then belongs to it necessarily, but not according to hypothetical necessity, as Cooper has seen well. We cannot say, "If a blooded animal is to exist, it is hypothetically necessary that it have blood," or "if a bird is to exist, it is hypothetically necessary that it have wings," because the blood or the wing is *that which makes* an animal blooded or a bird. Thus, the fact that human beings, for example, necessarily have eyes, might seem to be a consequence of a hypothetical necessity: if such and such a being is an animal, that is, a living thing provided with perception, then it is necessary that it have organs of perception, notably, in the case of the human being, eyes. In fact, the presence of organs of perception in an animal is necessary, not hypothetically (i.e., in taking into account an end or an essence), but from the fact that the capacity of perception is part of the essence of the animal, and is even definitory of the animal: an animal is a living thing (i.e., a being able to nourish itself and reproduce, two functions of the "nutritive soul") that also has the faculty of perception (function of the "sensitive soul"). To be sure, there are animals that are not endowed with vision, like shellfish and those that perceive by touch alone, the basic sense that all animals possess. But a human being necessarily has eyes, except in the case of a monstrosity, to which we will return.

That the presence of organs of perception in an animal does not derive from a hypothetical necessity means that the organs of perception are given with the animal, and we are not in a position of having to figure out how nature discovered that she was obligated (had the necessity) to dedicate an organ to a function, for example, to tie down the large blood vessel, or the smith the obligation to find a material sufficiently rigid to make a saw. However, one might ask oneself whether the fact that a human being has eyes does not derive from a hypothetical necessity. It seems to me that it does not, because, given that a living thing cannot have more than five senses, as the first chapter of *De Anima* 3 demonstrates, and that the human being is one of those animals sufficiently perfect to have all five senses, it follows necessarily from the human essence that human beings have eyes and that the fact that certain do not have eyes is a monstrosity. There is no hypothetical necessity there.

If the example of the eyes seems to be controversial, one can invoke others, for example that of the bipedal nature of birds. We read in the *Parts of Animals* (4.12, 693b5): "The bird is necessarily bipedal, because the essence (οὐσία) of the birds is that of blooded, and at the same time winged. But blooded animals move at no more than four points." We have seen that being blooded and being winged are not hypothetically necessary characteristics for a bird, but are parts of its essence: one cannot say, "If there is going to be a bird, it is hypothetically necessary that it have wings," because "winged" belongs to the definition (essence) of the bird. But the rationale by which Aristotle attributes bipedalism to birds rests on the general theory of animal movement: every blooded animal has four support points, and that can be verified in birds, because the two wings correspond to the front legs of quadrupeds, so that a bird necessarily has the sort of bipedalism that we observe. This is not a hypothetical necessity, which is now at issue, but a material necessity. But "winged" and "biped" are not of the same status, because, contrary to the fact of having wings, bipedal is not part of the definition of the bird, but derives from its essence via the general theory of animal movement. It therefore makes sense to say that the bird is necessarily a bipedal animal, but that necessity is not subsumed under hypothetical necessity, because for bipedalism as well as for the presence of wings, nature has yet to find the means necessary for the realization of an end: if the morphological structure of the bird is given, it will necessarily be biped. In an identical perspective, a remarkable passage in the *Parts of Animals* says:

> And as the sensory, locomotive, and nutritive faculties are found in the same part of the body, as has already been said elsewhere, it is necessary that the part that immediately possesses such principles, on the one hand in that it can receive all the sensibles must be a simple part, on the other hand, in that it is moving and active, it needs to be anomoiomerous. That is why in nonblooded animals it is the analogue of the heart that is the part in question, and in blooded animals, the heart. (2.1, 647a24)

It is necessary that the "perceptive center" of an animal be a part that is simultaneously simple and anomoiomerous, because as a perceptive organ it must be simple. It is doubtful that we have here a hypothetical necessity because one cannot say that, given the hypothesis that an animal should exist, it is necessary that it have the sort of command center that controls its basic animal faculties, that is, nutrition, sensation, and movement. Rather, it's a matter of a necessity that follows from the very nature of these functions: if a part must receive all sensibles, it must necessarily be simple; if it is the seat of local movement, it has to be anomoiomerous. Some explanations are in order. At 647a5, Aristotle explains that "perception occurs in all the homoiomerous parts because whichever of the senses is a sensation of a unique kind, and for each of the sensibles there is an organ of sense that can receive it. But that which is potentially affected by that which is actually, so that this and that are generically one and the same thing." It is a fundamental thesis of the Aristotelian theory of perception that the perceiver and perceived are generically (γένει, 647a9) one and the same thing. But, in each perception, each sense perceives one single sensible quality, or more precisely, one single contrariety (cf. *De Anima* 2.11, 422a23; 3.1, 425a19). Thus, the part that perceives must be simple. A fortiori, then, the organ that is at the origin of all the sensations, that is, the heart or its analogue, must be simple. That does not mean that the perceptual organs, for example the eye, are simple (they are obviously anomoiomeries), but that the affected part (impressed) by the sensible be homoiomerous. But "the heart, in fact, is divisible into homoiomeries like each of the other viscera, while as a consequence of its external configuration it is anomoiomerous" (*Parts of Animals* 2.1, 647a31), that is, it is simultaneously simple and anomoiomerous. It is therefore able to be the perceptive-motor center. The very terms used by Aristotle seem to signify that this necessity for the heart to be the seat of these functions is not dictated by an end: this part, the heart or its analogue, "possesses immediately" (τὸ ἔχον πρῶτον μόριον, 647b27) the principles of these functions.

The exercise of sense perception is thus framed by the set of necessary conditions that depend primarily on that which may be called a "law of compatibility," which imposes an appropriateness, in fact a form of identity between the perceiver and the perceived. It's not a matter of hypothetical necessity. We will return a little later to the relationship between the homoiomerous and anomoiomerous parts.

To return to the passage cited from *Generation of Animals* 5.1, the two other cases in which there is an intervention of the end are described thus: "that which either is for the sake of some end, or is an end for which something exists," that is, those concerned with finalist causal explanation, other than the attributes that belong to the essence of the animal (mentioned first in the passage), the properties that are for an end, and those that represent an end. One may understand better what Aristotle means if one reads this text in parallel with this well-known one from *Parts of Animals* 1: "That's why one must say especially that since this is the essence of a human being, this is why it has these characteristics, because it cannot exist without these parts. If this is not possible, one must say the next best thing that in a general way it cannot be otherwise, or at least that it is good thus" (1.1, 640a33). What can we get from this passage about hypothetical necessity? *By definition*, the human being, qua animal, possesses perceptual organs, as we have noted. The second case, "it is impossible that it be otherwise," could have been expressed by Aristotle in a positive form, "it is necessary that it be thus," for example, if a human being is going to be alive, it is necessary that it have an organ for concocting food (*Parts of Animals* 3.7, 670b25), but also to anchor its large blood vessel toward the front of the body (*Parts of Animals* 3.7, 670a13), this organ being, for it, the liver. Possibly Aristotle did not use this positive formulation here in order to refute in advance certain interpretations that might have been able to see in the first two cases two expressions, each a little different, of the same thing,[30] which goes contrary to the letter of Aristotle's text that introduces the second explanation if the first "is not possible." But, after what has been said, if the two cases describe indeed one form of *necessity*, the first does not express, in contrast to the second, a hypothetical necessity. James Lennox carefully avoids blending these two first cases into one, while he says that "the explanation takes the form of conditional necessity."[31] Thus we say again that for Aristotle it does not amount to the same thing to say that "if X is a human being, then it is necessary that he have sense organs (including eyes),"[32] and saying "for a human being to live, it is necessary that its food be concocted and its large vessel be fixed, and thus a liver

30. As does, for example, Sorabji, *Necessity, Cause, and Blame*, 155.

31. In his commented edition of the *On the Parts of Animals*, 134. [Translator's note: quote verified.]

32. I repeat what I said earlier: the possession of perceptual organs is not hypothetically necessary, and the fact that a human being has eyes is not either, because this is due to the degree of perfection of a human animal. We will ask ourselves a little later whether, as Cooper thinks, the fact that his eye is composed of liquid is hypothetically necessary for the eye to provide a sufficiently sharp vision.

is hypothetically necessary for the performance of these two functions." The liver is not, in fact, part of the essence (definition) of a human being, and furthermore is not necessary by virtue of the general laws of animal economy, as is the case for the bipedalism of birds, but it is an organ that conveniently fulfills this function, although we cannot say that another organ could not have fulfilled this function.

As for the third case, the two lists, that of the *Generation of Animals* and that of the *Parts of Animals*, in contrast to what it seems from first glance, seem to refer to the same type of explanation. The part of the sentence "that which either is for the sake of some end, or is an end for the sake of which something exists" in the passage in the *Generation of Animals* introduces a contrast between a characteristic that is for the sake of something (the liver for the sake of concocting the food, which is the final cause) and a characteristic that is not for the sake of something else, but itself constitutes its own end. An interesting example is that of the spleen, discussed in *Parts of Animals* 3.7, if only because Aristotle seems to experience some difficulty in dealing with it, which appears in some expressions that are astonishing, to say the least. At 669b26, Aristotle contrasts animals "that necessarily have a spleen" and those "that do not have it necessarily and in which it is very small," obviously assimilating the absence of a spleen and the presence of a vestigial spleen incapable of performing the function of a spleen. The spleen is necessary from two points of view. From one point of view, morphological and structural first: from the fact of the "law" presented in the preceding paragraph according to which viscera have a double nature,[33] the spleen seems to be necessary as a dependent of the liver. This is a matter of what has been called earlier a law of animal economy. But since the liver is itself divided in two, the spleen "is in a certain way necessary, but not very (μὴ λίαν) necessary" (670a1), an expression that is, at the very least, curious. From a functional point of view, the spleen serves both to fix the large blood vessel (concurrently with the liver, in a way that is not perfectly clear in the text), but "the liver and spleen also assist in the concoction of food, for being full of blood, they have a hot nature" (670a19). In this last function too,

33. "But, in fact, all the viscera have a double nature" (*Parts of Animals* 3.7, 669b17), "in fact," that is, contrarily to what would seem to be the case for some of them, like the heart. This "law of symmetry" applies to the *fact* that animals, especially the human being, have the same three pairs of dimensions as the universe: up/down, right/left, front/back (cf. 669b19).

the spleen seems to supplement the liver, and anyway "it indeed seems that the spleen is like a false liver" (669b28). Then comes this astounding assertion: "It is by accident that the spleen belongs necessarily to those that have it" (ὁ δὲ σπλὴν κατὰ συμβεβηκὸς ἐξ ἀνάγκης ὑπάρχει τοῖς ἔχουσιν, 670a29). It is a bit provocative to translate here κατὰ συμβεβηκὸς by "by accident"; we are rather in a context that is also that of "per se accidents" (συμβεβηκότα καθ' αὑτά), an expression that Aristotle uses to designate properties that belong per se to a subject without being part of its essence. Doubtless one must understand this expression as signifying "being necessarily something concomitant that is not essential" for an animal.[34]

The spleen belongs necessarily to animals that have one, "but not very necessarily," and some animals lack this part. We have here a case in point that will become much more intelligible for us when we will examine another theoretical tool deployed by Aristotle, different from hypothetical necessity, that I have called "the two natures." Produced by the necessary interaction of material elements, the spleen is found to be produced in certain animals. It's not a matter of hypothetical necessity, because this necessity is not for the sake of an end, as is the case for the liver: the spleen, if one accepts Aristotle's texts with all their obscurities, their lacunas, and their bizarre expressions, seems to be used teleologically because it is there, and Nature knows, as we will see anon in detail, to use all available means, and that, according to a very celebrated formula, "nature, like a good housewife, has the habit of not throwing away anything that might prove useful" (*Generation of Animals* 2.6, 744b16). But it is present by virtue of an important law of animal organization that brings it about that viscera are paired. One may also presume that, in the case of the concoction of food, but also in that of the fixing of the blood vessels, these functions are best accomplished when the spleen collaborates with the liver, which corresponds to the third of the cases enumerated in the passage from the *Parts of Animals* cited earlier.

It's not a matter of a reality that, as is the case for bile, is produced by material necessity alone, and would be associated indirectly with a goal *via* an original hypothetical necessity, according to the analysis proposed by Cooper, but of an organ that has been produced by a material necessity (like everything that composes a living body), but which also participates in the goal, the survival of the animal, however, not as something that is

34. I do not believe that one may, like Michael of Ephesus (62.3ff.), understand that the spleen necessarily produces a residue, but that this residue is accidental because it does not have a final cause.

necessary for that survival. It is better to have a spleen, because the food will be better concocted, and the large vessel better fixed, but this is not hypothetically necessary, because it is not necessary at all. In any case many animals do not have a spleen, or have a nonfunctional spleen. From all that we can rediscover a convergence between the two descriptions of the third cases given in the *Generation of Animals* and *Parts of Animals*: the spleen has indeed its end in itself, and can thus be considered as "something for the sake of some end" and, without being necessary for the performance of a function, it facilitates that performance, and "is well thus." If, then, in the case of bile, there is indeed an *originating* hypothetical necessity, if one accepts Cooper's analysis (it is necessary that an organ, the liver, fixes the large vessel if one wants the passage of blood to be accomplished correctly, and this organ, by material necessity, produces bile), there's nothing like that for the spleen: it is not a by-product of an organ that is itself hypothetically necessary, and furthermore, the spleen is not necessary for the function to which it is *connected*, since that function could have been accomplished without it, even if it would have happened less well. But hypothetical necessity is a kind of necessity.

The end of the passage in the *Generation of Animals* that we are in the course of studying opposes two sorts of necessity, taking the example of the eye. The first seems to have the form of a hypothetical necessity: an animal "would have an eye, because it is assumed that it is an animal of this kind"; but this hypothetical *form*, stressed by Lennox in relation to the first two cases, can express a necessity following from the essence of the animal, as we have seen. The second sort of necessity is a material necessity ("because there is produced by nature such or such an action and passion"). The contrast is thus between the necessity linked to finality, which would be hypothetical or following directly from the essence of an animal, and material necessity. It is the latter that brings it about that the eye is as it is, notably in terms of color. This case does not fully conform to the schema proposed by Cooper, because the color of the eyes, which is produced by a purely material necessity, is related to an end, but not by the intermediacy of a hypothetical necessity, but by a necessity following from the very essence of the animal, if our analysis of the necessity of the eyes (with which Cooper agrees) is correct.

It is sometimes difficult to decide what, according to Aristotle, belongs to the essence of an animal, that which, in the passage of the *Generation of Animals* cited earlier, he characterizes as "that which belongs to the definition of each being" (778b12). To the point that the necessity following from the essence and hypothetical necessity borrow, as we saw in the parallel passage in the *Parts of Animals*, very similar formulations. In any case, it is not the universal character of a function that makes it follow from the essence of an animal. A passage from the treatise *On Sleep* proposes an interesting form of hypothetical necessity in that it is a matter of a trait that belongs to *all* animals ("each animal," ἑκάστῳ τῶν ζῴων, 2, 455b25) and that, nevertheless, is not part of their essence, because it is explicable by hypothetical necessity. It is a precious passage also because it is almost the only one to be found in Aristotle's biological writings dealing explicitly with hypothetical necessity, without being a methodological text, but on "actual" zoology: "Sleep belongs of necessity to each animal. I use the term 'necessity' in its conditional sense, meaning that if an animal is to exist and have its own proper nature, it must have certain endowments; and, if these are to belong to it, certain others likewise must belong to it" (2, 455b25). It is not always easy, either, to decide whether Aristotle is appealing to hypothetical necessity or not. For example, in the case of the eyes, to which Cooper makes extensive reference. We have noted that the fact of having eyes, for example for a human being, does not arise from hypothetical necessity. From the fact that certain animals, "human beings, birds, viviparous and oviparous quadrupeds" (*Parts of Animals* 2.13, 657a25), have fluid eyes it follows that nature has endowed them with eyelids to protect them. The eyelid seems to be an incontestable example of hypothetical necessity: the fluidity of the eyes makes it necessary that they be protected. But should one consider, as does Cooper, that the fact of having fluid eyes is hypothetically necessary for the animals that have them? That is not sure. Perhaps it would be better to think that, given that this is a human being, for example, it is better that it have fluid eyes rather than hard eyes, like those of insects. But that does not make fluid eyes *necessary*: some animals that have fluid eyes could surely live with dry eyes, even if they would live less well, and Aristotle, as we will see, defines a threshold of acceptability for disadvantageous characteristics that an animal can have. We thus rediscover here a case close to that of the spleen.

We can from that make a little more precise what hypothetical necessity is by considering some of its characteristics, starting by coming back to a point already discussed. A characteristic is said to belong to, or to be formed in, an animal by hypothetical necessity if, given the essence of that animal, the characteristic in question is necessary for its existence, or for the carrying out of the functions of that animal. But, as the text of *Physics* 2.9 says, "The necessary resides in the matter, while that for the sake of which is in the idea" (200a14). No matter how tied to the end it may be, hypothetical necessity thus is applied to the matter of the animal under consideration, as is also the case for the saw: if there should be a saw, it must be made of iron or bronze, that is, that its matter be iron or bronze. There's a fundamental point that has escaped many commentators who have believed that this "new" material necessity abolishes "mechanical" material necessity as described by the Presocratics, and such as Aristotle saw at work both in nonliving and living things. The finality at work in living beings does not work miracles, if one means by "miracle" something that happens or exists in contravention of the laws of nature. The matter that composes living beings can only obey what we call the laws of nature: Aristotle's finalism does not carry him back to a magico-mythic level, as we have seen, and Aristotle, as we will see again later, was not a vitalist of exception as was Bichat. In a sense, then, and Cooper has adopted this position contrary to many others, there is only, in the animal world, necessity in the sense of material necessity, but it becomes hypothetical when it contributes to the realization of an end.

One must not be misled concerning the sense of this last assertion, for material necessity cannot by itself *produce* a living being. When a material element intervenes in the production of an animal, for example during the development of the embryo in utero, it indeed does so according to its own properties, but if it is not directed by an end, which Aristotle defines as form or soul, there will not be generation. This is true both at the level of the formation of the parts and at that of entire animals.

Among the passages that present this doctrine, the most famous, to which we will soon return, is found in the *Generation of Animals*:

> If then the hot and the cold can make something hard or soft, sticky or brittle, and all the other properties of this sort that belong to animated parts, on the other hand, the reason for which this becomes flesh, that becomes bone, the hot and cold can no longer produce: it is the movement of the generator that is in actuality (entelechy) what is potentially that from which there is a generation, as in the case of that which is generated by art, for hot and cold make the iron hard or soft, but what makes the sword is the movement of the tools, a movement that possesses the principle of the art. (2.1, 734b31)

Thus, hypothetical necessity does not replace Democritean necessity in living things: just as much in the production concerning the structure of an animal, given that the animal is what it is, this generation or structure requires a certain material disposed in a certain way. In the image of the wall presented at *Physics* 2.9, it is necessary that the stones be located beneath the earth, which is below the wood, for if the stones were in the upper part of the wall it would collapse. But it is the mason who carries out the project of a wall that will be a part of a house, and it is the mason who can arrange the constituent parts in a way to fulfill that function. But the stones, earth, and wood, once arranged in a "viable" wall, keep their material properties, notably that some are heavier than others, and thus continue to be subject to Democritean necessity alone. Cooper, as we have seen, expresses this very correctly when he says that "Aristotle subsumes Democritean necessity under hypothetical necessity." Reciprocally, hypothetical necessity is not substituted for the end, nor does it indicate the end, unless an Empedoclean position is adopted (as described by Aristotle). Or, to put things in yet another way, if it's going to saw, the saw must hypothetically be made of iron, but if the saw needs to be made of iron, it is not hypothetically necessary that it be able to saw.

Ultimately, what is it that makes something necessary, and in this instance, hypothetically necessary, that human beings have eyelids, and eyelids of a certain kind, and one may pose the same question for bones, organs, and so on? It is the fact that human beings have always existed, and will continue to exist forever, with the kind of eyelid that they have, that this sort of eyelid accomplishes a function that brings it about that, integrated into the whole set of functions of the human body, it permits, or facilitates, the everlasting survival of the human species. The human eyelid cannot be other than it is, because it has never been, and will never be otherwise; this is due to the fact that, during the conception of a human being, the female matter is informed in a *specific* way by the movements transmitted by the male sperm, and also by the movements proper to this matter. The matter is not an inert reality, as we will recall later, and this information, being applied successively to all the parts during the formation of the embryo, results in eyelids of a certain kind being formed. We must note that this necessity that comes from the form is a *natural* necessity, which is not always exercised, but "most of the time," since animal reproduction is not without hazards: not only are there small variations that bring it about that, for example, the eyelid of Socrates is not that of Coriscus, but there sometimes are real *deviations* that give birth to monsters. These deviations are mostly due to the fact that the female material is incompletely or badly formed by semen brought to it by the male.

There are cases of hypothetical necessity that may be called "doubtful," because they are in some way "negative." Thus, at *Parts of Animals* 3.1, 661b26, Aristotle, while applying a finalist principle according to which nature provides the means to those able to use them, writes:

> One must grasp a general fact that will be found applicable not only in this case but also in many others that will be dealt with later on. Nature allots each weapon, offensive and defensive alike, to those animals alone that can use it; or, if not to them alone, to them in a more marked degree; and she allots it in its most perfect state to those that can use it best; and this whether it be a sting, or a spur, or horns, or tusks, or what it may be of that kind. Thus as males are stronger and more courageous than females, it is in males that such parts as those are found, either exclusively, as in some species, or more fully developed, as in others. For though females are of course provided with such parts as they are necessary for them, the parts, for instance, that serve their nutrition, they have even these in an inferior degree, and the parts that answer no such necessary purpose they do not possess at all. This explains why stags have antlers, while does have none.

We need to make several remarks about this passage. The two mentions of the "necessary" (ἀναγκαῖον) at 661b34 and 36 (the parts that are necessary even for females and those that have "no necessary purpose") do not appeal to the same kind of necessity. The necessity of having a digestive tract follows directly from the very essence of the animal such as it is and, as we have seen, is not a hypothetical necessity. As we will see, to say that antlers are useless for stags is only approximately true, and to say that antlers are not necessary for the females means that their possession has such inconveniences that they would put the survival of the does in danger, as we will also see. This case is different from that of the ears of birds, which we will discuss, because it would certainly have been advantageous for birds to have ears, while here we can doubtless apply the schema of hypothetical necessity: if the does are going to survive, a sine qua non condition is that they do not have antlers. But that an absence be hypothetically necessary doubtless constitutes a limiting case of hypothetical necessity.

Certain absences, on the other hand, cannot be explained by hypothetical necessity. The short chapter 2.12 of the *Parts of Animals* explains why birds do not have ears. In fact, they do not have enough material to make ears because the supply of material available in these animals was used up in making feathers, a situation that we will find again concerning horns. This absence of ears is a necessary fact (if there is no matter, necessarily there will not be the organ), but it is difficult to see a hypothetical necessity or to derive this absence from a fundamental hypothetical necessity. It is interesting to compare this with the passage in *Generation of Animals* 5.2, 781b22, that discusses the seal, which also does not have ears. But the reason for that is that, from the fact of living primarily in the water, not only would the ears be useless, but they would even be harmful by accumulating water. It is nevertheless not certain that one may find here the traces of a hypothetical necessity, even though negative. The absence of ears, though it is assuredly advantageous to the seal, is doubtless not necessary in the sense that one may imagine that a seal could survive with ears. Otters and beavers, mentioned by Aristotle several times, live in the water and have ears, a fact that can be difficult to ignore.

There is, finally, a case that goes contrary to the pretention of commentators like Balme and Cooper to cover the entire field of zoology with hypothetical necessity. There is the example, particularly interesting and often invoked, of the formation of anomoiomerous parts from homoiomerous parts. It indeed seems hypothetically necessary that homoiomeries have the properties that they do:

> The homoiomerous parts are for the sake of the anomoiomeries. From these latter in fact come functions and actions, for example, an eye, a nose, a whole face, a finger, a hand, a whole arm. But given that the actions and movements that belong as much to whole animals as to parts of this kind are present in many different forms, it is necessary that their components have different powers. In fact, in some cases softness is needed, and in others hardness, in some cases the ability to extend is needed, and in others flexing. To the homoiomeries have been distributed the powers of this kind individually (one of them is soft, another hard, one humid, another dry, one sticky, another brittle). (*Parts of Animals* 2.1, 646b11)

Thus, we have here a schema of the type: if bones need to support flesh, they have to be hard. But that's not where Aristotle goes, because a few lines farther along, he writes:

> These different powers serve the hand for pressing and grasping. That is why the instrumental parts (τὰ ὀργανικὰ τῶν μορίων) are composed of bone, tendon, flesh, and other parts of this kind, but the latter are not composed of the former. Thus, as they are for the sake of something, by virtue of that cause (the final cause), these parts (homoiomeries and anomoiomeries) are arranged as we have said; but when one tries to find out also how it is necessarily thus, it is obvious that there existed already among them (the homoiomerous parts) by necessity the mutual relationships that they have. (*Parts of Animals* 2.1, 646b24)

It's a matter here of a situation purely Democritean in which the components that constitute more complex beings have among them preexisting affinities of forms that explain their aggregations. We are in a schema, one that we will study in more detail, different from that of hypothetical necessity, a schema according to which the form makes use of a state of affairs in which it does not participate: nature has to *profit* from these properties of homoiomerous parts, which they owe to the relationships that they have between each other since forever, to make them serve for the construction of anomoiomerous parts. It is because flesh and bone had between them determinate relationships that they could allow the flexion of members in a manner appropriate to each species. One cannot help finding in a passage like this an echo of Democritus's "hooked atoms."

In order to complete our discussion of Aristotle's hypothetical necessity, it is useful and very important to notice in what context Aristotle appeals to it. One must note that the explicit references to this doctrine in the presentation of the properties of animals are rare (we have presented just one, concerning sleep), the other references are found in methodological passages, but also in polemics against the mechanism of the Presocratics. But this mechanism puts into operation two theoretical schemas that are closely linked. The first affirms that there are no goals in nature, the second that structures are explained by how they came to be, and not how they came to be by what they are, two positions absolutely opposite to those adopted by Aristotle. Doubtless this relationship between structure and genesis is precisely the main point of disagreement between Aristotle and people like Democritus, since for them, structure is constituted simply by the play of material necessity, while for Aristotle, a new structure cannot be constituted unless it is posited by an already existent structure, that the new structure *reproduces*, with individual variations, which in turn depend on material necessity. It is necessary that a being exist before it comes into being. If one does not want to abandon the major achievement of Presocratic philosophy, namely determinism (we repeat, Aristotle has no intention of adhering to a miraculous conception of nature), it is necessary to maintain that necessity, which is material necessity in that it depends on the properties of matter, can be in some way regulated by finality.

I would therefore be ultimately inclined to argue that the doctrine of hypothetical necessity is above all *polemical*, meant to attack philosophers like Democritus on their cardinal concept, that of necessity, and show that the very notion of necessity can be rescued from an Aristotelian point of view, if only one gives it the form of hypothetical necessity. That it is more a matter of polemic than of science can be seen from at least two indications: First, as we have already noted, in many cases it is undecidable whether or not one may take them as cases of hypothetical necessity. Second, Aristotle appeals to hypothetical necessity primarily, or for the most part, to explain processes of *generation*, and that is easy to understand if one remembers that for the Presocratics, generation is the main starting point of the explanation of natural realities. But when he is himself explaining the generation of animals or their parts, Aristotle makes little or no appeal to hypothetical necessity. Thus, it is remarkable that in the work where Aristotle deals scientifically with animal generation, the *Generation of Animals*, he makes no explicit appeal to hypothetical necessity. Of course, Aristotle shows that by use of his concept of hypothetical necessity Presocratic mechanism can be attached to Aristotelian finalism, because controlled by it, but Aristotle gives only a rather weak heuristic value to hypothetical necessity. Everything happens as if this concept, useful in his anti-mechanist polemic, has been subsumed under a more effective schema, one that we will call, though this name has not become usual, "the two natures."

The Two Natures

In the chapter of *Parts of Animals* 3 dedicated to horns, Aristotle gives us a kind of summary of his entire zoology; that is the main reason why commentators have so often leaned on this text. Horns are made of earthy matter contained in the animals, which "necessarily flows upward, to be distributed in some to the teeth and defenses, in others to the horns" (*Parts of Animals* 3.2, 663b34, after PP). Aristotle conceives this earthy matter—without explaining why it moves upward (since earthy matter ought necessarily to tend downward[35]), like a residue of the digestion of food. Digestion is a necessary process that Aristotle, like many among the ancients, assimilated to concoction, which results in different sorts of substances, of which the most important is blood, produced in the heart as a consequence of its heat. The other parts are formed under the influence, notably, of hot and cold, for example the rigid parts, like bones, teeth, horns, and cartilage. We will come back to that with more details.

What Aristotle calls the parts of animals, whether they are homoiomeries (blood, flesh), anomoiomeries (face, limbs), or intermediate like the heart, which is indeed composed of a unique material but in a form that cannot be divided into hearts, are not formed by a single play of the properties that compose them (earth, water, air, sometimes fire), during the generation of the animal. They are constituted under the control of the end, which is the essence of the animal under consideration. Some of them, following necessarily from the essence of the animal, like blood; others, as hypothetically necessary for the existence and functioning of the organism in question. We have seen that for bones. Aristotle has explained to us that it is the form of the parent animals that in the last resort determine the successive appearance of all the parts with the properties that they have. This form is transmitted by the movements impressed by the semen of the male on the female material. That's what the famous passage of *Generation of Animals* cited previously says, without actually mentioning "form":

> If then the hot and the cold can make something hard or soft, sticky or brittle, and all the other properties of this sort that belong to animated parts, on the other hand, the reason for which this becomes flesh, that becomes bone, the hot and cold can no longer produce: it is the movement of the generator that is an actuality (entelechy) what is potentially that from which there is a generation. (2.1, 734b31)

35. J. Lennox is right to say that this movement upward cannot depend on the earthy nature of the horns per se, and needs to have an efficient cause for the process. Aristotle, *Aristotle: On the Parts of Animals*, 249. But Aristotle does not tell us which. I will propose a possible explanation in the next chapter.

What the movements of the generator introduce is a proportion, a *ratio* in the mathematical sense (the Greek term *logos* covers the same semantic territory as the French term *raison*) between the constituents of the matter, for even when Aristotle says that the matter of horns is "earthy," that in no way means that they are composed only of earth.

First we must add a clarification whose importance will be grasped in the next chapter: that which has just been said is valid only in the case of sexual reproduction, and concerns only indirectly, or with important amendments, all the living things, vegetable and animal, that reproduce in other ways, notably spontaneously. Next, to avoid the impression of schematism that arises from this conception of generation, we can make two remarks, repeating things that have already been said. First, the system of movements is proper to each species; human flesh is not the same thing as dog flesh. Aristotle even goes further by studying the resemblances between offspring and their ancestors over several generations, for the system of movements of the father of Socrates is not the same as that of the father of Coriscus. Next, Aristotle does not conceive of the female matter as pure passivity: not only does it resist the imposition of a form brought by the movements of the male semen (thus transmitting to the offspring certain maternal traits), but it is itself provided with a form that, in certain cases, leads the female to produce living beings, certainly imperfect, but alive none the less, as in the case of wind eggs.[36] We will need to talk about that again.

A second moment in the generation of an animal also depends on its specific form and the individual form of members of its lineage. In fact, once flesh and bone have been formed it is still necessary to assemble them and other homoiomeries to make limbs and organs. This process of composition is also regulated by the form of the animal. But what especially interests Aristotle, in his "eternalist" perspective, is first to shine a light on the hylomorphic relationship that makes an animal exist and be everlastingly viable. Then it is necessary to see how each part is constituted under the control of the form. We have there one of the basic schemas of Aristotelian biology. This situation is perfectly described by a remark that Aristotle makes about horns: "Since necessary nature exists, let us say how rational Nature[37] uses that which exists by necessity

36. Cf. Connell, *Aristotle on Female Animals*; and Pellegrin, "What Is Aristotle's *Generation of Animals* About?"

37. From here on we will capitalize "Nature" as the principle of organization.

for the sake of something" (Πῶς δὲ τῆς ἀναγκαίας φύσεως ἐχούσης τοῖς ὑπάρχουσιν ἐξ ἀνάγκης ἡ κατὰ τὸν λόγον φύσις ἕνεκά του κατακέχρηται λέγωμεν, *Parts of Animals* 3.2, 663b22). The "necessary nature" of an animal is what the first chapter of *Parts of Animals* book 1 calls its "material nature" (ὑλικὴ φύσις, *Parts of Animals* 1.1, 640b29), the same passage comparing it to the material nature of a bed, the wood or the brass. This material is endowed with natural properties, like heaviness/lightness, hardness/softness, that determine certain movements. Thus earth can't go anywhere but down. But we must remember that the organs of animals are never made of one single pure element, earth for example, but are always a mixture of elements. As for "rational Nature," that is what book 1 calls "Nature according to form" (ἡ κατὰ τὴν μορφὴν φύσις, 640b28), or sometimes "Nature as substance" (*ousia*, 641a25),[38] or "soul" (641a1).

One of the major differences between this schema and that of hypothetical necessity is that the necessary nature is a *given* (the genitive absolute τῆς ἀναγκαίας φύσεως ἐχούσης probably has a causal value: "since necessary nature exists") and that the Nature uses "that which exists by necessity." Material nature recovers an ontological autonomy that the relationship of hypothetical necessity had masked, since in that relationship the end "dictated" the manner in which the material had to be arranged: if there must be a saw, or lungs, or horns, the matter must be used thus and so and functions in such and such a way. In the new schema, the Nature finds before herself before a certain quantity of earthy matter and she possesses the potentiality, by introducing into it a proportion, of transforming it into various "parts," "for Nature takes it from there [the teeth] and gives it to the horns, that is, that the food destined to form teeth is used to grow the horns" (*Parts of Animals* 3.2, 664a1). For horns are big consumers of earthy matter. Aristotle notes elsewhere that this problem does not arise for animals that have a sufficient quantity of such and such matter, that is large animals, and "we know of no very small animals that have horns: the smallest, in fact, that is known is the gazelle" (663b27). From forever, then, Nature has *decided* that it was better to give teeth or horns to cattle, for example. Why don't the animals with horns have enough earthy matter to allow rational Nature to give them both plenty of teeth and horns? The answer, which Aristotle does

38. "Since nature is said, and exists, in two ways, as matter and as substance" (τῆς φύσεως διχῶς λεγομένης καὶ οὔσης τῆς μὲν ὡς ὕλης τῆς δ' ὡς οὐσίας (641a25).

not give explicitly, doubtless relies on the fact that teeth and horns are formed from the food, and that the animal can eat only so much food.[39]

So, Aristotle offers us an amazing construction that, by putting into operation the law of organic correlation and the law of the subordination of characters, is at the same time a remarkable example of *biological thought*. One had to wait twenty-three centuries to find again, in Cuvier, a structure of the same kind. We said some of that when we cited his *Research on the Fossil Bones of Quadrupeds*, and we will see it again when we deal with the diversity of animals. Nature having chosen to make horns, because they represent a vital advantage for the animals that have them, found herself short of earthy matter and could not provide these animals with two ranks of teeth. Actually, the food may come to their stomach poorly refined, which would be a big disadvantage. Digestion is, in fact, a basic function of living things, and it is even the case that the faculty of digesting distinguishes living from nonliving. Nature mitigates that inconvenience by giving animals with horns a system with several stomachs that allows them to ruminate and sufficiently elaborate their food. One may add a detail to this picture, one that concerns feet. Aristotle counts three kinds of feet according to the number of their divisions: solid-hooved, like the horse; polydactylous, like human beings and bears; and cloven-hooved, with the foot divided in two. Aristotle appeals to several principles to provide an account of how Nature manages the plurality of forms of animal feet. Nature does not give horns to polydactylous animals because they do not need them; because they can have claws, teeth, or other means of defense; and according to a major Aristotelian teleological principle, Nature does not (generally) provide redundant organs. But, on the other hand, Nature has made a cloven hoof for most animals with horns (Aristotle provides only one exception, the "Indian ass," which may be the Indian rhinoceros) in order to economize the earthy matter of which the horns are made, and also the hooves (3.2, 663a31). A cloven hoof in fact uses up less matter than a solid hoof, and that is why equids do not have horns. It is also a question of a finalized action by rational Nature, one consisting of managing the material provided that she cannot change.

Before returning to the horns, an inexhaustible example, we can clarify the relationship between necessary nature and rational Nature by

---

39. Balme proposes this hypothesis: "Since nutrition is limited, few animals can grow both horns and solid hoofs," this being valid for the teeth. Balme, "Teleology and Necessity," 300.

considering some other examples from the *Parts of Animals*, first that of the brain. Book 2 chapter 7, dedicated to the brain, is a little disorganized and some more knowledgeable, and more daring, than I, could conclude that it is put together out of chronologically distinct layers. This chapter begins by comparing the brain to the spinal marrow, in noting that the brain is the beginning point of the spinal marrow, confirmed by perception ("one sees," ὁρᾶν, 652a26); although the marrow is warm, while the brain is cold, but we should not be surprised about that since "Nature always manages (μηχανᾶται), as a remedy for any excess to set beside it, its contrary for the sake of equalizing (ἀνισάζῃ) the excess of the other" (652a31). Which implies that the brain is the "principle" (ἀρχή) of the spinal marrow in the sense of "origin" and not in a strong sense of existential basis. Following considerations of the relationships between the brain and perception (there aren't any) and the brain and the soul (there aren't any of them either, because if the soul is somewhere, it is in a warm place).

Then Aristotle repeats in a more universal form what would be for modern biology what could be called the law of equilibrium of organs:

> But since absolutely everything needs a counterbalance to achieve measure and the mean (for the mean has the substance and proportion that the two extremes, taken separately, do not) for this reason to balance the location of the heart and the heat to be found there, Nature has devised (μεμαηχάνηται) the brain, and for the sake of this, this part that combines the nature of water and of earth is present in animals. And for this reason, all blooded animals have a brain, while pretty much none of the others have one, unless by analogy, for example, for the octopus. (652b16)

Here too the chronological interpreters have run wild. David Balme, for example, as we have seen, using the pretext that the *History of Animals* (1.16, 494b28, and 4.1, 534b4) says that octopuses, like all mollusks, have a small brain, concludes that this last treatise, being more "accurate," is later.[40]

40. In fact, the words "by analogy" ought to allow removing the contradiction that one ordinarily finds between these two texts: the *History of Animals* distinguishes brains by their size, the *Parts of Animals* by their function, in conformity with the functional distinction between these two treatises noted in the previous chapter.

We see clearly here that nature, which here is rational Nature (the word λόγος is here translated as "proportion"), is the maker of order and *ousia*, like the Mind of the *Timaeus*, but with a fundamental difference. Aristotle's repeated use of the verb μηχανᾶσθαι,[41] translated "to construct," "to devise," shows the distance that separates him from Plato. Of course, μηχανᾶσθαι does mean "to artfully imagine or bring into being," but often it is taken, pejoratively, to mean "to scheme or plot," just as the noun μηχανή means an ingenious invention, but also an expedient, a machination, an artifice. When the word designates a "machine," it is often a theatrical machine, the deus ex machina that fools the audience completely. The machination put into operation by the Aristotelian Nature is not a demonstration of power, but the mark of a weakness that obliges her to take detours. Aristotle's Nature is a sly Nature that tries to get the best of a situation, though ruled by necessary laws that she cannot modify. We get this also in the repeated use of the verb καταχρῆσθαι, "to use, to apply," in the passage speaking of the two natures (necessary and rational): "Nature uses the excess of the residue of a body of this kind, which is present in the larger animals, for their defense and benefit" (663b32). We are definitely far from the dominant Mind of Plato. We will see, in the next chapter, the consequences of this at the level of nature as a whole. We are also far from the hypothetical necessity in which the form dictates conditions to the matter. But, returning for a moment to the horns, we can complete our picture of the relationships between mechanism and finality in Aristotle.

There are limiting cases, like that of deer, which Aristotle discusses at some length, or that of antelopes, goats, or bison. Like other animals with horns, stags make theirs with the earthy matter that Nature has not dedicated to their teeth. But in their case, the negotiation between necessary nature and rational Nature goes badly, "because the large size of their antlers and the multiple ramifications are more harmful to them than useful" (663a10). Aristotle does not say that their antlers are useless, and everyone knows their role in combat between males during rutting season. As in other cases, Nature compensates for this disadvantageous trait by giving the stag speed, and the limbs that make that possible, allowing the stag to escape from its predators. But one cannot help thinking that it would have been more expedient to give them antlers a little smaller

41. Cf. *Parts of Animals* 3.3, 664b32, 665a8; 3.4, 665b13; *Generation of Animals* 2.6, 745a31.

and use the matter thus economized for a more complete dentition. As for antelopes and goats, they are, with stags, animals "in which the excess of development of the horns makes them naturally useless" (663a8), but in them, the lesser development of the horns brings it about that they can use them *in certain cases* to defend themselves (doubtless one may think that in that case the enemies are small enough), and that in other cases (doubtless, large predators) Nature has given speed to them too. The teleological principle of sufficiency, according to which "Nature has not given to the same animals the means of defense that would be simultaneously sufficient and plural" (663a17) is thus not violated.

The case of bison demands a closer look. The text of *Parts of Animals* 3.2 says, "In bison (for their horns are by nature curved toward each other) Nature has added the ability of projecting their excrement" (663a14). Thus, Aristotle does not envisage bison as aggressive animals capable of charging their enemies, but as fearful animals. Since "its size is equal to that of a bull and more massive than that of an ox" (*History of Animals* 9.45, 630a20), the bison cannot be ranked among the animals whose large size assures safety, since the ox and bull are not either. Doubtless we need to be sure of exactly what Aristotle calls a βονάσος, which we translate "bison." He thinks that the disposition of their horns, turned toward each other, makes them useless for defense or offense. And, he says, they are disposed thus "by nature" (πέφυκε, 663a14). But that can't be rational Nature, which is obligated to take a corrective measure (the projection of excrements) to make up for the inefficiency of the horns as organs of defense: thus, this is a matter of "necessary nature." Aristotle does not say about these horns, as he does of those of the stag, that "they are more harmful than helpful," and one might presume that, for him, the horns of bison are useless to these fearful animals. The best solution is thus to consider the horns of bison as the result of necessary movements that affect the material of which bison are composed. Nonetheless, they are as they are necessarily, but not by a hypothetical necessity. This situation is reminiscent of that of bile, discussed earlier. It seems to contradict the teleological principle that "nature does nothing in vain." In fact, the case of the bison, like that also of bile, can help us determine the exact significance of this principle, and also the principle according to which "nature always brings about the best." Some of the material ruled by necessary nature escapes the action of rational Nature.

We can take another step thanks to the does. They do not have antlers, but then why don't they have two ranks of teeth, since the matter that would have gone to make their antlers is still available? "Antlers have been refused to the females from the fact that are not useful even for the males, but they are less harmful to the males because of their strength" (3.2, 664a6). In other words, males, due to their strength, are able to support the inconvenience of their large antlers, and to survive everlastingly as a species. But we may dare say, "just barely," and animals similar but less robust would not succeed. That's the case with does. But "as for the fact that does do not have antlers, though in terms of teeth they are like the males, the cause is that the two sexes have the same nature, namely that of animals with horns" (664a4). In the strategy that Nature puts into operation to assure the perpetuity of species, whatever relates to reproduction is obviously crucial. But, at least if we leave to one side the case of nonsexual generation, like spontaneous generation,[42] the rule that directs this reproduction, and the possibility of hybridity does not challenge it, is that it takes place between animals of the same species. Stags (and does) could, in fact, reproduce with females (and males) of other species, provided, says Aristotle, that the other species are of a comparable size with deer, and that they would have identical durations of gestation.[43] But this kind of event remains for Aristotle exceptional, and thus lacks one of the characteristics of a *natural* event, namely, to happen "always or for the most part."[44] Furthermore, Aristotle seems to have thought that the offspring of these trans-specific unions would not be fertile. We will come back to that point later. Thus, it is necessary that the stag and the doe be animals with horns, which implies a certain number of characteristics, among which one might mention: having horns, for sure, having just one rank of teeth, having several stomachs, having a cloven hoof. But certain animals may lack some of these characteristics and nevertheless be part of the family of horned animals, like the Indian ass, which does not have a cloven hoof, and the oryx, which (according to Aristotle) has just one horn. Thus it is necessary that the doe be a horned animal without horns, because it would be fatal for her to have horns, which shows that the expression "animal with horns" is descriptive, but in the same manner as the fact of having a cloven hoof, which does not reveal the essence of these animals.

---

42. Which are only an apparent exception to the rule. We will talk again about spontaneous generation in the next chapter.

43. Cf. *Generation of Animals* 2.7, 746a31; *History of Animals* 8.28, 606b22.

44. Cf. Groisard, "Hybridity and Sterility in Aristotle's Generation of Animals." Cf. *Physics* 2.8, 198b35.

We can then answer the question: Why has Nature not used the earthy matter made available by the absence of horns to give does a double rank of teeth? We can give two answers to that. The first is partial and insufficient. It consists of noticing that the problem is less serious for the does than for the stags, because they have less earthy matter from the fact that they are smaller than the stags.[45] The second is that the presence of a double rank of teeth would lead to a profound reorganization of the doe's organism, particularly its digestive system, because the presence of several stomachs would then become useless, so the does would violate an important teleological principle often posited by Aristotle, according to which Nature does not provide two different organs for the same function,[46] because "Nature does nothing in vain." A doe with two ranks of teeth would thus not be an animal of the same species as the stag, and would have acquired a secondary characteristic (having two ranks of teeth) while losing a fundamental characteristic (being able to reproduce within its species). We thus have here an example of the application of what Cuvier called the "law of the subordination of characters," which Aristotle had already formulated, in his way.

Ultimately, one has the impression of being present at a kind of reversal of relations in the schema of hypothetical necessity. For a necessity directed by a form that is simultaneously an end is substituted the image of a sly Nature, cleverly getting the best of a situation, more precisely the material givens, which she has not chosen. That constrains rational Nature to have a remarkable "flexibility." We can see that with the example of the camel. The camel too, like the doe, has the remarkable characteristic of having a digestive system like that of horned animals, although it does not have horns. At least four elements combine to explain this situation: (i) The camel does not need a defensive organ such as horns, because its large size is defense enough. (ii) The camel needs a digestive system with several stomachs because the food available to it from its environment is "difficult to concoct, spiny and woody" (*Parts of Animals* 3.14, 674a29), and doubtless that is also why "it is more necessary to have its stomach thus [i.e., multiple] than to have front teeth" (674a33). (iii) So that, since it has several stomachs, a complete dentition would be useless for it (674b2). (iv) With the material saved from incomplete dentition, which would usually

45. "The earthy bodily element is found in larger quantity in larger animals" (663b24), and thus in smaller quantity in smaller animals.

46. Cf. for example *Parts of Animals* 3.2, 663a17, previously cited.

go to horns, Nature has reinforced its tongue and palate: "Since its food is spiny and it is necessary that its tongue be fleshy, Nature has used the earthy matter from the teeth to make the palate hard" (674b2). There is therefore a remarkable inversion of the order of causes between the case of horned animals and that of the camel. For the first, it is because they have horns to defend themselves that they do not have a complete dentition, and it is because they lack a rank of teeth that they have several stomachs; for the camel, on the other hand, it is because its food is as it is that it has several stomachs, which makes a complete dentition supererogatory, and liberates the material that would have been used for teeth for a more useful application, the reinforcement of the palate. In the first case then we have the sequencer: fabrication of horns, limitation of dentition, creation of several stomachs; in the second: creation of several stomachs, incomplete dentition, reinforcement of the palate. Evidently, from the fact that both horned animals and camels are everlasting species, these differences in logical order do not translate into chronological order because these different operations did not occur successively in time. If Aristotle had appealed to hypothetical necessity, he could have shown that it applies in both cases, for it is necessary to have an incomplete dentition if one wants to have, in the one case, horns, and in the other, a reinforced palate. But then things would become more difficult, because in animals with horns, the plurality of stomachs is hypothetically necessary in order to palliate the inconvenience of an incomplete dentition, while in the camel, the plurality of stomachs is sufficient for the working up of food, and thus makes a complete dentition superfluous, and thus impossible in Aristotle's finalized Nature. We would then fall back into the difficulties noted earlier concerning negative hypothetical necessity: in order that the camel not have a useless part, it is hypothetically necessary that it not have a complete dentition . . .

A last remarkable example of Aristotle's exercise of several finalist principles accompanied by a conflict between several explanatory levels. Among animals not born by copulation but spontaneously, Aristotle says in the *History of Animals* (5.1, 539b7), that some of them produce both males and females, but, when these have intercourse, that results in offspring that don't have the characteristics of either of their parents; they are imperfect, and do not themselves have offspring. Thus, lice generate nits, flies and fleas larvae; Aristotle says of these, turning out to be a poor observer in these cases, that they remain as they are for the rest of their lives. Nature does not violate the principle of sufficiency, previously noted, in this case either: these animals reproduce by spontaneous generation; it would have been supererogatory if they also reproduced by copulation; but they produce, spontaneously, both males and females that are naturally destined to have intercourse. The best solution then is that this intercourse not result in a lineage. There remains a question: as we will see in the next chapter, sexual reproduction is more perfect than spontaneous generation, so one may ask why Nature privileged spontaneous generation. That comes from the fact that these animals are born "from liquids in putrefaction" (*Generation of Animals* 1.16, 721a7).

Nature's Excellence

Since it is everlasting, nature needs to be perfect. And Aristotle does not simply maintain the thesis of the global perfection of the universe, but considers all its parts as perfect. How is a position like that tenable by one who knows the animal world so well, with all its surprising irregularities? We can begin by returning to the formulas according to which Nature "always achieves the best," and "does nothing in vain." First, by citing a surprising passage from the *Parts of Animals*:

> The differences that exist in these parts between each other are for the sake of the better: one example, among others, the differences in blood from one to the other. . . . The thicker and hotter blood conduces to strength, while that which is thinner and colder contributes to perception and intellection. The same differences can be found among the substances that are analogous to blood. That's why bees and other animals of this kind are naturally more intelligent than many blooded animals, and among the blooded animals those that have cooler and thinner blood are more intelligent than those with the contrary qualities. But those with hot, thin, and pure blood are the best, because animals of this kind are well disposed both for courage and for prudence. That's why the same difference exists between the upper and lower parts, and even for the male in comparison with the female, and for the right in comparison with the left side of the body. It goes the same way for the other parts, both homoiomeries and anomoiomeries. It must be thought that they differ among each other, some for the better

> and worse, others relative to the function and essence (οὐσία) of each animal; for example, two species that have eyes, one has them hard and the other fluid, one does not have eyelids, the other does, in order that vision be sharper. (2.2, 647b29)

While considering the parts of different animals, Aristotle here opens the question of perfection from two different angles. The manuscript tradition is not solid in terms of the order of these points of view, one group, greater in number, reverses the two parts of the sentence "some . . . others" at 648a15. Perhaps that allows supposing that neither of the two takes precedence over the other. I have chosen the order of manuscripts E (the famous *Parisinus graecus* 1853) and P (the *Vaticanus graecus* 1339, whose dating by specialists varies between the twelfth and fifteenth centuries, but which, even though recent, reproduces an ancient copy), a minority tradition but more ancient. In the first distinction, the parts are considered in themselves, which allows arranging them on a scale that goes from better to worse. Thus, a hot blood, thin and pure, is better than that which does not have all these qualities, and a cold blood, thick and impure, is worse than all the others. From the second point of view, such a distinction does not make sense, because the perfection of a part is considered from the point of view of the functions that it carries out in a given organism. Thus, at the end of the passage, it seems that one should understand, apropos of the differences noted among the various sorts of eyes, "so that vision should be sharper *in every case, taking account of the circumstances*," for eyelids would be at the very least useless for fish, and it is better for some to have the eyes fluid and for others dry eyes, and it would not make sense to say that dry eyes are "worse" than fluid eyes. We have here the articulation of two concepts of excellence, general and specific, which we meet elsewhere in other domains,[47] and to which we will need to return.

Nevertheless, we still have a question that remains unanswered: Why is it that not all animals, at least the blooded ones, have warm, thin, and pure blood? That would have been a proof of the excellence of Nature. In the same way, "the best," in an absolute sense, would have been, for the stags, that they have smaller antlers and two complete ranks of teeth, and for the bison, that they have no horns at all, or very small ones, with, at

47. Thus, in politics, on the question of constitutional excellence. Cf. my *Endangered Excellence*, 293–303.

the same time, more functional teeth. Similarly, Aristotle recognizes that human beings have a sense of smell "less good than that of many animals" (*De Anima* 2.9, 421a10). It cannot be denied that a better sense of smell would have been beneficial for us, a question all the more intriguing given that Aristotle recognizes that the human being, and more precisely the man, is the most perfect of animals, as we will see again in detail.

Thus, we must reconcile the fact that human beings are inferior to dogs in terms of sense of smell, and that in human beings, as well as in dogs in fact, Nature has nevertheless *brought* about the best; this could lead us to adopt a *relative* interpretation of this principle. We need to read this principle of the excellence of Nature as a principle of perfection by attaching to it a principle of sufficiency. In Aristotle's world, as we have described it, the essential task of rational Nature has been to arrive, with the material available, so to speak, at hand, at constructing organisms capable of surviving forever. Bison can do it, with their stunted horns, since Nature has given them sufficient means of defense. Aristotle's world is a world in equilibrium. It is this global lack of historicity that permits it to be everlasting and globally perfect, which sometimes leads Aristotle to say that Nature is divine.[48] From that, "bringing about the best," for Nature, is to give herself the means to conserve this everlastingness by preventing the introduction into nature of corrupting elements. In the animal world, this takes a precise form: species must be constructed in such a way that they are *perfect*, that is, that their everlasting survival may be assured, and that, we repeat, without possibility of correction or adaptation. Aristotle doubtless would not maintain that perfection and everlastingness are mutually convertible, nor that the everlasting is the cause of perfection, but perhaps it is Aristotelian to say that everlastingness is a sign of perfection, as milk is a sign, and not the cause, that a woman has had a child (*Prior Analytics* 2.27, 70a14).

Hence the intervention of the principle of sufficiency, which Aristotle develops mainly under another form, that of a Nature that "does nothing in vain." It is useless to give horns to the camel, since its large size is enough to defend it against its possible enemies. One may reasonably assume that, for Aristotle, a human being *does not need* the developed

48. In the *De Caelo* (1.4, 271a33) we read, "god and Nature do nothing in vain," an expression to which we will return. We can doubtless consider the expression "god and Nature" as a kind of hendiadys meaning "Nature which is a god," to indicate that it is everlasting and globally perfect.

sense of smell that dogs have in order to survive forever, all the more so since the human being compensates this perceptive deficiency by a certain sensory superiority, since "he is the most sensitive to tactile impressions" (*Parts of Animals* 2.16, 660a13).[49] The sense of smell is thus not a very important sense for the survival of the human species, and we will see, when we study animal pleasure, how olfaction has a vital importance for the dog. One may thus think that a dog would have a hard time surviving forever with a sense of smell as weak as that of human beings. Special perfection carries the day against general perfection, because the absolute perfection of the sense of smell, or of the blood, in one case having the acuity of the dog's sense of smell, in the other, warm, fine, and pure, only makes sense "relatively to the functions and the essence of each animal" (*Parts of Animals* 2.2, 648a16, cited earlier). It would be *in vain* for a bull to have fine and pure blood, as for a human being to have *in vain* a very effective sense of smell.

An analysis like that allows us to think the animal world in its diversity without resorting to a Pollyanna teleology like that of Bernardin de Saint-Pierre, who held that every trait of every living thing is good both for itself and for the entire universe. Of course, "in each animal there is something naturally beautiful" (*Parts of Animals* 1.5, 645a22), if only the ability to survive in conditions that are sometimes difficult. But Aristotle put himself in the position, not only of being able to recognize that certain characteristics of living things could have been better (human beings could have had a dog's nose), but also that others are, at best, useless and others truly harmful, and he can go so far as considering some species of animals as *naturally* constituted as real monsters. Thus, the triton (axolotl) is provided with gills and not lungs, but looks for its food on solid ground: "It seems that the nature of these animals has undergone something like a deviation, like some males that become

49. Cf. *History of Animals* 1.15, 494b16: the passage "touch is in human beings the most acute sense, second comes taste. For the other senses the human being yields to many other animals," means that the other human senses yield to touch, but also, from the fact that this remark is located in the context of a comparison with other animals, that human touch wins over all the other touches, something not obvious in this translation. Cf. *De Anima* 2.9, 421a20; *Sense* 4, 441a2, this latter passage saying explicitly that human touch surpasses all others "in precision": human touch is ἀκριβεστάτη τῶν ἄλλων ζῴων. We will need to come back to this point in the last chapter, dealing with pleasure.

females and some females that become males" (*History of Animals* 8.2, 589b28). But this "deviation" (Aristotle uses the verb διαστρέφειν) is not really one, since the triton reproduces completely regularly, and does not *deviate* at all. There are also animals that Aristotle considers badly made. Thus "some have a completely defective liver, as if their body had inherited a bad mixture, for example the toad, the tortoise, and other animals of this kind" (*Parts of Animals* 3.12, 673b29). Aristotle thus reconciles his conception of a perfect world, especially a perfect living world, with a grasp of animal diversity that he was the only one, and for many centuries, to have grasped. His presentation is thus not apologetic, like that of most finalists. That is thanks to his subject-less theology, because it is not certain that he could have attributed to a deity, or to a nature, to be the organizer of the faults of the triton or the toad.

So that from Aristotle's texts it is possible to develop an animal typology based on the advantage and disadvantage. For each species, Nature has "done the best," that is, Nature has brought it about that, for each species, the sum total of the advantages and disadvantages are positive, and enough to assure its permanent survival. If one sets to one side neutral traits (if they exist, they probably show Aristotle's ignorance of some vital role, notably in reproduction, of some properties, for example the color of birds), there are three sorts of traits, from the point of view of advantage versus disadvantage: (i) advantageous traits (for bulls, having horns); (ii) disadvantageous (having only one rank of teeth for horned animals): on this disadvantageous trait, Aristotle goes so far as to say, in the *Parts of Animals* passage just cited, that they bring it about that certain parts are "for the worse"; (iii) both advantageous and disadvantageous, in two forms: (a) more advantageous than disadvantageous (having curved talons for raptors, which makes walking difficult, but capturing prey easy), (b) more disadvantageous than advantageous (the large antlers of stags "are more harmful to them than helpful"). Most disadvantageous traits are compensated for by an advantageous trait (having several stomachs for horned animals, fecundity for small fish). But not all. Thus, it is advantageous for octopuses to have two ranks of suckers on each tentacle; but there exists a variety of octopus that has only one, because the tentacles are too narrow to accommodate two. The only thing one may say is that "this characteristic does not belong to them because it is best, but because it is necessary because of the definition of their substance" (ἀλλ' ὡς ἀναγκαῖον διὰ τὸν ἴδιον λόγον τῆς οὐσίας, *Parts of Animals* 4.9, 685b14). "The definition of the substance" of this octopus, which is a redundant way of designating its form, has been given forever, and it has as one of its characteristics to be that of an animal of small size, which thus does not have room on a tentacle for two ranks of suckers. As this species has always existed and will exist forever, that shows that this disadvantage is not sufficiently important to put its survival at risk. Possibly, after all, it is compensated for by a characteristic that Aristotle does not mention: its small size means that perhaps it needs less food and thus has a less perfect tool for catching food. In the case of this small octopus, as in that of all animals, Nature has "brought about the best," the perfection proper to each species manifested by its permanent survival, as he says in one of the most famous passages of the *Generation of Animals*, cited twice earlier: living things have specific everlastingness (2.1, 731b31).

It would have been definitely more difficult to explain such situations making use of the schema of hypothetical necessity than by deriving them by a kind of negotiation between necessary nature and rational Nature, applying the structure of advantage/disadvantage. Such a structure permits, more than any other, to find the homology that Gotthelf has established between Aristotle and Darwin.

All of that directs us toward a grasp of the whole of Aristotelianism as principally founded on *autonomy*. We will see that again in several domains and from several viewpoints, but we can recognize it down in relation to perfection in the animal world. Animal species are not without relationships between each other, and some of those relationships are necessary, for how would predatory carnivores survive without their prey? At the same time, we will have occasion to examine in more detail the relations, both amicable and hostile, that the various species have among each other. Nevertheless, it remains that the horizon of Aristotelian biology is the species and the means that Nature has put in place so that each living species can survive forever. Aristotelian zoology is not aimed at the idea of an ecosystem through which modern zoology proposes a relational approach to animal species. At the end of the day, we have to keep together several approaches that are only apparently incompatible. On the one hand, the focus on species does not transform Aristotle's biology into a natural history like that of Buffon, who described animals species by species: it is indeed a study of organic, functional, and behavioral resemblances and differences. On the other hand, animal diversity does not have meaning by itself: we will strive to establish this crucially important thesis for the interpretation of Aristotle's zoology, and of his natural philosophy in general, and for the evaluation of Aristotle's place in the spectrum of Greek thinkers.

Chapter 3

A Philosophy of Life?

The question of the passage from the nonliving to living has haunted modern biology at least since the eighteenth century—did Aristotle himself ask that question? What we said in the previous chapter would seem to show that he did not. Since living species are everlasting, movement from nonliving matter to living matter to form the *first* individual of an animal or vegetable species cannot occur. On the other hand, change from living to inert is an experience shared by all sublunary living things—it is death. But we will see that the question of the passage from nonliving to living evidently does arise for the generation of living things, animals and plants. We will need to see, in particular, at what moment life appears in the fundamental process occurring within living things, the development of food that, in animals with blood, results in the blood from which all the viscera, the "parts," and the semen are made, a role played, in non-blooded animals, by the analogue of blood.

Aristotle was certainly aware that, as in many other issues, when he argued that life did not have any first appearance, he was in opposition to all cosmogonists, whether philosophers or mythologists. His constant position, whether about myths or philosophical systems, was to affirm that they contained a partial truth, no matter how distorted and disguised it may be, to be reinterpreted within an Aristotelian framework. That's what he does with the myth of autochthony, which as a remarkable little book[1] shows, is far from being limited to Athenians: the Aristotelian theory of spontaneous generation can give meaning to a myth like that. If one insists

1. Guthrie, *In the Beginning*.

on holding on to this myth according to which "one day" (ποτέ) human beings and quadrupeds were born from the earth, that could only be by coming from eggs or larvae; but since we notice that no animal is born from an egg by spontaneous generation, it remains that it would have to be from a larva (*Generation of Animals* 3.11, 762b28). But it is clear that the Aristotelian theory of spontaneous generation has no room for a possible emergence of a larva or animal of a new species.

The question of the difference between living and nonliving is, on the contrary, an Aristotelian question, posed explicitly by Aristotle. But the first obstacle raised to addressing this question is that of the plurality of senses of the word "life," starting from a passage of the *De Anima*: "Since 'living' is said in several ways, let us say that a being is alive if any of these functions appears in it: thought, perception, local movement, and rest, as well as the movement of nutrition, decrease, and growth" (2.2, 413a22). In this passage, it must be assumed that the functions in question belong to living beings per se, because Aristotle does not mean to say that a saw, animated in local movement when it is used by a carpenter, is alive. However, this passage allows taking account of different kinds of life, which some of the passages to be examined later do not. In fact there are for Aristotle several sorts of life that may be considered homonyms, that is, equivocal, in the moderately strong sense of "homonym," since homonymy may be more or less strict.[2] In addition to the organic life of animals and plants, this passage allows room for the life of the first Unmoved Mover, since it is "thought thinking thought," as well as of the whole universe and of the celestial bodies that make it up, that are animated only by regular and everlasting circular local movement. But there is also the "practical life" as when Aristotle speaks of a "life of leisure" or the "happy life" of virtuous citizens. In this case the equivocation is more obvious, and we don't need to concern ourselves with it. Several attempts have been made to establish a terminological difference between the sorts of life, since Aristotle uses two different words to designate life, *bios* and *zoē*, but they have not led to persuasive results. At best one may say that these two words are both applied to the organic life of animals and plants, while

2. At *Nicomachean Ethics* 5.1, 1129a26, Aristotle says that injustice and justice are said in several ways, in fact two ways, that "because of their closeness, their equivocity remains hidden from us." These are what the tradition calls "general" and "particular" justice. But particular justice is a part of general justice, which thus gives a "weak" equivocity. In any case homonymy (equivocity) has a variable force.

the ethico-political sense is usually, though not always, *bios*. But, taken literally, this passage in the *De Anima* could lead us to believe that all natural beings are alive, since a stone, for example, is indeed animated by an internal local movement, not provoked by an agent, that transports it toward the center of the earth. But it is not alive, as the following passage affirms: "Among natural beings, some are alive, and others are not. We call life the ability to feed oneself, grow, and die, by virtue of oneself" (τῶν δὲ φυσικῶν τὰ μὲν ἔχει ζωήν, τὰ δ' οὐκ ἔχει: ζωὴν δὲ λέγομεν τὴν δι' αὐτοῦ τροφήν τε καὶ αὔξησιν καὶ φθίσιν, *De Anima* 2.1, 412a13). This passage too is approximate in its expression, since, by defining *natural* living beings by what he elsewhere calls the nutritive soul (we will have a lot to say about that), Aristotle sets to one side celestial bodies, doubtless as well the cosmos as a whole, which are *natural* beings, but do not have a nutritive soul. But at least this text covers the domain that interests us, that of the organic life of sublunary living beings, which have as one of their main characteristics that of existing within stable species constituted of mortal beings. These are all natural bodily beings, composed in the last instance of the four elements distinguished by Aristotle: earth, water, air, fire, a list doubtless gotten from Empedocles.

We have already said that Aristotle was not what Georges Canguilhem has called a "vitalist of exception," that is, one who thinks, like certain vitalists of the end of the eighteenth and beginning of the nineteenth centuries, that the living world escapes the physiochemical laws governing the nonliving universe. For Aristotle, as we saw in the preceding chapter, the matter composing animals (we are concerned primarily with animals, most of the time considering plants only in contrast to animals) cannot annul what we call—and Aristotle does not—the laws of nonliving nature. This state of affairs, which no interpreter can ignore, has given rise to a famous dispute that has resulted in a significant literature, in which we will intervene without too much justification. Some, in fact, have argued that all natural realities, whether living or nonliving, may be explained completely and solely by recourse to physiochemical laws, teleological explanation being just another way of formulating the explanation of life phenomena. This sort of interpretation culminated with the work of Wolfgang Wieland on Aristotle's natural philosophy, already mentioned,[3] in which the author claimed that the teleological approach to living things was nothing but a way of describing the phenomena that would be more

3. Wieland, *Die aristotelische Physik*.

useful; he followed, as we have seen, the pathway indicated by the Kantian expression, *als ob*: everything happens *as if* the feet of ducks had been webbed *in order to* facilitate their passage in water. There is, in this reading, a fundamentally true element, which is to refuse all intention to nature, but it is unacceptable because it attributes to Aristotle a teleology that is *apparent* and not real.

Other interpreters, who are in fact right, think on the contrary that teleology defines a proper level of explanation irreducible to any other, which agrees with Aristotle's statements about "rational Nature," which is strongly active, using, making, balancing, while never derogating from the laws of "necessary" nature. But not derogating is not explaining: teleological explanation refers to a *supplementary* intervention by Nature that takes part in the natural processes governed by the laws of nonliving nature. We will see, later in this chapter, still on the example of horns, that derogation is sometimes not far off.

We will find this dispute again concerning the generation of living things, something that particularly interests us. In fact, although the priority of essence to generation is a characteristic, possibly the fundamental characteristic, of the Aristotelian approach to living nature, nothing prevents us from posing the problem of the relationship between nonliving and living when we look at the process of producing natural beings, whether nonliving or living. For nonliving beings, one might speak of "production," while for living things it's "generation." Two questions need to be formulated immediately; they will be explored more precisely later: in what sense can one say that *reproduction* is a privilege of living things? Does Aristotle intend to construct a general theory of the production of all natural beings, and more precisely, of that most simple state that homoiomeries represent both for living things and for nonliving realities like minerals and metals?

To begin with the second question, one has the impression that Aristotle bridges the gap between nonliving and living, for example in a text like this one:

> I call "homoiomeries" for example things mined like copper, gold, silver, brass, iron, stone, and other materials of this kind, as well as materials derived from them; also the parts of animals and plants, for example flesh, bone, tendon, skin, viscera, hair, fibers, blood vessels, from which the anomoiomeries are constituted, like face, hand, foot, and all the parts of this kind,

> and in plants the homoiomeries are wood, bark, leaf, root, and every part of this kind. (*Meteorologica* 4.10, 388a13)

In a simple reading of this passage, when one sees the list that it gives, one gets the impression that Aristotle has here, as a matter of fact, a vision of a general unified theory of homoiomeries and their formation, unless one takes the easy way out by disputing the authenticity of *Meteorologica* book 4.

But the homoiomeries are, in Aristotle's biology, the basic bricks of the living things, as is shown in the often-cited beginning of *Parts of Animals* 2. Aristotle there explains that the parts of animals, and he could have said the same thing for plants, at least to some degree, present three "compositions" (συνθέσεις, 2.1, 646a12), a possibly heterogeneous list because only the two latter are really *compositions*. At the first level we find the four elements, earth, water, air, and fire, which combine with each other to yield the homoiomerous parts of animals, like flesh and bone, which constitute the second level. Aristotle makes clear at this point that it is not so much the elements that we should take account of, as their "powers" (δυνάμεις), namely the four elementary qualities hot/cold, dry/wet, a doctrine that is presented in detail in *On Generation and Corruption*. Perhaps it is because they are combinations of these qualities that Aristotle thinks of the elements as composite. The homoiomerous parts combine to constitute the third level, that of anomoiomerous parts like face and hand. Aristotle says, as we have seen, the "the homoiomeries are for the sake of the anomoiomeries" (*Parts of Animals* 2.1, 646b11).

This doctrine of homoiomeries and anomoiomeries is more complex than it appears, because this simple distinction is insufficient. Some anomoiomeries are composed of several homoiomeries, for example, the arm is composed of skin, flesh (Aristotle does not have a concept of muscle), and bone. But the face, an example of an anomoiomere often given by Aristotle, is a different matter, since it is composed of nose, brow, and so on, which are themselves also anomoiomeries. Furthermore, "the anomoiomeries can be composed of homoiomeries, and some are composed of a single homoiomere, like certain viscera; for they are in various shapes while being formed from a homoiomerous body that is so to speak simple" (*Parts of Animals* 2.1, 646b30).

A little later Aristotle gives the example of the heart, coming to the point of saying about it and some other viscera that "they are from one point of view homoiomeries and from another anomoiomeries" (647b8).

In a subtle article, Marwan Rashed[4] even distinguishes a third kind of homoiomere, designating a part composed of small identical particles, like semen. Ultimately what is important for us here is to notice that animals are composed completely of anomoiomerous parts, never directly of *pure* homoiomeries, because the biceps is not simply flesh, nor the tibia simply bone, because the biceps and the tibia, even if the biceps is homogeneous (because in most cases, but not all, skeletal bones of an animal are composed of bone and marrow), are not formless, but are in the same situation as the heart.

Finally, last comment, homoiomeries are not simply mixes of the elements, but mixes in a given proportion, and that is more obvious for the homoiomeries that are included in the composition of living things, because one may make mud with variable proportions of earth and water, but for living things, we can rely on the passage, already cited twice, from the *Generation of Animals*, that explains the difference between making something soft or sticky and transforming a material into flesh or bone (2.1, 734b31): in the second case it is necessary to introduce a proportion between the components, earth, water, and perhaps air. A passage like that is not completely intelligible except to one who knows, at least in outline, Aristotle's physiology and theory of generation.

To return to the *Generation of Animals* passage, doubtless we need to make the text say a little more than it says explicitly. The material and efficient causes alone (hot, cold, earth, water) can produce a material with the properties of flesh, bone, blood, and so forth, but, as we will see several times in detail, that which the form transported in the sperm of the generator brings as a supplement is that this material will really be flesh or bone, that is, flesh or bone of a particular animal, dog or human, for each animal species has its own flesh, something that Aristotle has shown precisely for blood (leaving to one side the question of knowing whether we ought to say "human" or "of Socrates"), and flesh and bone able to fulfill the functions of flesh and bone. The action of the form that is contained in the semen of the generator, and that acts by means of a system of movements contained in the semen, which thus fashion the various organs (with the assistance of the "innate breath," a problem here set aside), consists of introducing a relationship (a reason, *logos*) between the properties of each homoiomere. But a comparison with natural beings that (almost) no one, and certainly not Aristotle, considers to be living

4. Rashed, "A Latent Difficulty in Aristotle's Theory of Semen."

shows us that this introduction of a relationship, possibly and doubtless ideally, numerical, is not that which characterizes the living thing. We have an instance of this with minerals and metals, as we will see.

As for the question about reproduction, that is closely tied to that of the criteria of life noted earlier. By defining organic life as the capacity "of feeding oneself, growing and dying of oneself," Aristotle defines the functions that he elsewhere calls the "nutritive soul" (ἡ θρεπτικὴ ψυχή, for example at *De Anima* 2.4, 415a23). If one adds to that nutritive soul a "sensitive soul," which gives to living things the faculty of discriminating, then the living thing is an animal. If other faculties are added, like the fact of moving oneself, desiring, imagining, thinking, one has animals that are more and more perfect.

This split between beings that have and beings that do not have the nutritive soul is so clear that it forces us to give a "disengaged" reading of two passages that are generally called upon in order to attribute to Aristotle a *scala naturae* establishing a gray zone between living and nonliving:

> Thus, nature passes little by little from inanimate beings to animals in such a way that, as a consequence of this continuity, one hardly notices the border between them (λανθάνει τὸ μεθόριον αὐτῶν) nor to which of the two groups an intermediate form belongs. For following the class of inanimate beings comes first that of plants. (*History of Animals* 8.1, 588b4)

> Nature in fact passes in a continuous way from inanimate things to animals, going by way of living beings that are not animals, in such a way that one has the impression that they differ very little from one to the other (ὥστε δοκεῖν πάμπαν μικρὸν διαφέρειν θατέρου θάτερον) from the fact of their reciprocal proximity. (*Parts of Animals* 4.5, 681a12)

It seems clear that this continuity between nonliving and living is a matter of the *impression* of people little accustomed to observing or little acquainted with Aristotelian distinctions, but in no case would he allow the existence of bodies intermediate between living and nonliving. Seven lines after our *History of Animals* passage, in contrast, Aristotle can assert that "the passage from plants to animals is continuous" (μετάβασις ἐξ αὐτῶν εἰς τὰ ζῷα συνεχής ἐστιν, 588b11). That is true in a slightly extended sense of "continuous." In the most proper sense, two quantities are continuous "whose limits are one" (*Physics* 6.1, 231a22). So, then, in no case could an inanimate being and a living being be in continuity, but a plant (which possesses only a nutritive soul) and an animal (which has in addition the sensitive soul) can be considered continuous from the fact that the animal also has a nutritive soul. Thus, we need to stop a moment on this "nutritive soul."

The Nutritive Soul

Aristotle is aware of the degree to which the expression "nutritive soul" is approximate and inadequate, also the case for the expression "nutritive part of the soul," which was more or less imposed on him by the Platonic tradition.[5] The most Aristotelian description is in terms of "power" or "faculty" (δύναμις). In qualifying it as "nutritive" (θρεπτική), Aristotle characterizes it very badly or at the very least extremely incompletely. The nutritive soul, in fact, is in charge of the growth of the living thing (and it is similarly concerned with its death, which would deserve explanation) and also of its reproduction. We can see that in several statements like these:

> The nutritive soul belongs to all living things, and it is the primary and most common faculty of the soul: it is that by which life belongs to all. Its functions are generation and carrying out nutrition. (*De Animal* 2.4, 415a23)

> In all living things, whether plant or animal, the nutritive soul exists in the same way, that is, it is able to generate another being like itself. (*Generation of Animals* 2.1, 735a17)

> How animals nourish themselves and with what, in what way they assimilate the food from the stomach, is more appropriate to consider and talk about in a treatise on generation. (*Parts of Animals* 3.5, 668a6)

---

5. Cf. *De Anima* 3.10, 433b1: τὰ μέρη τῆς ψυχῦς; for the use of *dynamis*, cf. apropos of plants, "they have no other faculty of the soul" than the nutritive (οὐδεμία γὰρ αὐτοῖς ὑπάρχει δύναμις ἄλλη ψυχῆς, 2.2, 413a33).

Obviously, Aristotle does not argue that nutrition and generation are one and the same thing,[6] but rather that the same *dynamis* allows the animal (Aristotle is thinking especially of them, as we will see) to assimilate food for achieving the three functions in question: preservation, growth (often grouped together under nutrition), and generation. It is thus not surprising that Aristotle asks himself about the priority of one of the functions, nutrition or generation, in relation to the other, a question to which he gives seemingly contradictory answers, just a few lines apart:

> Since the same power (faculty, δύναμις) of the soul is both nutritive and generative, we must deal with nutrition first, for the distinction in relationship to other powers is made by this function. (*De Anima* 2.4, 416a18)

> As the correct appellation is always taken from the end, and the end consists in the engendering of a being similar to itself, the first soul will be the generating principle of such a being. (*De Anima* 2.4, 416b2)

We have here a distinction between two kinds of priority, that which is first because it is more "basic" (nutrition), and that which is first because it is an end. But reproduction is, for living things, an end because only it can assure the permanent survival of species, as we saw when discussing teleology.

To grasp why nutrition, development, and generation all depend on the same *dynamis*, which is at the same time the definitory capacity of the living thing, is one of the best ways of grasping also how the living thing functions as an *organism*, while the nonliving does not. The first reason that makes it that the nutritive soul provides these three functions is that it is also the case that semen and the "menstrual fluid," which is taken to be the female material acted on by the semen, comes from the

6. Cf. for example *Parts of Animals* 2.7, 653b13: "But the study of the residues of food is to be found in the works that consider and study food (there one studies which animals have them and for what reasons), while that of sperm and milk are found in the works on generation, for the first is the principle of generation, and the second exists because of generation." Cf. Pellegrin, "What Is Aristotle's *Generation of Animals* About?"

elaboration of the food. But there are texts that go further in the exposition of these reasons. We may cite, and analyze, two remarkable passages from the *Generation of Animals*:

> It's the female that furnishes the matter, but the male provides the source of movement. Just as the product of art is fabricated by the use of tools (or rather the movement of tools), as this movement is the activity of the art, and the art is the form of that which is produced in something other, in the same way the power of the nutritive soul, just as it later produces in animals and plants growth from nutrition, using as tools hot and cold (for it is in them that it finds its [i.e., of power of nutritive soul] movement and each living thing forms itself according to a certain relationship), in the same way it constitutes at the beginning that which is engendered by nature [i.e., the embryo]. In fact, the matter in which growth takes place, and that by which at the beginning that which is brought into being by nature is formed, are identical, so that the productive power is also identical right from the beginning. But that [the power bringing about growth] is more important. If that is the nutritive soul, then that is also generative. This [nutritive soul] is the nature of each living thing, present in plants as well as all animals. (2.4, 740b24)

> It doesn't make any difference if one speaks of semen or of the movement that makes each of the parts grow, or again the movement that brings about growth or that which constitutes at the beginning. In fact, the reason (*logos*) of the movement is the same. (4.3, 767b18)

A few comments on the first passage. The comparison between the operations of art and those of nature occurs frequently in Aristotle, but what does it do here? As the tools achieve the form of the art present in the mind of the artist, by their movements, in the same way the power of the nutritive soul carries out two operations: the second is the growth of the embryo, and the first is the formation of that embryo. The tool, immanent and not separate as in the case of art, of these operations is the cold and the hot (and doubtless certain other qualities), "for it is in them that its movement is to be found" (740b32): "its movement" (ἡ κίνησεις ἐκείνης) cannot be understood otherwise than as signifying "the movement of the nutritive soul." But, since the beginning of the passage recalls that it is the male that is the source of the movement, it must be understood that it is the nutritive soul of the male that, by making the hot and cold work, forms the embryo in the female material, without, we remember, contributing materially to that embryo.[7] But this matter that furnishes the material for the growth of the embryo is also that from which the embryo is originally formed.

But, as we see in the second passage, it's not only that the matter of generation and that of the growth of the embryo are one and the same material, but also that the movement that the semen of the male imprints to constitute the embryo and the movement that makes it grow are one and the same movement. Here we must understand "semen" (γονή) as designating the sperm, and the "reason" (λόγος) as indicating the relationship that the movements carried by the semen introduce among the components of the female matter, as the first passage says. The generation and growth of an animal are thus one and the same thing, even though different in their being (it's a frequent formula in Aristotle to say that A and B are *tauto*, but that their *einai* is different), and it is thus not surprising that both depend on the same "soul."

Let us then envisage in parallel, as proclaimed, the process of production of natural beings, by considering first the sexual generation of living things, then their spontaneous generation, and finally the formation of nonliving natural realities like metals and minerals. First, we need to make a few preliminary remarks on natural beings themselves, which will allow us to justify this order of exposition.

For us, the split between living and nonliving is reflected in the difference between physical sciences and sciences of life, although it is true that life sciences are also concerned with the physiochemical makeup of living beings. It's not the same for Aristotle, for whom the study of living things, at least living things that exist in the sublunary world, belongs to physics, provided that we understand the word "physics" in the ancient sense of "science of nature," the science of those whom Aristotle calls *physikoi* or *physiologoi*, as we saw in the preceding chapter. The fact that

7. This absence of male material in the embryo has been a contested point since antiquity, including by those who generally agreed with the Aristotelian theory of the reproduction of living things. One of the most remarkable examples is Galen's *De Semine*.

the difference between living and nonliving plays out *partially* within the class of natural (physical) beings is not without consequences.

Natural beings are those, says Aristotle, that have "in themselves a source of movement and rest" (ἐν ἑαυτῷ ἀρχὴν ἔχει κινήσεως καὶ στάσεως, *Physics* 2.1, 192b14). This famous definition of natural beings is more difficult to understand than it may appear and, something hardly surprising, has given rise to many commentaries. Themistius and Simplicius thought that here Aristotle wanted to establish a contrast between natural beings and those of art, chance, or deliberate choice. In fact, Aristotle, in what follows, shows that he is thinking uniquely or primarily about the opposition between that which is by nature and that which is by art (technique). The passage deserves one more look:

> In contrast, a bed, a cloak, and whatever else would be of that kind, in that they have been given these names and to the degree that each of them is the product of an art, do not have any inner impulse[8] for change; but, on the other hand, to the degree that things are accidentally made of stone, earth, or a mixture of the two, they have that impulse, and to that degree, because nature is a certain principle, namely a cause of being moved and being at rest in that to which it belongs immediately per se and not by accident. (192b16)

At least two things ought to attract our attention. First, the products of art having been made ultimately of natural materials (the matter of the bed was wood), they have an indirect relationship to nature. That means that everything that surrounds us is natural in some way. But Aristotle's example goes a little further. In the third line of *Physics* 2.1, Aristotle gives as examples of natural beings "the simple bodies, namely earth, fire, air, water," that is, the elements. But, in considering as natural by accident objects made "of stone, of earth, or a mixture of the two," and whatever the word "earth" designates here (the element earth or empirical earth that we run into in our everyday experience), Aristotle considers stone, and thus very likely all minerals, and no less probably metals, as natural

8. At 192b18 instead of "impulse" (ὁρμή) Themistius reads "principle" (ἀρχή), which Simplicius notes as an alternative reading (265.15). This reading would slightly ameliorate the sense, notably the relationship with the proposition beginning "since nature is a certain principle." But the majority reading gives an acceptable sense.

beings. In fact, it's because it's "of stone" that a statue is said to be natural, even if accidentally. We will speak later about the origin of minerals and metals. The "parts" of living things cited just after these as natural beings do not have in themselves the source of their movement except when they are part of a living organism. Next, the ancient commentators have not failed to notice that the definition here proposed excludes some beings that Aristotle had to think of as natural. By talking about a "principle of movement *and rest*," he in effect seems to exclude the beings perpetually in motion, like the celestial bodies; for that reason, Alexander of Aphrodisias proposes to read "or" instead of "and." Perhaps we may think that Aristotle means by "natural beings" only the natural beings found in the sublunary region. But neither this passage in the *Physics* nor those presenting the same doctrine establish a distinction among natural beings between living and nonliving.

It cannot be left unnoticed that Aristotle begins, in *Physics* 2.1 as elsewhere, his lists of natural beings with living things and that most of the time whether he says so or not, he thinks of living things when he speaks of natural beings. In fact, living things are not the only natural beings, but the relationship between nonliving and living natural beings cannot be simply exclusionary as it is for us. In fact, it is a well-known Aristotelian practice to group together realities in the same category X because they share certain properties of this category X, but positing that some of them are more X, or more completely X, than others. That happens in all domains, as we have previously seen: courage in the full sense is that of a citizen, male, Greek, adult, and free, who is also sufficiently virtuous, and not of a woman, child, barbarian, slave, who can have a sort of courage, but "incomplete."[9] An Aristotelian would not say that a man is more courageous than a woman, but that a woman is less courageous than a man, since the courage of a (male) citizen serves as a paradigm for all other forms of courage. But living is more perfect than nonliving, for "the soul is better than the body, the ensouled better than the soul-less because of the soul, being than nonbeing, living than nonliving" (*Generation of Animals* 2.1, 731b28, cited earlier).

Somewhat in the manner of the Anselm-Descartes proof for the existence of God, which rests on the fact that rejecting existence to a perfect being is rejecting perfection, Aristotle attributes life to that which

9. Cf. *Politics* 1.13, 1260a20.

is perfect, or almost perfect: the First Unmoved Mover, that is, God, the universe as a whole, and the celestial bodies are everlasting living things. In the sublunary world, one must understand a nonliving thing as an imperfect living thing rather than the living thing as a nonliving thing plus something: since, as we have seen, as Foucault says, "Life is the root of all existence, and the non-living, nature in its inert form, is merely spent life; mere being is the non-being of life."[10] In other words, if Aristotle has a problem about reductionism, it is not posed going up, that is, in asking the simpler to furnish rules of intelligibility for the more complex, but in going down, that is, in showing that simpler states cannot be understood except by reference to more complex states. This explanatory schema runs from one end to the other throughout Aristotle's entire work, and is of crucial importance in the study of living things; it brings it about that the embryonic stage is understood in relation to the developed stage—this is a variant of the doctrine of the explanatory preeminence of essence to genesis. This "descendant" schema opposes Aristotle to all his Presocratic predecessors.

In one fell swoop, the order of exposition, sexual reproduction, spontaneous generation, the production of metals and minerals, is all more or less justified: we will have hardly any problem showing that spontaneous generation is an imperfect form of sexual generation, from the fact that both are really *generations*: we would next have to grasp that the formation of minerals and metals becomes intelligible when one refers it to the spontaneous generation of living things, and thus, transitively, to sexual generation.

Sexual Generation and the Female Material

The general Aristotelian schema of fecundation in living things that have separate sexes is well known. Aristotle repeats throughout the *Generation of Animals* that in conception the female provides the body and the male the soul (2.4, 738b26), that the female provides the matter, and the male puts it in motion and shapes it (b20),[11] and that by means of movements contained in the semen of the male, which works on the female material

10. Foucault, *The Order of Things*, 278.

11. Cf. κινοῦν πρῶτον καὶ δημιουργοῦν, *Generation of Animals* 2.1, 735a27.

as the plane or drill works on the wood of a future table. And Aristotle makes clear, in a very important passage cited earlier, that it is "the movement of the generator that is potentially that which is generated" (*Generation of Animals* 2.1, 734b34) that brings about this formation. One can, therefore, ask whether Aristotle is not partly responsible for some of the bad readings that have been made of his theory of sexual reproduction. Thus, when describing the female material, assimilated to the "menses," as a "residue" (περίττωμα)[12] of food, Aristotle seems to consider it as an ignoble product. Actually, Aristotle distinguishes useless residues, like urine and feces, from useful residues such as semen. In a truly remarkable work, Sophia Connell[13] has notably denounced what one might call the "authorized feminist version" of Aristotle's theory of generation. Authors like Luce Irigaray, Julia Sissa, and Suzanne Saïd, none of them, it must be emphasized, a specialist on Aristotle, go so far as to argue that according to Aristotle the female provides an indeterminate prime matter, which is a hermeneutic monstrosity. From a perspective like that, only the male really has a *genetic* role, in other words, there is only one sex, the female being content to parasitize the genetic process from the fact that the material that she brings resists the information coming from the male, which would be sufficient to explain the resemblances of offspring to their maternal lineage, and especially the most important of those resemblances, the birth of females, but also the birth of monsters and hybrids. A very inadequate explanation, as we will see and Sophia Connell has definitively shown, which means that we don't need to do it again, that females have a real genetic role, at the same time not exonerating Aristotle completely from a phallocentric approach, for he thinks it's better to be male than female, because a male is a more complete animal. On the other hand, there are possible exceptions to that rule, since, for example, as we have already noted and will see again, the females of the bear and panther are more courageous than the males.

We need to say a bit more about the matter by which the female contributes to generation in animals in which the two sexes exist separately. As always, we need to conform to the Aristotelian explanatory schema in which the most developed in any series is paradigmatic. But, among animals that reproduce via a relation between female and male sexes, some

12. A word used 140 times in the *Generation of Animals*.

13. Connell, *Aristotle on Female Animals*.

of them have the sexes in different individuals, others are hermaphrodites. Aristotle gives a general justification that sexual reproduction is better than other ways, and that in which the sexes are separated is even better; this can only add weight to the case file that has been put together against his phallocratism:

> But since they [animals and plants in the sublunary world] have as principles female and male, it is for the sake of generation that the female and male exist in beings that have male and female.[14] But as the first moving cause (to which belongs the definition and the form) is by nature better and more divine than the matter, it is also better that that which is worth more be divided from that which is worth less. That is why, in all those in which it is possible, and to the degree that it is possible, the male is separate from the female, for it is as something better and more divine that the principle of movement belongs, as male, to generated beings, while the female is matter. But the male unites and mixes with the female for the sake of the process that constitutes generation, for this process is common to both. (*Generation of Animals* 2.1, 732a1)

But it's not any random matter that the movements from the male are able to put into operation; at least in the case of "canonical" sexual reproduction within a species yielding offspring that are themselves fertile, it's the matter of a female of the same species as the male. But the case of wind eggs[15] of birds, discussed in *Generation of Animals* 3.1, shows that this material is, at least in certain cases and to a certain degree, alive. Of course, Aristotle says, "not in the sense that fertilized eggs are alive (for from them is born an actual living being), but not like wood or stone" (2.5, 741a19), in other words, of nonliving material. The proof that these eggs are alive is that they rot, but rotting is one of the characteristics of a living thing.[16] But if they are alive, they have a nutritive soul. Aristo-

14. At 732a3 ἐν τοῖς [ἔχ]ουσιν is a correction by Drossaart Lulofs of the ἐν τοῖς οὖσιν of the manuscripts, "among beings." Those who do not adopt the emendation (Peck, Louis) are forced to add ἑκάτερον τούτων ("beings that have two sexes"), which appears in only one manuscript, although in fact the oldest (Z).

15. In Greek, τὰ ὑπηνέμια καὶ ζεφύρια (3.1, 749b1)—whence "wind eggs."

16. It is useful to distinguish decay (φθορά), which happens to living things, from putrefaction (σῆψις). Cf. Meteorologica 4.1, 379a16; *Generation of Animals* 5.4, 784b8.

tle says that they have it "potentially" (741a23), which demands some explanation. One must not forget that, in the case of wind eggs, they are produced by *animals*, in this case birds, and as we will see, some fish, that is, living beings with a sensitive soul. But a passage in the *De Anima*, famous for its difficulty, compares the interlocking of psychic powers (nutritive, sensitive, locomotive, etc.) to a sequence that says that some figures are included within others, a sequence in which "the later term always contains the previous potentially" (2.3, 414b19). Aristotle gives as an example the rectangle that contains, it is implied "potentially," the triangle. Should we then understand that the sensitive soul contains the nutritive soul "potentially"? We will need to return to this question when we deal with the various powers of the soul.

The crucial consequence of all this is that, at least as observed in the case of certain birds and certain fish for which *History of Animals* 5.1, 539a31 tells us that they have wind eggs, the female is able to produce living material by herself. But it's the same matter, or as Aristotle puts it, the same "residue," that constitutes both wind eggs and fertile eggs. So, the homoiomerous parts of animals are not constructed by the action of the vital movements (those contained in the semen) on nonliving matter (earth or water), but by the action of these movements on living matter. But that can be said of all animals that reproduce via sexual union, as Sophia Connell has very well demonstrated.

In the history of biology, the mysteries of sexual reproduction have taken a long time to dissipate, and there still remain a few. For possibly various reasons, the spermatozoa were observed two centuries before the ova, that are larger, discovered only in 1827 by Karl Ernst von Baer. This does not mean that Antonie van Leeuwenhoek understood the role of spermatozoa just from having perfected the microscope that allowed him to see them. As for Aristotle, he was completely ignorant of the existence of what were called, from the seventeenth century onward, "animalcules." As for the "menses," this "bloody secretion produced in the female" (*Generation of Animals* 1.19, 726b35), Aristotle thought that they were produced in the same way as semen, by elaboration of food thanks to the internal heat of the animal, but elaborated to a lesser degree, given that females have less heat. However, thanks to his analysis of wind eggs, we can doubtless think that Aristotle came pretty close to the notion of ovulation, as close in any case as William Harvey and his "ovist" successors.

In fact, according to Aristotle, only certain birds and certain fish produce wind eggs, and it is important to understand why. Wind eggs are produced only in birds that "are prolific because they have a great deal of residue" (*Generation of Animals* 3.1, 749b2), understood as the residue due to the concoction of food. This processing of food is in fact the central phenomenon of animal economy, since it is at the origin of the formation of all parts of the living thing: blood in the heart, which will solidify into flesh, the various organs, semen, and right up to the most peripheral parts such as nails and hair. Why, for example, do raptors have less residue than chickens? Because that which is produced as residue in chickens is used by raptors for feathers for long, strong wings, a necessary characteristic for their permanent survival given their way of life. Thus, birds with a surplus of residue use it for wind eggs. Then comes this passage, crucial for us: "Wind eggs, as was said before, are produced because spermatic material is indeed found in the female, but in birds the menstrual secretion is not produced as it is in blooded vivipara. In the latter, in fact, some produce more than others, some in a quantity that it is barely perceptible" (*Generation of Animals* 3.1,750b3).

It is necessary to understand "as in blooded vivipara" (ὥσπερ τοῖς ζῳοτόκοις τοῖς ἐναίμοις, 750b6) as signifying, not that the secretion is produced in birds, but in a different way than it is in blooded vivipara, but that it is not produced at all in birds, while it is found in blooded vivipara. In the females of blooded vivipara, women for example, the blood of the menses is an overflow of the residue due to the elaboration of the food, which is found in their uterus and serves as material for the embryo that should be shaped by the movements brought by the semen of a male. Consequently, despite very different appearances (eggs have a shell, a white, and a yolk; menses are a bloody flow), wind eggs and the menstrual blood are *the same thing*, namely, a material resulting from the elaboration of food, material from which the embryo is made (since, we repeat, none of the material of the male semen remains in the embryo), but this part of the material that has not been fertilized, that is to say, enformed by the movements of the semen. These wind eggs are thus a "diminished" version of fertilized eggs: they are smaller because incomplete, and they are less tasty because less concocted, but because they are smaller there are more of them (750b21–26). Unfortunately, Aristotle does not explain why, instead of taking place in the form of an effusion, as is the case for the menses of women, this expulsion of superabundant residue takes place in the form of eggs.

Wind eggs, then, and menses, share major characteristics, notably that of being a living material at least in a certain way. We see evidence of that in the case of wind eggs in the fact that they have an internal principle of growth and death, since they both grow and rot. But they would not be alive if the material of which they are a part was not itself also alive, for it's not by leaving the animal that they acquire life. But the most remarkable point in this passage is perhaps that the female material is called "spermatic" (σπερματική), that is, seminal.

Let us return, then, to the nutritive soul from the angle of a problem that Aristotle raises in *Generation of Animals* 2.5: If she is able to produce living matter, why does the female need a male to engender a living animal? We should look closely at this text:

> One might ask oneself why [the female is not able to generate alone]. Since the female possesses the same soul [as the male] and the material [of the embryo] is the residue in the female, why does the female need the male in addition, and why doesn't she generate herself from herself? The difference is that an animal differs from a plant in respect of perception. But it is impossible that a face, a hand, flesh, or any other part exist unless it has, either actually or potentially, relationally or absolutely sensitive soul, for otherwise it would be a corpse or a part of a corpse. If then it is the male that is able to produce such a soul, and it is that in respect of which the female and male are distinct, it is impossible for the female to generate herself, from herself, an animal. What we have just been talking about [the sensitive soul] in fact constitutes the being of the male. Nevertheless, the difficulty raised does have a point, as is clear from the case of birds that make wind eggs, because it is certainly the case that the female may generate, up to a point. But that too poses in addition the difficulty of knowing in what sense one may say that their eggs are alive, for it is not possible that it would be in the sense that fertile eggs are alive (because it is from these that a new actual animal is born), nor in the sense that it is a question of wood or stone. In fact, there is, even for these eggs, a kind of rotting as if that had previously participated in a kind of life. It is thus evident that they possess potentially some soul. Then which is it? Necessarily the last: nutritive. It is present in all animals as well as in plants. Then why does it not bring about the parts of animals? Because it has need of the sensitive soul, because the parts of animals are not like those of plants: there is need

> of the cooperation of the male, because in animals the male is distinct [from the female]. (*Generation of Animals* 2.5, 741a6, partially cited earlier)

We need to comment on several points in this passage. It is clear that it is talking only about the "dominant" mode of sexual reproduction, which, in virtue of an Aristotelian schema previously mentioned, serves as a model for the other kinds: that can be seen from the last sentence of the passage, because there do exist animals in which male and female are not distinct. This passage also says clearly, if one brings together the first and last lines, that the female residue possesses nutritive soul. It has it "potentially," because the residue, as is more evident, once again, in the case of wind eggs, does not exercise the developed functions of a nutritive soul, nutrition and reproduction, functions that it would exercise if its nutritive soul were actual. Wind eggs are a living *animal* material to the extent that they are produced by an animal, but they are not actual animals because they lack the sensitive soul, which is brought by the male, Aristotle going so far as to say that the sensitive soul is "the being of the male" (741a15). But an animal part like the hand or flesh cannot be found in a living thing if it does not possess the sensitive soul "in a certain relation or absolutely" (ἤ πῃ ἤ ἁπλῶς, 741a12), which doubtless must be understood as signifying that a part like the hand depends absolutely on the sensitive soul in that it is in itself endowed with sensation, and that a part like bone depends mediately in that it participates in the structure of an organism endowed with sensation, although not having sensation itself. We will return, in the next chapter, to this particular relationship between the nutritive and sensitive souls.

The female residue, and especially wind eggs, have nutritive soul only potentially, because, as we will see again, *in animals* the nutritive soul cannot really exist actually, that is, fulfill its proper functions, without the sensitive soul. That is why, as Aristotle says elsewhere, "in certain animals, for example [female] birds, their nature is able to generate up to a certain point, because they indeed produce what are called wind eggs, but that which they make is not complete" (*Generation of Animals* 1.21, 730a30). But to be incomplete (ἀτελής) is another way of being potentially.

There is at least one case that says a great deal about the Aristotelian concept of the vitality of the female material:

> But if there exists a certain class that would be female without having a separate male, it is possible (ἐνδέχεται) that it would

> generate an animal from itself. Even if, up to now in any case, this has not been observed in a reliable way, one may hesitate about the family of fish, because in the fish called red mullets no male has yet been seen, but there are females, and plenty of embryos. But we still don't have any observation worthy of credence; and in the family of fish there are some like eels and certain gray mullets living in marshes that are neither female nor male. (*Generation of Animals* 2.5, 741a32)

We have here a remarkable situation: it's a matter of states of affairs that Aristotle deems *possible*, that is, in agreement with his conception of generation, but that have not been observed, at least not clearly. It is not sure that it is appropriate to translate ἐρυθρῖνος as "red mullet," but it doesn't matter much here.[17] Aristotle insists on the fact that these red mullets are among animals, which, on the one hand, have sexual organs, and on the other hand, have distinct sexes, differentiating them from animals without gender such as eels, and hermaphroditic, which have both sexes. It's a matter, in the case of the red mullets, of *females* that generate alone and generate animals, that is, living beings that have a sensitive soul, which would then necessarily have been transmitted to them by their mother, since they do not have a father. Aristotle makes clear that the matter has *not yet* been observed, but it is clear that he hopes that it will be, and that other females are, as one says, "within an inch" of producing offspring truly alive, even if males remain normally the providers of the sensitive soul.

There remains at least one question that Aristotle leaves unanswered, that of knowing why males cannot generate by themselves. They in fact have the sensitive soul, which is contained in and carried by the semen, but also the same kind of material as that from which the female makes the fetus. This material, which Aristotle describes as "menses," is fabricated by the female from blood, which is food developed by the process of digestion—an assimilation that we have discussed. The male, in fact, makes something "better" than the menses, namely the semen, because he is warmer than the female. But whoever can do the more can also do the less, so the male ought also be able to produce the matter from which the fetus is made. The semen is not destined to be matter but to transmit movements. By not granting semen any material function, Aristotle turns his back on hylozoism.

17. According to D'Arcy Thompson in his *Glossary of Greek Fishes*, they belong to the family of Serranidae or that of Sparidae.

If, in fact, the semen were to furnish also the material for the embryo, the semen would, properly speaking, be alive, without the intervention of any other principle, which would be in conformity with the definition of hylozoism. Did Aristotle refuse a material role to the semen *in order* not to fall into hylozoism? An insolvable problem. But the female material has to be animated by an external principle, the semen from the male. The Stoics, in contrast, were clearly hylozoists, because they posited that the primordial matter is alive. The Aristotelian schema, already examined, which explains the inferior (nonliving) by the superior (living) is an efficacious defense against hylozoism, because the material composing living beings has to *receive* life from outside itself. The Presocratic systems that explain the superior by the inferior have, in contrast, a hylozoist tropism. All the same, there is room for hylozoism in Aristotle, because the immortal material beings, celestial bodies and the celestial vault, are presented by him as alive, and they have not received life from anything else than themselves, and have received it without any process of animation.

The female material, even though it is like every material in the world in the last instance composed of the four elements or of some of them, is not an inert matter, that is, a material that would obey only its own laws. Certainly wind eggs are, like everything else in the sublunary world, made of the four elements, but, as the long passage that we have just partially analyzed says, even if they are not like fertilized eggs of real living animals, neither are they nonliving entities like wood (Aristotle obviously thinking here of dead wood of which furniture is made) or stone (741a21), which would thus act only as dead wood or stone acts.

A famous dispute, one to which we have already alluded and which is another version of the dispute noted earlier concerning teleological explanation, opposes interpreters who have claimed that the material elements left to themselves would be able to form living parts from the sole fact of their physiochemical properties (which contradicts, for example, the passage from the *Generation of Animals* 2.2, 734b31, cited earlier), against those who think, like David Balme, that "the sublunary elements, air earth fire and water, act teleologically only when they are part a living being."[18] It is the latter group that is right, because it is not in the nature of water and earth to make bone, horns, or flesh. Balme is obviously right, because when the soul of the male undertakes to form an embryo, via a system of movements, the material on which that soul works is already prepared

18. Balme, "Teleology and Necessity," 277.

by an elaboration by the nutritive soul of the female. Thus the female, as Sophia Connell has shown, plays a real genetic role, and contributes to generation by producing a *seed*.[19]Thus, among many other passages, right from the beginning of the *Generation of Animals*, Aristotle affirms that the *sperma* that results in the offspring comes from both male and female, which can be understood very well when one knows that for him male semen and menses are the same reality, food worked up by the internal heat of the animal, but to different degrees of elaboration as a consequence of the diminished heat of the female:

> [We must] consider how and where seed is produced. In fact, it is from seed that beings generated by nature are formed, and the way in which seed is found to be produced in the female and male should not remain in obscurity. It is, in fact, because one part of this kind is secreted by the female and one part by the male, and because this secretion is to be found in them and comes from them that the female and male are principles of generation, for we call male an animal that generates in another, and female that which generates in herself. (1.2, 716a7)

Even if, in the rest of the *Generation of Animals*, Aristotle tends to reserve the word *sperma* to the male seed (what we call sperm), its use here to designate the generating contributions of both sexes shows clearly that he grants to both males and females a real generative role.

Aristotle is not very forthcoming about this preparation by the female of a material that will be used teleologically by the male. But since this preparation happens in and by a living being, it is not surprising that this material would be as it were torn away from its pure material nature and that this preparation would take the form of a "teleologization." There is a very precious passage that shows this, again about horns. At *Parts of Animals* 3.2, 663b29, Aristotle explains that the "bony element is earthy," and that therefore the horns are earthy, and that there is more of it, as we have seen, in larger animals:

> It is certain that Nature uses the excess of residue of a body of this kind, which is present in larger animals, for their defense and their benefit, and that this residue, necessarily flowing

19. Cf. Connell, *Aristotle on Female Animals*, 101ff.

> upward, is found to be distributed in some to teeth and tusks, in others to horns. That is why no animal with horns has two ranks of teeth. In fact, they do not have front teeth on top. (663b31, partially cited earlier)

As I mentioned earlier, this "necessary flow upward" should have attracted the attention of the Aristotelian commentators, because the nature of earth and things composed of earth naturally carries them downward. Thus, there needs to be another kind of "necessity" to direct the earthy element upward.

If Nature means to give animals with horns both means of defense and other useful attributes (cf. for their defense and interest, ἐπὶ βοήθειαν καὶ τὸ συμφέρον, 663b32), that is, horns for their defense, and also teeth, it is necessary that the material of the horns flow upward. Because, as Aristotle shows a few lines earlier, horns have their maximum usefulness on the head.[20] It is thus indeed the *form* of the animal that requires that the earthy material go upward, there where, and only where, it could accomplish a function advantageously. That is why it is tempting to interpret the "by necessity" (ἐξ ἀνάγκης, 663b34) as indicating not an absolute necessity, but a hypothetical necessity: if it is necessary that the earthy matter serve the ends that Nature assigns to it, then it is necessary that it flow upward. In fact, this example shows, better than many others, the difference, slight but real, that exists between explanation by hypothetical necessity and that by the "two natures": it seems closer to Aristotle's intention to describe the process in question as being teleologically neutral and even possibly bringing disadvantage and to say that it is up to rational Nature to "do something," rather than saying that it is necessary that the earthy material go upward if one wants the animal to have horns. The very vocabulary used by Aristotle seems to show that: Nature "uses"

20. "If the horns had been naturally located elsewhere on the body, they would add weight without any other usefulness, and would even obstruct many of their functions; even if they added weight without any other usefulness, they would even be a natural obstacle on the shoulders. In fact, it is necessary to think not only how the body would be stronger, but also where the horns would go otherwise. So that, since bulls do not have hands, it is impossible for them to have horns on their feet, and if they were on the knees they would get in the way of bending, it is necessary that, as is actually the case, that they have them on the head" (66eb3). At 663a35, Aristotle cites the Momus of Aesop. [Translator's note: Momus complained that bulls did not have eyes on their horns, to help them aim at things they were goring.]

(καταχρῆται ἡ φύσις, 663b33), which shows that she does not *decide* to direct the material upward as the smith decides to take iron to make a saw; furthermore, Nature makes the best of a bad situation, because she turns to advantage an originally negative situation, an "excess" (ὑπερβολή, 663b32), the result being marked by a certain indeterminacy, since Nature could, with this material, make teeth, tusks, or horns. Aristotle, on the other hand, does not say why this earthy material goes upward, while it seems that it ought to go downward. But the two approaches raise the same difficulty because earthy matter cannot act in contravention of the laws of its material nature, which commands it to go downward because of its weight.

Possibly the solution to this problem is indicated by the mention of the "excess" (ὑπερβολή) of the residue indicated at 663b32. In his commented translation of the *Parts of Animals*, James Lennox has correctly remarked that τὸ σωματῶδες καὶ γεῶδες πλεῖον ὑπάρχει τοῖς μείζοσι τῶν ζῴων at 663b24 ("the bodily and earthen is present in greater amounts in larger animals") could mean, not that it is found in smaller quantity in smaller animals, but that it is found in excess in larger animals, which the word "excess" at 663b32 would confirm.[21] In fact, the two readings can coexist. Since the overabundant material cannot accumulate any more downward, it flows upward.

In a way, all the interventions of rational Nature, which looks to get the best in a situation, bring about the better, even by trickery, are done at the price of a modification of that which would be the action of a given material according to the laws that are purely its own. Thus, it is not in the nature of earth to accumulate to fortify the tongue and palate of the camel, nor to extend into a plane surface to make the web feet of ducks. But neither of these two actions is done contrary to the laws of material nature.

Another aspect of the relation between the sexes in reproduction, which undermines a little more the image of the female as pure matter, is revealed by the Aristotelian analysis of the resemblances between parents and offspring, a doctrine that has fascinated commentators. The result of fertilization is presented by Aristotle as the result of a contest, a result that depends in the first place on the degree of heat. The movements contained in the semen are the essential tool for the transformation of the food into blood or flesh, but its heat is the principal condition, male semen benefiting

21. Cf. Lennox, Aristotle, *On the Parts of Animals*, 249.

from higher heat than that of the female seed. The major principle that will direct the mechanism of the formation of the fetus is that "it is necessary that that which is not dominated by that which fashions it change into its contrary" (τὸ μὴ κρατούμενον ὑπὸ τοῦ δημιουργοῦντος ἀνάγκη μεταβάλλειν εἰς τοὐναντίον, *Generation of Animals* 4.1, 766a15), but, as Aristotle repeats several times, the contrary of male is female. This notion of "domination" is crucial for the questions that interest us here. If the semen is not hot enough, it does not dominate the female material, and everything happens as if it were the female who would become, at least to a certain degree, the "fashioner." Males who are too young or too old, those who have a body that is rather more humid than dry, and more feminine, have a tendency to father females: "all that happens because of a lack of natural heat" (4.2, 766b32). Any cooling element thus encourages the birth of females, for example, the fact of absorbing "hard, cold water" (4.2, 767a34). In fact, it is not so much the internal heat that determines the sex, as the more or less great ability to cook the food to make of it a residue serviceable for generation, but heat is indeed the principal cause of this ability. But the abundance of the residue also plays a crucial role, because the more abundant it is, the more difficult to concoct (4.2, 766b35). Thus, exposure to south winds, which we would automatically think would produce males because they are warm winds, actually favors the production of females because they are humid and also because they bring about a surplus of residue.

But this tendency toward the contrary due to the fact of the absence of domination is not the only one to intervene. It coexists with a tendency to go toward a similar, but which is inferior to that to which it is similar, an inferiority that Aristotle describes in terms of "relapsing":

> On the one hand, each movement when it degenerates thus changes into its opposites and, on the other hand, the fashioning movements relapse (λύονται) toward that which is nearest: for example, if the movement of the generator relapses, by the smallest difference it changes toward that of his father, and next toward that of his grandfather; and it goes in this way both for males and for females: the movement of the female generator changes toward that of her mother, and if not her, then toward that of her grandmother. And so on for their other ancestors. (*Generation of Animals* 4.3, 768a14)

The word that is here translated "degenerates" is the participle ἐξιστάμε-νος from the verb ἐξίστημι, which means the act of going out from, but with a pejorative nuance: when used transitively it can mean "to drive out of one's senses," by anger, indignation, or another movement of that kind, and intransitively to be expelled involuntarily. Theophrastus has that notable use when speaking of the destruction of plant grains: "All other destructions are contrary to nature (παρὰ φύσιν), namely those in which the grains are infested with worms, those in which they become too wet, and those in which they degenerate (ἐξίσταται) in some other way" (*Causes of Plants* 5.18.1). This duality of the destiny of spermatic movements is even better presented by Aristotle a few lines farther along:

> But, in a general way, here's what one must take as basic principles: first, that among movements, some are actual, some are potential; and two others according to which, if a movement is dominated it degenerated toward its opposite, while if it relapses, it is toward the next movement in line; if it relapses a little, it is toward the nearest movement, while if it relapses more, it is toward a movement farther along. Ultimately the movements are so scrambled (συγχέονται) that [the offspring] does not resemble anyone in the family or relations, and nothing remains in common except that it is a human being. (*Generation of Animals* 4.3, 768b5)

By losing their power of domination, the movements contained in the semen become distant from their genus (degenerate) and go toward their opposite, primarily that which is male toward that which is female. These "degenerations" are therefore described as a sort of abandonment of the nature of the individual in whom they occur. And Aristotle adds, a few lines farther along, that that which is "natural" (πέφυκεν, 768a21) and happens "most frequently" (ὡς ἐπὶ τὸ πολὺ, 768a24) is that the son resembles his father, and the daughter her mother. But if the movement issuing from the father does not dominate, one may have a son who resembles his mother or a daughter who resembles her father. On the other hand, by relapsing, that is, by persevering in its nature but in a weaker form, this movement goes toward the neighboring forms in going progressively farther from the original form. In the first instance, that yields females, in the second offspring that resemble less and less their father until they become just animals. Sometimes, in fact, the offspring resembles no one in the family and then it is the movements related to its *human* form that carry the day, and even sometimes it's the movements related to its *animal* form, and then one gets a monstrous baby, for example a sort of elephant man. Because Socrates is not only the son of his father and grandson of his grandfather, and so on, but he is also a human being and an animal, all these characteristics passed on to the embryo by the means of inherited and transmitted spermatic movements. These processes have been, in recent years, relatively well studied.

But what interests us most of all is the role of the female lineage. If one describes the intervention of this lineage into the reproductive process in terms of *resistance* to information, as interpreters often do, one has a simple parasitism of the process in question. A resistance of that kind exists, which leads Aristotle to argue, among other things, that, since it is the role of the male to transmit the type of which he is the carrier, and that the female material can make that transmission *deviate*, one may define monsters as deviations far from the type; thus monsters cannot be made by the male parent and "one ought, in general, rather think that their cause is in the material and in the embryos in the course of their development" (4.3, 770a6).[22] But that gives a very incomplete image of the role of the female.

One gains a better grasp of the nature of that role if one returns to a passage already cited, in which Aristotle distinguishes the movements that "degenerate" toward their opposite, and those that "relapse" toward movements that are close to them: the passage explains that the *movement* of the female generator can also regress toward that of her mother, and then her grandmother (768a18).[23] The female is thus the carrier of *movements*, giving her an active role in generation.

---

22. But the passage (ὅλως δὲ μᾶλλον τὴν αἰτίον οἰητέον ἐν τῇ ὕλῃ καὶ τοῖς συνισταμένοις κυήμασιν εἶναι) can be read in a relative sense: the ὅλως leaves room for cases where this is not true, and the μᾶλλον can have a comparative value ("the cause is found more in the material [than in the formal cause]").

23. At 768a18, Drossaart Lulofs corrects the text "and it goes this way both for males and for females" to "it also goes this way for females," a useless correction and one without paleographic basis, but it shows that Drossart Lulofs understood the passage very well. The words ἐπί τῶν ἄνωθεν at 768a21 should be understood as concerning just the ancestors, and not the female ancestors of both sexes. Aristotle, *Aristotelis De Generatione Animalium.*

Nevertheless, we should not respond to the excesses of "feminist" interpretations of Aristotle's biology by making him a paragon of sexual egalitarianism. In that debate, which does not interest us directly here, one must at least notice the following points. Aristotle clearly affirms the ethical and political inferiority of women, giving them virtues less well developed than those of men, also depriving them of access to citizenship, a circumstance that they share with infants, barbarians, and slaves. We must note that this ethical inferiority is not a universal law of nature, since "females are all less courageous than males, except the bear and the panther: the females of these species seem to be more courageous" (*History of Animals* 9.1, 608a34), as we have seen several times. On the other hand, from a strictly biological point of view, women are inferior to men because females are inferior to males. But no text that has come down to us establishes a link between the ethical inferiority and the biological inferiority of women. As for a possible intellectual inferiority of women, Aristotle says nothing.

Nevertheless it remains that for Aristotle "the female is like a disabled male" (*Generation of Animals* 2.3, 737a27), and that he thinks that males are superior to females, ultimately because of their superior heat, which, notably, allows them to work up the food better, and one of the signs of this superiority is incontestably the preponderant role that the male plays in reproduction. The female cannot hope to outweigh that superiority except in the case of a weakness of the male, but if, as one says, everything goes well, a male resembling his father is born:

> The offspring that does not resemble its parents is already in a way a monster, because, in a way, in these cases nature has already strayed from the lineage. The very beginning of that is the birth of a female rather than a male, but that is naturally necessary, because it is required for maintaining the genus of beings in which male and female are separate; but since it is possible that the male sometimes is not dominant, because he is young, or old, or some other reason of this kind, the birth of females is necessary in these animals. (*Generation of Animals* 4.3, 767b5)

What can we say about this necessity of the birth of females? Perhaps we could think of it as a hypothetical necessity, for it is hypothetically necessary that there be females if the species is going to survive, but Aristotle seems to correct this initial impression by proposing a second kind of necessity for the birth of females: "it is possible that the male not dominate" or, adds Aristotle, "some other reason of this kind" (767b10), these two forms of necessity having no relation to finality, the second form of necessity taking the form, "if certain conditions are given, then necessarily a female is born." This would be a form of absolute necessity à la Democritus, which rational Nature uses to attain an end, the reproduction of animals. The status of this natural necessity of the birth of females remains slightly obscure, and perhaps one may think that Aristotle calls it "necessary" especially because it is not accidental, given the frequency of the birth of females.

The difference between genesis and essence, a fundamental point of disagreement between Aristotle and the Presocratics, includes another aspect that interests us here because it concerns the relationship between nonliving and living. As far as animal species are concerned, the passage from nonliving to living never happened, nor could it happen as an *event*, since living things have never had a first member of their line that would have appeared one day. As David Balme says in a text cited earlier, nonliving elements left to themselves, even those that constitute food any more than the others, would never form animal "parts." For the atomists, in contrast, the soul is a specific disposition of atoms that, when it happens as a result of random atomic contacts, makes a body in which the result is alive. But in the sexual generation of living things, there is a passage from nonliving to living, since food is transformed into semen and menses, which are the means employed by the living being to produce life. But two things must be added. First, once fertilization has taken place, the mother that carries the embryo must be alive for it to develop. Next, that fertilization itself, described as the information of a material, must put in operation two vital agents. Thus, female passivity is an "active passivity."

In fact, as we have just said, this passage from nonliving to living in the case of the fetus is often described, according to a technical paradigm used by Aristotle himself, as the informing of the female material by the movements contained in the male seed. But, as we have seen, this material is alive. The complex reproductive operation described in this chapter never leaves the living realm. It's only in the case of the formation of semen and its equivalent in the female that food, most of the time nonliving material, becomes a living matter. But, as we have seen, vital realities like that are not constituted simply by the action of hot and cold, but by the movements transmitted to this nonliving material by a generator, or more precisely both generators, who exist *in actuality.* If an animal, then, depends on its environment, for animals that breathe would not survive without ambient air, nor aquatic animals without the water of rivers or the sea, it still keeps what may be called its "vital initiative," which brings it about that it is the animal who chooses objects of interest in its environment, whether in terms of material or in terms of process. The environment is thus limited to a status of a *condition of existence.* What the Aristotelian theory of sexual generation reveals is that on this point too, Aristotle shows himself to be a real *biologist.*

It's a real shame that, as some of his students can attest, Georges Canguilhem did not know Aristotle better, because he would have found in him a remarkable prefiguration of his own analyses, notably of that which he proposed in his very celebrated article, "Le Vivant et son milieu" (The Living and Its Milieu), which was first a lecture presented in 1947 at the Collège Philosophique.[24] It's worth citing a few passages of this article that, in twenty-five pages, proposed the bases of a biological thought, first of all by delimiting clearly, as the title indicates, the relationships of the living thing with that which is not itself. Relying on studies like those of Kurt Goldstein and Jakob von Uexküll,[25] Canguilhem put himself in opposition, as is well known, to an approach uniquely physiochemical of the living thing, but he did that in a form that perhaps was not adequately understood by his readers, by way of a critique of the theory of the environment that he thought to be like "the positive translation, apparently verifiable, of Condillac's fable about the statue":

> The theory of milieu was at first the positive and apparently verifiable translation of Condillac's fable of the statue. [footnote omitted] When the air smells like roses, a statue is rose-scented. In the same way, the living, within the physical milieu, is light and heat, carbon and oxygen, calcium and weight. It responds by muscular contractions to sensory excitations; it responds

24. Translator's note: Collège Philosophique was an association founded in 1946 by Jean Wahl in the Latin Quarter, in Paris. Wahl created it because he felt the lack of an alternative to the Sorbonne (University of Paris), where it would be possible to give voice to nonacademic discourses; it became the place where nonconformist intellectuals—and those believing themselves to be so—were tolerated and given consideration. Information from Wikipedia.

25. Canguilhem, *La connaissance de la vie*, 144ff, *Knowledge of Life* 110.

> with a scratch to an itch, with flight to an explosion. But one can and must ask: Where is the living? We see individuals, but these are objects; we see gestures, but these are displacements; centers, but these are environments; machinists, but these are machines. The milieu of behavior coincides with the geographical milieu; the geographical milieu, with the physical milieu.[26]

The living thing, says Canguilhem, is a center of reference, a bearer of values: "The milieu on which the organism depends is structured, organized, by the organism itself. . . . It is for this reason that, within what appears to man as a single milieu, various living beings carve out their specific and singular milieus in incomparable ways."[27] The living thing is thus an organism that functions according to its own laws, which "brings its own proper norms of appreciating situations," Canguilhem says,[28] and which distinguishes self from not-self. In describing the milieu constructed by the living thing (*Umwelt*) like an "elective extraction" in the environment (*Umgebung*),[29] Canguilhem really puts to work a procedure that seems to be very close to that which Aristotle uses in his doctrine of the "two natures": the living thing selects in its environment the elements (tolerable thresholds in the chemical composition of the milieu, nutritive materials, perceptible signals, etc.) that are significant for it, which permit it to make these elements function for its benefit for survival and development. That's a properly Aristotelian approach.

But, to come back to the question of the generation of living beings, there does exist a process of formation of living things in which this formation of the living being seems indeed to demolish what we have just been saying in that it fundamentally does not depend on a vital initiative of the living thing itself, but is imposed on it by its nonliving environment, and by furnishing, at the same time, a remarkable example

26. Canguilhem, *Knowledge of Life*, 108. Cf. the citation of Edouard Claparede in his preface to Frederick Buytendijk's *Animal Psychology*: "What distinguishes the animal is the fact that it is a center in relation to ambient forces that are, in relation to it, no more than stimulants or signals; a center, that is to say, a system with internal regulation, whose reactions are determined by an internal cause: momentary need." Cited from Canguilhem, *Knowledge of Life*, 118.

27. Canguilhem, *Knowledge of Life*, 118.

28. Canguilhem, *Knowledge of Life*, 113.

29. Canguilhem, *Knowledge of Life*, 112.

of change from nonliving to living: it's spontaneous generation. We have to see what that is.

Spontaneous Generation

Most interpreters think that the doctrine of spontaneous generation constitutes a marginal part of the Aristotelian theory of reproduction, one that Aristotle is not easy about. Thus, Sophia Connell talks about the "obvious discomfort with the phenomenon."[30] Others, on the other hand, like James Lennox or John Dudley, think that spontaneous generation is a phenomenon that Aristotle has well under control, and that this theory plays a role in the general study of the generation of living things. Their position is more correct than that of the others.[31]

Two preliminary remarks. First, we have to respond to the first group that, if Aristotle seems hesitant and imprecise about the spontaneous generation of certain animals, that is doubtless not due to any uncertainty on his part about the process of spontaneous generation and its causes themselves, but about the difficulty that he has of observing it in many cases. That's an example of a kind of *topos* in Aristotle's biology, particularly in the *History of Animals*, of saying that a phenomenon difficult to observe requires additional study; we saw that in the case of females giving birth without males. That lack of certainty can go pretty far, for example to the point of not seeing what the cause (or causes) of spontaneous generation are in a particular species, or even if the generation of a species in question is actually spontaneous. Thus, one can see clearly that, among animals with shells, snails couple, but it is impossible to say if that coupling results in reproduction (*Generation of Animals* 3.11, 762a32). That definitely shows that Aristotle sometimes has a hard time grasping the causes at work in a particular case, but not that he has only an imprecise idea of the process of spontaneous generation itself. Next, we must also reject the idea, also pretty common, that this generation is a marginal phenomenon. It is one of the two, or three, great modes of animal and vegetable reproduction. Spontaneous generation concerns a great number of animals—rodents,

30. Cf. Connell, *Aristotle on Female Animals*, 261, n. 70: "The entire discussion displays Aristotle's perplexity and consternation rather than any confident theorizing."

31. Lennox, Aristotle, *On the Parts of Animals*, 229–249; Dudley, *Aristotle's Concept of Chance*, 329ff.

fish, and especially shellfish—and Aristotle knows how numerous those species are, and we will come back to that.

To open the question of the spontaneous generation of living beings, there is, in fact, a lot to get from a little-used passage, doubtless because it is thought to be more descriptive than explanatory. It's at the beginning of *History of Animals* book 5. It would be necessary to look at the first chapter almost line by line. Here are a few remarks. An introductory paragraph lays down, as is habitual for Aristotle, how far we have come, and what we have left to study: the various parts of various animals have been considered, including parts devoted to reproduction, as well as the functions like perception, voice, and sleep; "it remains to define that which relates to their generation" (περὶ δὲ τῶν γενέσεων αὐτῶν λοιπὸν διελθεῖν, 5.1, 539a1), that is, their modes of reproduction. But "these modes of reproduction are numerous and very varied and, if from one point of view they are dissimilar, from another point of view, they are similar" (539a2). That's a typical connecting passage in Aristotle, especially in the *History of Animals*, between similar and dissimilar, which almost always functions in the same way: it's a matter of discovering, behind a very large variety (which is the case here) and sometimes enormous differences, a unity that is at first invisible, that this unity would be specific (and in this case the difference would be slim), generic, or analogical, a unity that rests on function. Thus, the lungs and gills are the same by way of the function of cooling the organism they serve, even though they look very different. Thus, one should be able to expect something of the sort for the modes of reproduction. And, in fact, one of the great tasks of Aristotelian science, and particularly in the study of living things, is this hunt for identity. We must note that Aristotle does not outline this conceptual matrix as he usually does: he does not argue, as he does in other cases, that the reality (the genital organs, for example) that he examines are "in one sense the same, in another different," but that the modes of generation, despite their diversity, "resemble each other." He uses the verb προσεικέναι, rather rare for him, that he uses elsewhere to indicate the resemblance of infants to their parents (*Generation of Animals* 1.17, 721b29). It is difficult to decide what to make of this textual fact, but it could signify that the similarities between the modes of reproduction are greater than those to be found between lungs and gills. Thus, then, as we have seen, the *History of Animals* mainly offers an accounting of the diversity of living things (in contrast to the *Parts of Animals*, especially based on the underlying unity of that diversity), the study of the modes of reproduction (re)introduces the primacy of unity. But this is not the same unity as that of the *Parts of Animals*, because this one is a unity of function (cooling, for the lungs and gills), while in the case of the modes of generation, and even, we will see, in the modes of production of nonliving natural beings, it's a matter of bringing into evidence the resemblances of the explanatory processes.

Next comes a methodological remark that has taxed the ingenuity of commentators—one should stop taking the supreme animal (human being) as a model and point of departure; it will be better to go up the modes of reproduction from the most rudimentary to human reproduction: "Let us begin with the animals with shells, then it will be necessary to deal with the crustacea and then the other families in the same way. It's a question of mollusks, insects, and then the family of fish" (5.1, 539a8).[32] But a good many of these animals, including almost all animals with shells, reproduce by spontaneous generation. A passage like this makes affirmations like that of Jules Tricot insupportable, though he has actually the majority of commentators at his side; he claimed that "a natural irregularity . . . spontaneous generation seems to be of the same kind as the numerous aberrations that constitute *monstrosities*."[33] And anyway, at 539b15, Aristotle says a few words about this diversity in the modes of reproduction, from which one might draw several conclusions, and first, that the great dichotomy that divides the living world, that is, just as much animals as plants, is that which opposes reproduction that comes from parents of the same form via a seed and spontaneous generation. Faithful to his chiasmic method of exposition, Aristotle begins then with the plants, to say that in those that do not reproduce by a seed, it is necessary that they be constituted by "a principle of this kind" (539a18),[34] that is, something that resembles a seed.

So even before beginning our study of the content of the Aristotelian

32. An inversion of the same kind is proposed at *Parts of Animals* 4.5, 682a30: "Since we have dealt with the internal parts of animals, we must return to the external parts that remain. And we must begin with the animals that we have just now been discussing [i.e., bloodless, especially insects], and not with those that we have put to one side, in order that, starting with those that demand less work, the presentation will dedicate more time to complete blooded animals."

33. In Tricot's French translation of the *History of Animals*, Aristotle, *Histoire des animaux*, 280.

34. Cf. *History of Animals* 5.1, 539a15: "There is in any case a common characteristic that belongs to animals as well as plants (κοινὸν μὲν οὖν συμβέβηκε καὶ ἐπὶ τῶν ζῴων, ὥσπερ καὶ ἐπὶ τῶν φυτῶν): these in fact either coming from the seed of other plants, or being born spontaneously, if a principle of this kind is formed."

doctrine of spontaneous generation, we can frame the question in terms of the following theses.

(i) Spontaneous generation is one of the two great modes of generation of living things (in fact there are three, if we add budding, discussed later in the text), which concerns a vast part of the vegetable and animal world. Cockles, clams, razor clams, scallops, ascidians, limpets, sea snails, sea anemones, sponges on rocks, hermit crabs, intestinal worms, fish lice, many insects, but also eels and certain mullet fish, as well as a number of rodents are born by spontaneous generation. If spontaneous generation is a marginal phenomenon, the margins are pretty large.

Furthermore, spontaneous generation is a multiform phenomenon as is shown, for example, in that same chapter 1 of *History of Animals* book 5. Because of that multiformity, the relationships between spontaneous generation and sexual generation are not always the same. Thus in certain fish the eggs are produced by spontaneous generation, but living animals emerge from them (*History of Animals* 5.1, 539b2). Here we are at the frontier between sexual and spontaneous generation, something that Aristotle notes when he says that this kind of generation "is rather similar to that which happens in birds" (539b6).

(ii) Spontaneous generation is a natural process if only because of its regularity. It would seem to be difficult to agree with Sophia Connell that "every single instance of testacean generation is an accident."[35]

(iii) Aristotle has made remarkable efforts to establish that spontaneous generation and generation by *sperma* emerge from a single causal schema, the two explanatory procedures putting to work analogous elements, at least in the Aristotelian sense of the word. It is this thesis that will serve as the foundation for our attempt at elucidating the nature of the process of spontaneous generation so that we can compare it with sexual reproduction and, later, with the formation of nonliving realities like metals and minerals. Thus, we do not adhere to the position of Sophia Connell who thinks that animals that reproduce by spontaneous generation are "so far away from the paradigm [of the reproductive function] that the usual causes are absent."[36]

The cardinal text for our topic, on which all interpreters base their analyses of spontaneous generation, is at *Generation of Animals* 3.11, 762a8–762b21. The first part of this passage, to 762a32, is aimed at

---

35. Connell, *Aristotle on Female Animals*, 262.

36. Connell, *Aristotle on Female Animals*, 260.

eliminating an error that Aristotle seems to think one might be tempted to commit. Although many animals reproducing by spontaneous generation develop in a context of putrefaction, one must be careful to avoid concluding that it is putrefaction that triggers the process of spontaneous generation. Passages that seem to support that idea, for example *History of Animals* 5.1, just cited, or 5.31, 556b25, which says that "fleas are born in minimal putrefaction," do not project the viewpoint of the explanation of the mechanism of spontaneous generation, and because of that, they can speak in a descriptive and approximate way, because, in fact, *one sees very well*, or at least Aristotle sees, that fleas are born in rotten materials. But to be born in a rotten context is not the same thing as being born because of rot. It would be unreasonable, and totally useless, to try to infer chronological conclusions from this. What gets the process going is an actual concoction of the milieu in which it happens, and the materials resulting from that putrefaction are the residue of that concoction, for "nothing is produced from the whole [of the matter]." (762a15), as is also the case in the arts, since the carpenter wastes a lot of wood when making a table.

At 762a35 the passage begins describing the process of spontaneous generation. But the whole passage rests on the comparison (analogy) between spontaneous generation and sexual generation, as was the case of the passage cited earlier, from the *History of Animals*. Thus, it is necessary, in animals that reproduce spontaneously, to find "the formation that corresponds to the material principle" (τὸ κατὰ τὴν ὑλικὴν ἀρχὴν συνιστάμενον, 762b1) of animals with sexes. Aristotle then sums up, in less than three lines, his theory of sexual generation: "In the females there is a certain animal residue that is potentially that which the animal from which it comes is [actually]; acted upon by a first mover from the male it produces an actual animal" (762b2).

We notice that this description confirms what was said earlier: the residue in the female is not characterized as an inert elementary composition, but as a material produced by the female herself, a material that is potentially that which will become the finished animal. The following sentence shows completely clearly Aristotle's analogical procedure: "In the present case [that of spontaneous generation] we must say what corresponds (τὸ τοιοῦτον) [to the female residue] and where the first mover from the male comes from, and what it is" (762b4). So spontaneous generation and sexual reproduction tell us the same story. As for the residue that forms the material of the embryo, then the animal, it comes from an elaboration of food: the internal heat of the body separates[37] and cooks this food, this process being controlled and directed by the nutritive soul of the female, which produces the residue in question. In the case of spontaneous generation, Aristotle describes things by carefully using the same words: in both cases there is a food, for some it is water and earth, for others it is something that comes from these two elements. It is probable that, always according to the chiasmic procedure of exposition, "the others" are those animals that reproduce sexually, while those that have earth and water for food are plants and animals that generate spontaneously, as was said at the beginning of the passage that we are in the course of examining.[38] In both cases there is heat, in one the vital head of the animal, in the other the climatic heat exercised at certain seasons on sea water and earth. So that one gets the same elaboration of food in both cases.[39]

We can say a little more about this residue produced by concoction of the earth, water, or a mixture of the two. As in the case of metals and minerals, to be discussed later, under the action of heat the residue produced by cooking coagulates, but in spontaneous generation the tendency is toward a rapid constitution (cf. ταχέως, 762a22) of small units, which Aristotle expresses with the verb συνίστημι and the translators render "form" or "take form." For that to happen, the heat in question must find itself "enclosed" (ἐμπεριληφθῇ, ἐμπεριλαμβάνεται, 762a22), which forms, Aristotle says, "like a foam bubble" (οἷον ἀφρώδης πομφόλυξ. 762a23). Sometimes, in the case of a shelled animal for example, it forms a hard envelope that protects the animal: in the sea, which is rich in earthy material, the earthy element solidifies around the animal and becomes "as hard as bone or horn" (762a30), which shows that Aristotle thinks of the shell as a part of the animal, just as much as bone for vertebrates, while modern physiology establishes a difference between bone, which is an organic reality, and the shell that is an envelope, certainly indispensable to the organism, but not a part of it.

37. That is, separates the useful part from the useless part, which the organism needs to get rid of, a process described by Aristotle several times.

38. Is Aristotle also thinking about plants that reproduce only by spontaneous generation? It's not clear.

39. "So that that which in the animals [that reproduce sexually] is elaborated by their internal heat out of the food, is [in the case of spontaneous generation] the heat of the ambient climate that unites by cooking and gives them form from the sea and earth" (762b13).

So as far as the formation of the residue that will serve as the material for the animal, we have, then, two parallel "accounts": in the case of sexual reproduction and in the case of spontaneous generation. But what about the moving cause that makes this material become, first an embryo, then an animal, at least in the case of sexual reproduction (Aristotle does not mention "embryo" in the case of spontaneous generation)? In this case too Aristotle tries a parallel exposition. His general thesis seems to be that in the case of living things (animals or plants) that are born by spontaneous generation, there is *pneuma* in the water most of the time, or in that which corresponds to the water, and which plays the role of an "animating breath." So that when this *pneuma* is enclosed in the foam bubble that we were talking about, the residue found in it also becomes ensouled.

We must be careful not to confuse the two situations. At 762b9, Aristotle, after having briefly recalled how the matter of the embryo is formed, writes: "It goes the same in plants (καὶ ἐν φυτοῖς) except that in plants and in some animals, there is no need for the male principle (for it is found mixed in themselves), but in most animals the residue does need it." Does Aristotle mean to speak of "plants" or of "(certain) plants"? In the second case, it would be a matter of plants that reproduce spontaneously, that is, without seed. In the first case, which seems to correspond with what he means to say,[40] Aristotle would be claiming that there are not two clearly distinct sexes in plants, each carrying in itself the female and male principles.[41] In fact, one must distinguish between plants that reproduce by a seed (grain) and those that reproduce by spontaneous generation. In the first, the male principle does exist, but it is not produced by a distinct individual. What are these "some animals"? Two readings are possible, parallel to those about the plants. Either it's about animals that reproduce by spontaneous generation, or, more likely, it's about certain animals that do not include males and females, like the sea bass or red mullet, whose species seem to be composed entirely of females, as we saw. But then they

40. It's hard to decide in the absence of the definite article in the text, but that is a weak argument for two reasons. First, if he had wanted to say, "in *some* plants," Aristotle could have used an expression like ἐν τισι τῶν φυτῶν; and second there is the minority reading (but that of manuscript E), chosen by Drossaart Lulofs, ἐν τοῖς φυτοῖς, instead of the majority reading, ἐν φυτοῖς.

41. Cf. *Generation of Animals* 1.23, 731a1: "In plants (ἐν δὲ τοῖς φυτοῖς) the powers [female and male] are mixed and the female is not distinct from the male."

need to have in themselves a male principle, that is, that they be hermaphrodites. In these animals, it cannot be said that they reproduce by spontaneous generation. In the present passage we thus have, in all probability, an *exception* to the rule, pronounced since the beginning of the treatise, according to which in animals able to move themselves "by swimming, flying, or walking" (*Generation of Animals* 1.1, 715a26), male and female are distinct, an exception in a way signaled in advance by the words ὡς δὲ κατὰ παντὸς εἰπεῖν, "generally," at 715a25:[42] "In general, concerning all the animals that move themselves, whether swimming, flying, or walking with their body, all have female and male, not only in the blooded animals, but also in some of the nonblooded." In fact, if one takes account of the fact that there are many insects that reproduce by spontaneous generation, and of the fact that some of them (to be sure not many) do not have distinct males and females (cf. *Generation of Animals* 1.1, 715b2), this exception is ultimately not that exceptional. But this does confirm that for Aristotle the typical case of spontaneous generation concerns shellfish.

It is indeed the absence of a fertilizing male principle, the principle coming from another individual or residing with the female principle in the same individual, that confronts us with a case of spontaneous generation. There is no system of movements that would be transported by the sperm via the innate breath or directly introduced into the female matter as in certain insects,[43] which imposes on the female residue a form that is simultaneously individual and specific, as is the case in sexual generation. On the other hand, there is, as has been said, in sea water (to stick with marine animals) *pneuma*, which is "full of psychic heat" (*Generation of Animals* 3.11, 762a20), that is, a head that can play the role of a soul, in

42. This expression does not seem to me to designate a universal relationship ("to speak of one fashion that concerns all cases"), an interpretation that has going for it the equivalence between κατὰ πᾶντος and καθόλου (cf. Bonitz, *Index Aristotelicus*, 571b6). It seems to me to be synonymous with another phrase, ὡς ἐπὶ τὸ πᾶν εἰπεῖν (cf. *Generation of Animals* 2.1, 732a20; *Length and Shortness of Life* 5, 466b14). Thus, it means "speaking generally" (Peck). Cf. *Generation of Animals* 1.23, 730b33, even more explicit: "in all animals that move themselves the female is distinct from the male," which does not envisage exceptions, but opposes the animals that move themselves to plants in which male and female are not distinct.

43. In certain insects, it is the female that introduces a part of herself into the male so that this part may be in contact with the warmest part of the male, close to the heart, which suffices for informing the material of the female (*Generation of Animals* 1.21, 729b22).

other words a floating principle of animation, permitting Aristotle to say, "In a way, all things are full of soul" (762a21). The crucial point seems to be that this floating psychic principle cannot play the role of the male principle in generation unless it is gathered and enclosed (ἐναπολαμβανόμενον ἢ ἀποκρινόμενον, 762b26) in the bubble that the residue worked up by the heat has prepared for it. In other words, the principal role in spontaneous generation is played by the residue, which, by preparing, by cooking and separating into small units (bubbles), the receptacle for the psychic *pneuma*, will permit this residue to become ensouled.

We seem to be quite close to hylozoist positions, all the more so since some passages in Aristotle seem to adhere to that position completely. Thus, at *Metaphysics* Z.9, 1034b4: "Beings born by spontaneous generation . . . are all those whose matter is able to move itself by the movement by which seed moves." Doubtless one might attack texts like this one by applying to them the remarks like that made earlier about passages in the *History of Animals* that seem to say that it is decomposition that generates living things. In both cases it's a matter of descriptive passages that are not at all interested in the cause of the phenomenon. But, in fact, if hylozoism is a doctrine that attributes vital capacities or properties to matter, as far as animals that reproduce sexually are concerned, Aristotle, we repeat, is not a hylozoist, because, even in the case of wind eggs, their material is not matter pure and simple, as we have seen. It is indeed from outside of itself, in this case by the action of the nutritive soul of the female, that the material residue receives life, and not from the existence of an internal tendency of this residue itself. In the case of spontaneous generation, things are less clear.

In sexual generation, that which establishes the permanence of a species is the form that the male transmits to its descendants by fertilizing a female, but doubtless also, according to what we have seen, the characteristics of the "residue" offered by the female to the forming process by the male. The major role is nonetheless played by the male. That does not mean that there are eternal Forms that would exist apart from the individuals that carry them, nor that the form is unchangeable.[44] Not only does the

44. Which does not prevent the form from being immobile, as the crucial passage cited earlier reminds us: "But the principles that naturally move things are two, of which one does not belong to the nature; in fact, it does not have in itself the principle of movement; that is the case of that which moves without being moved, like that which is completely immobile (i.e., the first being of all) and the essence and form, for it is an end and that for the sake of something" (*Physics* 2.7, 198a35).

form not remain intact in reproduction, but it is strongly influenced by the material that it enforms during fertilization, as we have seen. One proof, among others, of this, is the birth of hybrids via crossing of two different species.[45] On the other hand, even in intra-specific sexual generation the form is not transmitted without variations—sometimes those variations are very large, as in the case of monsters, or smaller, in the case of the birth of females or of offspring with characteristics of their maternal lineage. Nevertheless, it remains that by containing these variations within fairly narrow boundaries, sexual generation perpetuates animal species that exist forever, provided with globally stable characteristics.

But it is necessary to go further in the description of generation as a transmission of form. According to what was said earlier, we see that, if the female is a "lesser male," she has her role, even if less, in the transmission of the form. That's what follows from the fact that Aristotle attributes to menses *movements* of the same sort as those in the semen, since they can act in concert with them, or oppose themselves to them. The form, transmitted by the movements contained in the semen of the generators, can thus be rediscovered at all the levels and all the steps of the process of fertilization. But that *reproduction* of the form does not happen in spontaneous generation: "The most natural of functions of living things that are complete and not defective, and which do not reproduce themselves by spontaneous generation, is to produce another being like themselves" (*De Anima* 2.4, 415a26).

In the case of spontaneous generation, in fact, the process of transmission of the form undergoes a major transformation. The principle that corresponds to the male principle of sexual generation is, to speak in a somewhat simplistic way, the *pneuma* brought by the sea water in the case of marine shellfish, or by some other vehicle not precisely noted in the other cases of spontaneous generation; that is a universal provider of soul (as a consequence, as we have seen, "in a certain way all things are full of soul" [*Generation of Animals* 3.11, 762a21]). But the text of the *Generation of Animals* does not seem to attribute to this floating *pneuma* a role in differentiation, which would explain why in one case it forms mussels and in another oysters. Thus, it remains that it would be the material residue that, after concoction, determines the specific characteristics of oysters

---

45. As we have seen, Aristotle is not very demanding for two different species to be interfertile: it's enough that their size and periods of gestation be approximately similar (*History of Animals* 8.28, 606b22).

and other animals with shells. Which is exactly what Aristotle says: "In a general way all shellfish are born in the mud by spontaneous generation and differ according to the variations of the mud, oysters in slimy mud, conches and the others I mentioned [clams, razorshells, scallops] in sandy mud, in the crevasses of rocks the sea squirts, barnacles, and those that rise to the surface, like the limpets and nerites" (*History of Animals* 5.15, 547b19). The two processes of spontaneous generation and sexual generation are thus parallel, but at the cost of a remarkable inversion. Sexual generation occurs by the imposition of a form transmitted by the semen of the male to a material present in the female. It is that form that assures the global conformity of the future offspring to the general characteristics of its species. But the (female) material, far from being purely passive, intervenes, one may say, as a secondary creator in the process of forming the embryo, which thus inherits maternal traits, for fortunately offspring are not born strictly identical to their fathers, as we said. In the case of spontaneous generation, the psychic *pneuma* will animate the embryo formed from the residue of cooked matter. But this embryo already has what one may call its specific form due to the material from which it is formed, and as the passage from the *History of Animals* asserts, according to the place where it is formed. We will come back to that point. Thus, it is the material that is the principal element in the *formation* of the embryo, while the *pneuma*, which is the analogue of the male principle, has an efficient function, but not *formal*, or less formal than the material, since it is the material residue that gives the animal—or the plant (if it is one of the species that reproduce by spontaneous generation)—its specific form and its individuality.

In fact, it seems impossible that it would be the wander *pneuma* that differentiates individual animals in the residual material. Thus, it is necessary that it be this residue that, by means of physiochemical reactions undergone by the materials that make it up, constitute the units (bubbles) that the *pneuma* will animate. And it's the proportion of the elements that constitute the bubbles that determines the specific characteristics of the future animal. In other words, it is indeed the material that determines the *specific formal characteristics* as well as individual oysters, scallops, and other shellfish. That's what the passage from the *History of Animals*, just cited, says. One needs to be careful not to minimize the importance of this assertion. It's not a question of the influences of the matter or of the milieu on the final state of the offspring. These influences are proven and are often very important, including in animals that reproduce sexually—these go from the action of the wind on offspring to the color of sheep in relation to the water that they drink.[46] What is at stake here is the essential structure and character that make an oyster an oyster.

Among interpreters of Aristotelianism, notably on its metaphysical side, there has been a dispute on the question of individuation, some thinking that the individual form of a substance would be that which makes this substance unique and different from other substances, including those of the same species; others maintaining that it's the individual matter of each substance that assures its individuality within a specific form (Socrates and Callias share the same human form, but are differentiated by their material). It's not certain that the spontaneous generation of living things allows deciding this question, but it does offer a way of evading it: in many living things, the material is able, in a regular and continuous manner to give specific characteristics. One may perhaps assume that it is not impossible for it to play a major role in individuation.

Three more remarks finish up our discussion of spontaneous generation. Sexual generation is certainly, for Aristotle, a sign of a greater perfection than spontaneous generation. Similarly sexual generation in which the two sexes are carried by different individual animals is more perfect than that in which, as in plants and hermaphrodites, the same individual has both sexes. As we have seen, "In all those where it is possible and to the degree that it is possible, the male is separate from the female, for it is as something better and more divine that the principle of movement belongs, as male, to generated beings, while the female is the material" (*Generation of Animals* 2.1, 732a6). In a given situation, Nature does her best to allow living things that find themselves in specific conditions to live and reproduce themselves forever. We have seen what that means in terms of teleology. Nature thus chooses the best possible, the better being sexual generation carried out by distinct males and females. It is thus not surprising that this kind of generation has a paradigmatic function in relation to other modes of reproduction. So, it is necessary to understand correctly the fact that, in a well-known methodological passage in the *History of Animals*, cited earlier, Aristotle says that, in the case of the study of the generation of animals, it is necessary to reverse the order of exposition and start with "simple" animals to end up at human beings.

46. "On the island of Antandros there are two streams; one makes sheep white, the other black. It also seems that the Scamander River makes sheep yellow" (*History of Animals* 3.12, 519a16).

It's a matter here, in the logic of the *History of Animals*, of surveying and describing the genital parts, the modes of coupling, the times of gestation, and so on. In the extremely causal treatise, the *Generation of Animals*, on the contrary, it is in fact sexual generation where Aristotle starts, and it is even female and male that he defines first.

People have often noticed, and praised, the curiosity that pushed Aristotle to describe a considerable number of animals, while even though the study of the functions and organs that he achieved, which is doubtless the major goal of his research in natural history, had relatively little to gain. I have myself spoken of "theoretically unmotivated" curiosity.[47] In contrast, one may be astonished about the small amount of information that he gives us on certain subjects, and spontaneous generation is one of those. It is, for example, difficult, to the great despair of interpreters, to decide, in the texts dealing with spontaneous generation, and especially *Generation of Animals* 3.11, what really plays which roles of the different causes in the process of generation. Thus, our interpretation of spontaneous generation rests mainly on the case of marine animals because Aristotle is a bit more forthcoming (or less laconic) about them, but there are all the others born in the earth or in rotting wood, coming from the interior of animals or plants, from mud, and so forth. There are also animals born by spontaneous generation, but male and female, which, thus, couple and have offspring, but not of the same species as their parents. But Aristotle ultimately says very little about all these animals, and it is doubtless not by chance that, in his major theoretical text on spontaneous generation (*Generation of Animals* 3.11), he seems to speak mainly about the spontaneous generation of marine animals, with the characterization of the analogue of the female principle as a foamy "bubble" and the analogue of the male principle as *pneuma*.

I previously noted that the case of animals with shells was privileged in the study of spontaneous generation. And, in fact, everything happens as if—in the enormous investigation that would have demanded a precise knowledge of the modes of reproduction of all these species, work that Aristotle recognizes several times that it is far from having been accomplished, and not only about their reproduction—he has deliberately chosen the marine type of spontaneous generation, leaving us the task of applying the schema to the other types. But why did he choose this case? Possibly he knew it better than the others, for example as a consequence of his extended marine research, the strait of Pyrrha seeming to be rich in

47. Pellegrin, *La Classification des animaux chez Aristote*, 186. English translation, 152.

shellfish. Perhaps he worked at presenting a more readable analogy with sexual reproduction. But there is for this choice a reason stronger and more "material" that the *Generation of Animals* presents very well: "Sea water is much more bodily than fresh water, it is naturally warm, and it shares in all the parts, humid, *pneuma*, and earth" (3.11, 761b9). This means that in the spontaneous generation of an oyster, for example, one may see more clearly the generative role of the milieu: one understands how the body of these animals *takes form* in this environment rich in various material elements.

A modern biologist would doubtless have fastened on the reproductive mode of these amazing critters as well as on that of viviparous quadrupeds or human beings. What the analyses that precede it show us is that Aristotle does not conceive of nonsexual modes of reproduction as realities obeying their own logic, but as obscure and degraded versions of a more perfect mode. But if what has been said already is true, it is enough, at least provisionally, for Aristotle to show the analogy between the mechanism of spontaneous generation and that of sexual generation with the clearest or best known example of spontaneous generation, namely marine shellfish, in order that one of his main theoretical objectives may be achieved, that of showing that beyond their diversity, from the point of view of generation too, all animals are fundamentally the same despite their very great differences. One may therefore conclude from all this that in every spontaneous generation, that of shellfish as well as of mice, there is a fertilization of material unities formed from the material of the milieu by a wandering "male" principle.

Second remark: Aristotle himself makes the connection between his ideas about wind eggs and spontaneous generation. It's that some fish make a kind of wind egg. Here is the passage:

> However, in the fish family there are born some that are born neither male nor female, and, though being in the same family as other fish, they differ in form, and some are completely unique. Some have only females and no males, and give birth to what would, in birds, be wind eggs. In birds these eggs are all infertile, because their nature cannot go further than the formation of this egg, unless there were another way for them to have union with the males. . . . But in certain fish, when the eggs are produced by spontaneous generation (ὅταν αὐτόματα γεννήσωσιν ᾠά), it also happens that living animals are born

> from them, only some of them do this by themselves, while in others it does not happen without the intervention of a male. (*History of Animals* 5.1, 539a27)

The difference between birds and fish is that the wind eggs of birds are never fertile, while those of fish can sometimes be. This is connected to the difference in the modes of fertilization, since the male bird fertilizes the female by copulation, and the fish by sprinkling the eggs outside the female. But it is necessary to understand the ὅταν αὐτόματα γεννήσωσιν ᾠά of 539b3. It doesn't seem to mean something like "when the eggs are formed spontaneously,"[48] which leaves the impression that this could be the case, but also might not be. Aristotle here indicates, with the aorist subjunctive, that this is what happens normally: each time that these fish lay eggs, one may say that they are the equivalent of wind eggs of birds, and Aristotle says that they are produced "spontaneously," but not all result in a true spontaneous generation of living animals. In fact, the real spontaneous generation comes later, since some of these eggs are fertilized by the semen that the male sprinkles on them, while others give directly, that is, spontaneously, living animals. Therefore, it is not possible to qualify the generation from all these eggs of fish as "spontaneous."

Nevertheless, Aristotle indeed asserts that the birds' wind eggs form "spontaneously" (cf. the αὐτόματα of *Generation of Animals* 3.1, 749b35), which confirms that he thinks of them as living beings. They are generated spontaneously because, both, they are not the fruit of formation by the male semen, and they come from the elaboration, by the nutritive soul of the bird, of that nonliving reality that is the food. From that one may draw the conclusion, rather bold actually, that at the basis of every generation, sexual or not, there is a spontaneous generation. Oysters remain at that first phase, while vivipara submit the living thing born spontaneously to that additional elaboration that is the fertilization by a male principle. And one may interpret this "at the basis" as a chronological priority as well as a logical priority, since the female elaborates the material, which we have seen is alive, *before* the fertilization by the male semen occurs. This agrees with the thesis here defended that sexual reproduction, and that done by distinct sexes, is the perfected version of the process of which spontaneous generation is an incomplete copy. This also goes in the

48. Aristotle, *Histoire des animaux* (Louis, xi–xii).

direction of what was said earlier: Aristotle is not far from recognizing a function of ovulation in all females.

Third remark: The stability of species born by spontaneous generation shows that certain places are more propitious than others for the formation of certain fertilizable bubbles by some *pneuma*. This is surely due to a particular elementary composition. But one may imagine that it is with a certain variability: slime (ἰλύς), which is a variety of mud (Aristotle speaks of "muddy slime," ἰλύς βορβορώδης, *History of Animals* 5.15, 547b19), produces oysters, even, it seems, if the elementary composition of this mud is not everywhere the same. That perhaps explains why there are several varieties of oysters. That the material composition of the *place* where the generation occurs would be a determining factor in the specification of living things constitutes a large gap that separates sexual and spontaneous generation.

Here we must face a problem that is not only terminological. In spontaneous generation, then, a major role is played by the residue as it is elaborated by heat from decomposing materials. It is appropriate to return to the naturalness of this process of spontaneous generation, which raises the difficult problem of the unity of the Aristotelian doctrine of spontaneity. In fact, there is a real difficulty, one that commentators have perhaps too often ignored, of relating what Aristotle says about spontaneous generation in his zoological writings and his analyses of the concepts of chance (ἡ τύχη) and spontaneity (τὸ αὐτόματον), in any case difficult to interpret in detail, especially as it concerns the relationship of spontaneity (for Aristotle, chance is a concept that relates to practical activity of human beings, which is not our concern here) with the regularity and finality of phenomena that it concerns. The reference texts are to be found in *Physics* 2, 4–6. In these chapters, spontaneity is considered to be a *cause*. A passage in the *Metaphysics* sums up all of Aristotle's negative thoughts about spontaneity: "Things happen either by nature, by art, or spontaneously. But art has its starting point in something else, while nature has it in itself (because a human being generates a human being), the other causes being privations of these [i.e., art and nature]" (Lambda 3, 1070a6). In fact, the results of spontaneity are at the same time accidental, parasitic on other processes, themselves either art or nature, and irregular; thus, they do not comply with the rules of art or science. But spontaneous events can never violate the laws of nature, as *Parts of Animals* recognizes, saying that "the products of spontaneity (here called "chance") are produced in the same way as those of art" (1.1, 640a32). But if spontaneity (τὸ αὐτόματον) is an "accidental cause, in the realm of things that cannot happen always or for the most part," that is, is not natural (*Physics* 2.5, 197a33), it's hard to see how a definition like this can apply to the production of oysters or eels, which reproduce entirely regularly, specifically identical to each other.

It's necessary to try to respond, briefly, to this question of Aristotelian exegesis, which to be sure interests us only indirectly here. In the first place, we should say that one ought to not ask too much of etymology, all the more so since the term αὐτόματον is not entirely clear. Obviously, automaton is composed of two parts, *autos* (self) and *maton*, which Pierre Chantraine, in his incomparable *Dictionnaire étymologique de la langue grecque*, attaches to the perfect *memona* and to the noun *menos*, which both mark desire, will, passion. Thus, automaton is that which moves itself by its own will or passion (one often gives the Homeric example of the gates of Olympus that open by themselves) and that is thus not moved by something else. That can be read in two ways. Either one insists on the fact that that which moves is the cause of its own movement, or one insists on the fact that something else that moves it is absent, not accessible, or at least hidden. Anyway, Aristotle notes (*Physics* 2.4, 196a1) that there have been philosophers—perhaps he is thinking of Democritus—for whom spontaneity is simply an illusion: we have the impression that spontaneity exists in the world, but that's only because we don't perceive the causes. Automatic gates open by themselves in the sense that the mechanism that causes their opening remains hidden to us, but it exists all the same. That's a characteristic that is true of automata in the modern sense of the word.

It would seem that the analyses of the *Physics* rather belong to the first reading. There is a very interesting passage to this point, one that has brought about incomprehension among the best commentators. At the end of his study of chance and spontaneity, Aristotle says this about the difference between the two concepts: "But it is in the domain of that which happens by nature that the spontaneous is farthest removed from that which happens by chance; in fact, when something happens contrary to nature [understood: within a natural process], we don't say that it happened by chance, but rather spontaneously; and even in this case there is a difference: one has an exterior cause, the other an interior cause" (2.6, 197b32).

That means that when something contrary to nature intrudes in a natural process, one has spontaneity and not chance, because when you meet in the market the person who owes you money the question of naturalness does not arise. David Ross thinks that there is here an allusion to spontaneous generation. But that comes from Ross's mistaken conception of spontaneous generation, which he imagines, like many other commentators, as an *exceptional resemblance* to "normal" generation. Ross also criticizes Themistius, who saw in this text a reference to monsters, on the pretext that monsters are a snag in the teleological activity of nature, and that which happens spontaneously is, says Ross, an "end-like result." But it is Themistius who is right: in a natural process like that of sexual generation, the spontaneous that is here at issue is falsely opposed to the naturalness of this process (and that is why Aristotle speaks of "against nature"), without, obviously, going against the laws of material nature. That comes from the moved itself, more precisely always or most often monsters come from the resistance that the (female) material opposes to the (male) formal principal.

In spontaneous generation, however, the formation of new living beings is a natural process, which is in no way compromised or deflected by anything at all that is contrary to nature; it exhibits the regularity of natural processes. But this process *seems* to happen without a determining exterior cause like that which works in the case of sexual reproduction, namely the form transported by the semen of the male. In fact, every analysis that Aristotle gives of spontaneous generation, especially in the *Generation of Animals*, reinforces the idea that such a cause exists, as we will see again, but that it is not immediately obvious. That brings us closer to the second reading of the spontaneous distinguished earlier (that in which what moves is something not accessible, or hidden). That is why, when one *describes* the phenomenon of spontaneous generation, one even comes, and Aristotle himself comes, to say that it is the mud or rot that produces the animals in question, and, in the same way, this process of spontaneous generation seems to happen in contravention of all natural regularities.

For it is ultimately astonishing that Aristotle speaks of αὐτόματα or of birth ἀπὸ ταὐτομάτου for animals that are not born from fertile intercourse. *Perhaps* the reason is the following. In the *Physics*, when getting involved in the dispute about the causes at work in nature, a dispute that puts him into opposition particularly with the atomists and Empedocleans, Aristotle means to forge an operational concept of the automaton to perfect his causal theory. Faced by the atomists who attribute the construction of worlds to the automaton, he proposes his analysis of chance (spontaneity) as accidental causality relating above all to phenomena contrary to nature. But it turns out that those whom Aristotle calls "physiologists," most of whom, evidently, believed like Aristotle and even more than Aristotle in the spontaneous generation of certain living beings (or even of all), applied the word automaton also to this phenomenon.[49] One of Aristotle's goals was to show that Democritus and the other "mechanists" were unable, using their concept of chance, to account for regular phenomena like the generation of living things within their species. At *Physics* 2.4, 196a29, and *Parts of Animals* 1.1, 642a21, Aristotle mocks those who say that the generation of living things obeys the regularity of a "nature," while the universe as a whole is taken to be the result of chance. In other words, *Physics* 2.4–6 offers what Aristotle believes to be a correct analysis of *the automaton of the physiologists.* But that automaton, applied to various phenomena going from the constitution of living things to that of the whole universe, is thinkable in Aristotelian terms as an accidental cause, and thus not natural, intervening in the domain of natural things. More precisely, it is in the domain of finalized things that spontaneity reveals its true nature. Thus, in the example of the stone (*Physics* 2.6, 197b30): if a stone falls on a passerby because of purely material causality (because the mortar holding it split, or some other reason) that would have the same effect (kill or wound the passerby) as if it had been thrown intentionally, but this phenomenon comes under the automaton. We have then an accidental intersection of two causal chains, one making the stone fall, the other leading the passerby to this location. Furthermore, by saying that that which happens spontaneously has an *exterior* cause (197b36), always according to chiasmic structure so dear to him, Aristotle returns to the idea that the automaton is brought about by the intersection of two causal chains.

Thus, we find ourselves in a situation comparable to that which we came to concerning hypothetical necessity: the analysis of chance and of spontaneity in *Physics* book 2 are also part of the polemic that Aristotle aims at the Presocratics, with the proviso that hypothetical necessity can be used effectively in Aristotle's study of nature, especially zoology, while in the case of the concept of spontaneity we find ourselves faced by two different and incompatible approaches.

Thus, we doubtless must resolve, with heavy heart, to say that the analyses of spontaneity that we find in the *Physics* are not really applicable to spontaneous generation as it is presented in the zoological works. We

49. Cf. Theophrastus, *History of Plants* 3.1.4: "the spontaneous generation (of plants) of which the physiologies speak."

find in the Aristotelian corpus passages that describe the birth, including the birth of living beings, as exceptions to the regularity of the natural order. The triton, for example, transmits across generations the flawed form of its species, as we have seen. But we may take that transmission as natural, if only because of its regularity. One may say that it is a fortiori impossible that Aristotle the zoologist thinks that the spontaneous generation of animals in which it happens is not natural. This spontaneous generation is, in fact, a process with remarkable regularity, as we have said. Here too the parallelism between spontaneous generation and sexual generation is clear, since both processes produce living things that belong to stable species. The permanence of living species is a fundamental trait of Aristotle's universe, and one of the givens that any explanation in biology needs to account for, as we saw when we were talking about teleology. The permanence of the species "oyster" is no less established, and not less necessary in Aristotle's world, than that of the species "horse."

Ultimately, spontaneous generation poses the problem of passage from nonliving to living in a new way. It doesn't seem that there would be a mussel already present in actuality that works on the matter furnished by the environment for the sake of the generation of more mussels, in the way that an actually existing female works on the food to make it into the material for the future embryo. Maybe we have to turn toward the *pneuma*, which would then recover on one side what was lost on the other. In fact, we have seen that we cannot confer on this *pneuma*, even though it plays the role of the male principle, the function of transmitting a specific form. But maybe we can consider it as the *vital* reality that preexists the process of development. But this *pneuma* is described by Aristotle as a *psychic* reality, that is, if not living (because this *pneuma* is not a living being in the sense of being an individual animal), as at least *vital*. Thus, in the case of animals reproducing sexually as well as in those that do so by spontaneous generation, a vital reality always preexists the process of assimilation of nonliving materials, permitting the appearance of new animals.

There is another fundamental difference between sexual and spontaneous generation that can be seen clearly if one recalls the passage from *Physics* 2.7 cited earlier: "But the three last (formal, efficient, and final causes) often converge into one. In fact, the essence and that for the sake of which are the same thing, and the first source of movement is specifically identical, because a human being engenders a human being" (198a24). And, in fact, in sexual generation, the male parent is simultaneously, via the movements contained in his semen, the efficient cause of the embryo, the formal cause of the fact that this semen enforms the female material, and its final cause, because the process has for its goal the production of a being specifically identical to its generator.

But we do not find an identical situation in the case of spontaneous generation. However, if mussels reproduce themselves with such regularity, there has to be something that plays the role of a model, that is, a formal cause, the role played in sexual generation by the actual parent who is the formal and final cause of the offspring, for "human being generates human being." So, what corresponds to the paradigmatic function of the male, first of all, and of the female, secondarily? The product itself cannot play the role of the parent, because it is not a mussel that generates a mussel, and that's precisely because there are not ascendent mussels and descendent mussels that the generation of mussels is called "spontaneous." Nor can it be the *pneuma*, though that does indeed play the role of efficient cause, because it is much too undifferentiated to play the role of a model, since it has to form mussels in one place and oysters in another. Nor can it play the role of final cause, because it has to be mussels that come into being, and not more *pneuma*. Therefore, it remains, once again, that it cannot be anything other than the matter enclosed in the "bubbles," discussed earlier, which as we have seen inherited the role of specification played by semen in sexual generation. This material preserves the major characteristics of the place in which it is found, slime, for example, producing oysters. But even though one may say that the elementary composition of the material enclosed in a bubble, that is, the ratio (*logos*) of that composition, takes the place of the formal cause of the mussel, it seems difficult to make of that a model that could serve as a final cause, since, if a living dog serves well as a model for the puppies that he fathers and shows them at the same time the goal toward which the reproductive process is going, once again it isn't the same for a mussel or an oyster. Perhaps we have here an additional reason for calling this generation "spontaneous" (automatic): due to its elementary composition, slime produces oysters automatically in the manner that we would call a chemical reaction. Thus, there is no need for a preexistent exterior model. In any case, the beautiful unity of formal, final, and efficient causes exhibited by sexual generation explodes in pieces in the case of spontaneous generation.

In other words, in both cases, there is never a movement from nonliving to living except by the action of a living reality, or at least a vital reality. But there do exist nonliving realities, which are natural, as is seen notably in the fact that they produce themselves in a regular manner, for which there cannot be a chronological priority like that of actuality and potentiality. I'm talking about minerals and metals. To what degree does their production tell the same story as the generation of living things, whether sexual or spontaneous?

A General Theory of Homoiomeries?

When one reads the passage, cited at the beginning of this chapter, from book 4 of the *Meteorologica* (4.10, 388a13), one gets the impression, as I said, that Aristotle here has in mind a general unified theory of homoiomeries. Let's begin by citing this passage more completely:

> It is by these properties and differences that homoiomerous bodies differ from each other, as we have said, by touch, taste, odor, and color. I call "homoiomeries" for example things mined like copper, gold, silver, brass, iron, stone, and other materials of this kind, as well as materials derived from them; also the parts of animals and plants, for example flesh, bone, tendon, skin, viscera, hair, fibers, blood vessels, from which the anomoiomeries are constituted, like face, hand, foot, and all the parts of this kind, and in plants the homoiomeries are wood, bark, feather, root, and every part of this kind. Since these (ταῦτα) are constituted by another cause, but that out of which they are constituted has for matter the dry and the humid (consequently, earth and water, for these possess these two powers in the most obvious way) and those have for agent causes the hot and cold (for these are what produces condensation and solidification from the first ones), let us consider which of the homoiomeries are forms of earth, which of water, and which are both. (*Meteorologica* 4.10, 388a10)

Commentators (I. Düring, J. Tricot, P. Louis, J. Groisard)[50] agree in seeing in "the other cause" at 388a21 the formal cause. The ταῦτα at 388a20 ("since *these*") can only refer to the anomoiomeries; Aristotle lists three causes of anomoiomeries: the formal cause ("the other cause"), the material cause (the homoiomeries, which are themselves constituted of

50. Düring, *Aristotle's Chemical Treatise*; Aristotle, *Les Météorologiques* (Tricot); Aristotle, *Météorologiques* (Louis); Aristotle, *Météorologiques* (Groisard).

earth and water), the efficient cause (hot and cold). Which means that homoiomeries and anomoiomeries share the last two kinds of cause, since homoiomeries are themselves constituted out of earth and water by the action of hot and cold.

Metals, metalloids, minerals, and gasses are absolutely natural products. They are studied in the *Meteorologica*, which Aristotle calls, in the often-cited introduction, a physical treatise. One may imagine that we do not know a great deal about this question because Aristotle was not very interested in it. Unless, as was doubtless the case for plants, his writings on the subject were quickly replaced in the Lyceum by treatises of Theophrastus on the same subject. However, we need to remember that the treatises that we have by Theophrastus are far from concerning all the topics raised in Aristotle's *Meteorologica*: we have only the *De Lapidibus*, the *De Ventis*, the *De Signis Tempestatum*, and perhaps the *De Igne*.

There is only, so to speak, *Meteorologica* 3.6 that deals, rapidly, with the problem of the formation of minerals and metals. Minerals and metals depend, like all other phenomena studied in the first three books of the *Meteorologica*, on a theory that Aristotle claims as original with him (cf. *Meteorologica* 1.4, 341b7) of the two exhalations. In Greek, the term ἀναθυμίασις designates a hot and dry exhalation, opposed to ἀτμίς, a hot and humid exhalation. Aristotle breaks with this usage, but not with the customary distinction, by positing that there are two exhalations, both called ἀναθυμίασις and both caused by the heat of the Sun,[51] the one vaporizing the humidity that is in and on the earth, "the other being like a smoke coming from the earth itself, a smoke that is dry" (1.4, 341b10). This last exhalation, which Aristotle characterizes also as "more windy" (πνευματωδεστέραν, 341b9), being extremely flammable, is at the origin of often fiery phenomena, like shooting stars, lightning, thunder, earthquakes, while rain, frost, dew, and snow are attributed to the first exhalation. The treatise *Sense and Sensibilia* tells us, furthermore, that from the smoky exhalation, it is not water that forms, but "a kind of earth" (γῆς τι εἶδος, 5, 443a29). With the little information that we have, we will try, all the same, to understand what Aristotle has in mind. D. E. Eichholz[52] argues that it is false that the descriptions of the formation of minerals and metals are parallel.

51. We recall that, according to Aristotle, the Sun, like all other celestial bodies, is not hot per se, but rather its rotational movement produces heat (cf. *Meteorologica* 1.3, 341a12).

52. Eichholz, *Theophrastus, de Lapidibus*, 38ff.

"The dry exhalation produces all the minerals by consuming them" (ἡ μὲν οὖν ξηρὰ ἀναθυμίασίς ἐστιν ἥ τις ἐκπυροῦσα ποιεῖ τὰ ὀρυκτὰ πάντα, *Meteorologica* 3.6, 378a21): here the dry exhalation is the *efficient* cause of minerals. On the other hand, it is hard to decide what this exhalation "consumes." Jocelyn Groisard interprets the word ἐκπυροῦσα as a reflexive verb ("by consuming itself"), Olympiodorus in a passive sense (ἐκπυρουμένη καὶ περιφρυττομένη;[53] the last term is obscure, but doubtless related to φρύγω, "burn, roast"), but that seems difficult, because then, as Eichholz remarks,[54] the exhalation would be the *material* cause of minerals. The best is doubtless to understand that the dry exhalation produces minerals by being (or from the fact that it is) burning. Possibly Aristotle wants to say that, underground (ἐν τῇ γῇ, 378a20), this exhalation coagulates (agglomerates) the earth, which then changes into stone.[55] One would then have something like a coalescence of the humid by concoction.

On the other hand, for the formation of metals things seem to go as follows: the humid exhalation, unified with the dry exhalation, condenses on contact with a cold stone (one may suppose that it is exactly this coldness that makes it condense), which gives a metal. The cohesion of this type of body is thus not the result of concoction, but that of being seized by the cold. It is because they are solidified by cold that metals are liquified by heat. Also, metals contain earth (doubtless because they are formed in the heart of minerals that are earthy realities) and Aristotle explains that in a certain way they contain water:

> That vaporous exhalation produces all the metals when it is imprisoned, especially within minerals, and it is compressed and solidified because of the dryness, as is the case for dew or frost when it is separated [from the dry exhalation]; but, in the present case, metals are created before it is separated (ἐνταῦθα δὲ πρὶν ἀποκριθῆναι γεννᾶται ταῦτα). That's why they are in a sense water and in a sense not: in fact, the material was potentially water, but it is no longer. (*Meteorologica* 3.6, 378a28)

53. Olympiodorus, *Olympiodori in Aristotelis Meteora Commentaria*, 269.5.

54. Eichholz, *Theophrastus, de Lapidibus*, 44.

55. Cf. Tricot: "the heat of the dry exhalation is the cause of all the minerals." Aristotle, *Les Météorologiques*.

The contradiction of this with what is said at 4.10, 389a7: "Thus gold, silver, copper, tin, lead, and glass, as well as many nameless minerals come from water, because all these bodies are melted by heat," is perhaps not insurmountable. But there is here a difficulty that could in fact argue in favor of the inauthenticity of book 4, either as a whole or in part, or, at least, make its text suspect. But it is not unreasonable to think that we have here an Aristotelian teaching conveyed to us by a text that is not entirely from Aristotle's hand.

If, as we have seen, the dry exhalation cannot be the material cause of minerals, but is very likely their efficient cause, the situation is even more ambiguous in the case of metals. Thus, in this passage: "Thus, the homoiomerous bodies are constituted of earth and water, in plants and animals and metals, for example gold, silver, and all the other bodies of this kind, simultaneously from them and from the exhalation of each of the two bodies when it is imprisoned, as we have said elsewhere" (4.8, 384b30). Let us note first that the commentators are in agreement that the "elsewhere" cannot refer to anything else than the end of book 3 of this treatise (3.6, 378a15), which is a nonnegligible argument in favor, not only of the authenticity of book 4, but of the unity of the treatise. Jocelyn Groisard also remarks that this is the only allusion to the doctrine of exhalations in book 4. We also note, relative to our argument, that we have here one of the passages that connect nonliving homoiomeries with living homoiomeries.

The vocabulary used in the expression "both from them and from the exhalation" (ἐξ αὐτῶν [i.e., water and earth] τε καὶ ἐκ τῆς ἀναθυμιάσεως, 384b33) seems to indicate that the exhalation is, for metals, in the position to be a material cause. But maybe this "from" (ἐκ) is meant to point out causality in general, so that it could then indicate the efficient cause. Or, perhaps, it's the *humid* exhalation that is a material co-cause, because we have seen that the dry exhalation was equally present in metals even if they are especially produced by the humid exhalation. Maybe Aristotle thinks that the dry exhalation is the efficient cause of two kinds of bodies, minerals and metals, but we don't have enough textual basis to support that interpretation. In any case, the exact identification of the causes at work is tricky.

One may assume, but it is only an assumption, that it would be different proportions of the composing parts that bring it about that there are different kinds of minerals and metals. On the last page of *Meteorologica* 3, Aristotle gives an explicit resumé of the formation of metals, which we tried to understand earlier. He concludes: "Copper and gold are not formed like that [by the transformation of water], but each existed by the congealing of the exhalation before water was formed. That is why all metals are affected by fire, and contain earth, because all contain the dry exhalation. Only gold is not affected by fire" (3.6, 378b1). When we read these texts, we can see that they are too short and too ambiguous for us to be able to derive an interpretation that could gain a consensus. We have seen, for example, that it's difficult to decide which cause Aristotle is talking about—efficient or material? Some textual variants are loaded with consequences. Thus, at *Meteorologica* 4.6, one part of the manuscript tradition has Aristotle saying that earthy bodies, including metals, are formed because "all the humid evaporates at the same time" (383a30), and another part speaks of "all the hot."[56] We can invoke the intrinsic difficulty of Aristotle's texts and, as we have seen, the fact that these questions do not seem to have been of the greatest importance for him. Eichholz seems to adopt an acceptable position; he thinks that the greater or lesser proportion of earth explains the difference between the various metals, that proportion being at a minimum for gold and a maximum for iron, which has a great deal of earthy matter and when it melts leaves slag (cf. 388a33).[57]

But if Eichholz is right, then when metals form, and doubtless also minerals, there is a factor that plays the role assigned to the form transmitted by semen in the formation of animals, as described in the *Generation of Animals*. This factor seems to be tied to the "elementary" situation of the location where the metals or minerals form by the action of the "exhalations." Thus, it is more nearly the explanation of spontaneous generation than that of sexual generation that is closest to this explanation. But Aristotle is hardly loquacious on this topic, and one gets a rather fuzzy impression from all this; it seems to me that he could have avoided that without too much trouble. Thus, the temptation to treat this indeterminacy as a *symptom*: Aristotle's theory of the generation of minerals and metals, if it is that imprecise, will more easily fit into the project of which this study is a part, a general theory of homoiomeries.

Such a general theory, in order to be consistent, should be able to assume sufficient parallelism between the movements contained in the

56. Groisard explains in a long note why he chooses the first reading. Aristotle, *Météorologiques*.

57. Eichholz, *Theophrastus, de Lapidibus*, 41.

semen that act on the female material of the "menses" and the action of the environment on the combination of earth and water that is the material of metals and minerals, and that seems all the more possible because in both cases hot and cold, and probably also dry and wet, play an efficient role. Doubtless the parallelism is even closer between the formation of metals and minerals and the action of pneuma in the case of spontaneous generation. But there remains an important obstacle to the construction of such a general theory—there is a fundamental difference in the causes that bring about the construction of homoiomeries in the two (or three) cases. For the metals and minerals, it's a matter of what one could describe, in modern terms, as a chemical reaction, while in living things there has to be a living reality present, or at least a vital reality: in the case of sexual generation, there must be parents that are actual, that is, alive, that serve as formal and final cause for the offspring, in the case of marine shellfish, it is "the part of the psychic principle that is enclosed and separated in the breath that produces the embryo and introduces the movement" (*Generation of Animals* 3.11, 762b16, already cited). But the "Latin" equivalent of "psychic principle" is "animating principle." It ought to go the same way, but in a less clear fashion, in the other cases of spontaneous generation. Even if, as we have seen, local conditions play a formal role in the spontaneous generation of shellfish, they are not enough to bring about the appearance of a living being. From that point of view, we are in the same situation as that of sexual reproduction, in which the material elements of a living being, left to themselves, cannot produce a living being or the parts of that living being.

In fact, that question, previously noted, of knowing whether the elements left to their physiochemical properties would be able to produce flesh or bone, a question much debated among the interpreters of Aristotle's biology, could not be and never could be posed for the Stagirite, in the sense that this situation could not arise. It could, on the other hand, arise for a mechanism à la Democritus, or in a creationist finalism like that of Plato, but not for Aristotle's everlasting universe. This is always, as we have seen, because *essence* has explanatory priority over *genesis*. Earth and water are combined in a way to form the flesh of a dog because this combination is carried by a living dog that already exists that transmits it to the material of the bitch via his semen. Similarly, if, in the case of spontaneous generation, the environment offers a material able to receive life, again it is necessary that there be a vital movement, that of the *pneuma*, that brings about this passage.

The fact that a dog and an oyster are born and develop so to speak "inside life" and that it is not the case for a vein of iron ore, has important consequences; we will note at least two. First, the fate of homoiomeries like metals and stones seems to be to remain homoiomeries, while the biological texts tell us again and again that the fate of organic homoiomeries is to be transformed. Aristotelian Nature, in fact, *does nothing* with the iron and marble that she produces. Unless we attribute a providentialist notion of Nature to Aristotle, a nature that anticipates the use of iron to make sickles—but as we have seen, that is a sort of teleology that is impossible for Aristotle. The fate of the homoiomeries of living things, on the other hand, whether vegetable or animal, is *to be used by nature* for constituting anomoiomeries, which are then included in an *organism*: "The homoiomeries are for the sake of the anomoiomeries. These have functions and activities" (*Parts of Animals* 2.1, 646b10, passage already cited). And that is true both of dogs and oysters. That leads us to reaffirm and make more precise what was said before: these homoiomeries become either "true" anomoiomeries (the flesh and bone of a hand), or homogeneous anomoiomeries like the heart, and we have seen that what we call muscle is, for Aristotle, "flesh," but obviously the biceps of an animal are not "flesh pure and simple." Same for bone. Most are truly anomoiomeries that contain several homoiomeries (bone and marrow), but even bones without marrow (for "in the same animal some bones have marrow, others not," *History of Animals* 3.7, 516b5) are not bones "pure and simple." And, as several passages previously cited show clearly, and as Aristotle says many times, organic anomoiomeries are called to be integrated into larger groups, not by simple addition, but by functional integration, which is characteristic of an *organism*: "Since every tool is for the sake of something and every part of the body is for the sake of something, and that which they are for the sake of is a certain action, it is obvious that the whole body has been constructed for the sake of some complete action" (*Parts of Animals* 1.5, 645b14). Iron is not a tool for a function except from a human point of view, but not from the fact of its own nature. It would thus be *in vain* for it to possess properties like those that Aristotle attaches to life, namely, to assimilate food, reproduce, discriminate, imagine, and so forth. We can further clarify things by looking again at the passage at the beginning of book 2 of *Parts of Animals* (2.1, 646b10). The text adds up to saying that homoiomeries do not acquire a function until they are transformed into anomoiomeries, because only the later have "functions and actions." A passage in the *Generation of Animals* in which, for once, Aristotle does not employ chiasmic exposition, is even more clear and more complete: "Some parts are defined by a function (δυνάμει), others by properties (πάθεσι); anomoiomeries by the power to do something, for example the tongue or the hand, the homoiomeries by hardness and softness and other properties of this kind" (1.18, 722b30). That means that not every heterogeneous composition can be considered an anomoiomere. A vein of iron ore surrounded by minerals, for example, lacks the achievement of actions and functions to be considered a real anomoiomere. Rather it's a matter of an aggregate, and that goes the same for nonliving bodies that are mixed together whether by human industry (a metallic alloy, for example) or by natural conditions (during a volcanic eruption). So, translating ἀνομοιομερής by "nonuniform" (or "not homogeneous") is not a good idea. We can invent an example for which there is no explicit textual reference, while remaining within the Aristotelian lines: Aristotle asserts several times that the eye or the hand of a cadaver is an eye or a hand only homonymously, because it cannot perform the functions of an eye or a hand; but it seems that the flesh of a cadaver, to the extent that it has not decomposed, is not flesh by homonymy—it is a homoiomerous part endowed with "hardness, softness, and all the other properties of that kind." Possibly it would be necessary to invent, for realities of this kind, an additional kind of homoiomere, that of real homoiomeries, but unable to constitute anomoiomeries because there is no longer a soul able to direct that constitution.

Secondly, a characteristic that meshes readily with those we have just emphasized—metals and minerals do not form *individuals*. Individuality, as Canguilhem has shown clearly, is an essential characteristic of the living, in that it allows it to be a center of reference and a *subject*. A crafted object, like a table, has only pragmatic individuality, because it serves people who sit at it. It needs human art to "individualize" the wood and to make a table by simultaneously providing it with a function; it's always possible to make one large table out of two small ones. Although enclosed within its limits, a seam of iron ore is not an individual.[58]

58. The clearest presentation of the difference between a "crude or inorganic" body and a living body that could be adapted to an Aristotelian point of view is perhaps that which Lamarck proposes in his *Philosophie zoologique*. Following Anthelme Richerand in his *Nouveaux éléments de physiologie* (1801), Lamarck lists nine differences: a crude body (i) does not possess individuality "in its mass and volume"; (ii) is "truly homogeneous" or heterogeneous only by aggregation; (iii) can be completely dry,

Once again, if Aristotle had been a Presocratic, one might say that for him the production of homoiomeries like metals is a fundamental (basic) process in the production of all homoiomeries and, therefore, that the generation of living things, including human beings, is a production of stones, metals, and so on, continued by other means. But the reverse is true: there is a movement of explanatory decline that goes from sexual generation, to spontaneous generation, then to the production of nonliving natural bodies. But as the Aristotelian causes are not only ways of talking about reality, but descriptions of real situations (that's the famous Aristotelian "realism"), without really conferring finality on them, that in a way reintegrates nonliving bodies into a kind of cosmic structure. We know that according to Aristotle "inferior" realities do not participate in superior realities, and that he replaces the Platonic notion of participation by that of *imitation*. Thus, the uninterrupted series of animal generations imitates the everlasting circulation of the stars, so that the specific permanence of animals imitates the numerical permanence of celestial bodies. But can we say that iron, in being produced, without participating in life, imitates the reproduction of living things? We will see about that as we conclude this chapter.

To conclude, it really does seem that there is in Aristotle at least the project, or temptation, to build a general theory of homoiomeries, all the more so as living things that reproduce by spontaneous generation seem, in their reproduction, to approach realities like minerals and metals. There is, notably, an important point that the preceding analyses allow adding to the argument: when one compares the world of "superior" animals to that of minerals, one recognizes again that finality is the backbone of Aristotle's biology. But this difference seems to fade in the case of animals born by spontaneous generation, because it is not an individual actual animal that serves as a model and end for the generation of shellfish, since it is not an adult oyster that initiates in the matter for the future oyster the movements that will give it its form. It seems that there is a common conceptual matrix that accounts for the production of oysters and of iron: for the oysters as well as for the iron, there is not, properly

completely fluid, or completely gas; (iv) the molecules that make it up are independent of each other; (v) it does not need to move to persist; (vi) the increase in its volume or mass occurs by juxtaposition; (vii) it does not need to feed itself to persist; (viii) is made of "separate parts that are accidentally conjoined"; (ix) cannot die. Lamarck, *Philosophie zoologique*, 378–384.

speaking, a *reproduction*. But fading does not equal disappearance; finality is not absent in the generation of oysters, because the formation of anomoiomeries from homoiomeries is a finalized process, because the shell as well as the body of an oyster, products of the slime, are indeed functional anomoiomeries (homoiomeries are *for the sake of* anomoiomeries), which is not the case in the formation of a seam of iron ore, or the fusion of two metals in a volcano. This proximity between iron and oysters is thus not enough to eliminate the frontier that separates the living and the nonliving. Every animal possesses—we return to this point in the next chapter—a nutritive soul and a sensitive soul, and a mineral does not. The cases that seem doubtful or ambiguous really aren't, as we saw when we were talking about the *scala naturae*. We say it again: Aristotle maintains that the generation of living beings in the sphere of the living requires a preexisting vital reality prior to any physiochemical process of the constitution of any living thing. Thus, every attempt at reductionism is made impossible; that makes him, according to Canguilhem's criteria, definitely a biological thinker.

We have to come back once more to the coexistence in Aristotle of two great theoretical schemas, the one presented in this chapter, and the other in the preceding chapter. The first of these schemas is that the superior[59] furnishes a model of intelligibility for the inferior. The second is that of plural causality, which integrates mechanist approaches into the explanations of Aristotle as natural philosopher. Those two schemas may seem opposed to each other, because the first presents "inferior" realities as incomplete and/or degraded versions of "superior" realities that correspond to them (we have seen this in the case of the formation of living and nonliving beings), while the second establishes, as we will see, a kind of independence of the inferior in relation to the superior.

From the first schema, I will emphasize just two points. First, I reemphasize that the formation of the homoiomere that is iron is a truncated version of the formation of flesh in the living being, just as augmentation is a truncated version of growth, and the courage of a woman is a truncated version of that of a citizen. And, as a matter of fact, in *the formation* of minerals and metals, among other nonliving natural realities, the process put in operation is an incomplete version of the process of generation of living beings. But it is also a homothetic version, in that the agents and

59. Cf. *Generation of Animals* 2.1, 731b22: explanation by "the best, that is, the final cause, has its source from above (ἄνωθεν)."

phases of the production of nonliving natural bodies are analogues of the agents and phases of the generation of living beings. We must note that in practice, that is, to speak in Aristotelian terms, when one takes on the perspective of priority *for us*, this schema is especially valuable for clarifying the sense of the word "superior": to understand the complex process of growth, it is sometimes useful to grasp augmentation, and to grasp "true" courage, perhaps it is better to grasp first that fainthearted form of courage, that of a woman (if we can attribute to Aristotle an idea of this kind). We have, with this schema, a primary factor of the *unity* of Aristotle's natural world. But it is a matter of a hierarchical unity that we will need to clarify a bit more.

As for the second schema, interpreters of Aristotle know well that, instead of simply rejecting the systems of his predecessors, he replaces them within his own explanatory structure. We need to repeat what we said in the second chapter of this book: unlike Socrates in the *Phaedo*, Aristotle would have said that one of the causes of Socrates finding himself in prison would be because of his bones and muscles, because material and efficient causality, what one may call "Presocratic causalities," exhibit real causes. To make of material and efficient causes simply necessary conditions for the deployment of causality and not causes properly speaking, amounts to a rejection of natural philosophy, and that is exactly what Aristotle blames the "people at the time of Socrates" for doing. There is no need of chronological hypotheses for explaining the absence of Plato in the Aristotelian polemics about causality in works like the *Parts of Animals* or the *Physics*: in the *Parts of Animals*, Aristotle criticizes Platonists on a matter that one may call "logical," that of definition; and in the treatise *De Caelo*, he attacks the Platonic doctrine of the composition of the body as proposed in the *Timaeus*. But Plato is definitively not a natural philosopher: in his dialogues, the *physiologoi* are criticized and even ridiculed, they are not *aufgehoben* as they are in Aristotle.

This recuperation of the Presocratics in Aristotle's causal construction does seem to be brought about at a "lower" level than that of the formal and final causes, as is said for example in the passage from the *Generation of Animals* previously cited in a note (2.1, 731b22), and all the passages in which Aristotle affirms the preeminence of finality. In fact, this impression needs to be partly corrected, because, in that which I have called the theory of the "two natures," the last word returns to necessary nature, to which rational Nature must adapt, and necessary nature does not have to adapt to rational Nature. Robert Bolton, in a striking article, maintains that material-efficient causality is the last word of explanation in natural philosophy,[60] notably taking literally a passage in the *Physics*, in fact remarkable:

> The "why" ultimately comes down to either the essence, as in the case of unchanging entities (e.g., in mathematics it is in fact a definition of the straight, of the commensurable, or something else of this kind that leads to the ultimate "why"), or to that which has first started a movement (for example: Why did they make war? Because they had been pillaged), or for the sake of something (to rule), or in things that come into being, the matter. (2.7, 198a16)

I emphasized the ambiguities of this text in the notes to my introduction to my French translation of the *Physics*. For now, let us be satisfied to have shown that, according to Aristotle, without recourse to "matter," that is, the material cause, there is no complete explanation in natural philosophy.

But Aristotle goes further than simply integrating "Presocratic" causalities into the explanation of natural phenomena: he thinks that these causalities are sufficient for explaining a good number of natural phenomena, including vital realities. Thus, in this well-known passage, already cited: "Certainly, nature sometimes makes use even of residues; nevertheless, that is not a reason to look for a final cause in every case; because certain things are as they are, because of them, others happen necessarily" (*Parts of Animals* 4.2, 677a15). We have seen that a natural phenomenon like an eclipse is among those realities that do not have a final cause, and one may say the same about thunder, two cases often invoked in later years as examples of victories of science over superstition. In my opinion, we should include in the same category all the phenomena that Aristotle calls "meteorological" and, in fact, in the work where he deals with them, the *Meteorologica*, he proposes an exclusively *mechanistic* description of their coming to be. I think that rain is a phenomenon of this kind. Let's look again at the dispute between interpreters, noted earlier, about the passage in the *Physics* quoted at length in the chapter on teleology (2.8, 198b16). The question is knowing if Aristotle reports, in this chapter of the *Physics*, the concept that the mechanistic philosophers have of rain, adding that they mean this kind of explanation to apply to

60. Bolton, "The Material Cause."

all finalized processes, especially vital, or if, while obviously refusing to reduce finalized phenomena to purely mechanical phenomena, Aristotle nevertheless adheres to a mechanist explanation of rain. I think that the second alternative is true. John Cooper, who thinks that for any "Democritean" explanation to be acceptable to Aristotle it needs to be subsumed under hypothetical necessity, has to choose the first alternative.

In fact, rain is part of the environment for wheat, which cannot grow naturally in regions that are too dry, and it conforms to the schema of the two natures in that rational Nature endows this living being that is wheat with the organs that permit it to appropriate, conserve, and use water, and not that rational Nature sends rain to benefit the living thing. Similarly rational Nature has adapted the camel's palate to the spiky character of the plants in its environment; it would be wrong to say that rational Nature provides spiny plants to be eaten by an animal with a hard palate. I think that the production of iron is, for Aristotle, on the same order as the production of rain. In fact, we have seen that iron is not a homoiomere that assists in the construction of one or more anomoiomeries, while the production of flesh in sexual generation or the production of the hard material that forms the shell during the spontaneous generation of a shellfish are for the sake of the formation of anomoiomeries, and in any case are themselves already quasi-anomoiomeries, which is not at all the case for iron. If the formation of anomoiomeries from homoiomeries is not a guarantee of finality, since it can occur in the production of realities that do not have a final cause, like certain residues or bile, on the other hand the fact that some homoiomeries do not produce anomoiomeries is enough to maintain that the realities that these parts concern are outside finality. But we have seen that, in the case of minerals and metals, that they can sometimes produce what I have called "aggregates" (as in the case of a seam of iron ore enclosed by minerals), but not an anomoiomere part properly so called, that is, one able to perform an action or fulfill a function.

It is remarkable that in Aristotle's "everlasting" system, where a living adult dog and bitch are needed to set in motion that which will produce a puppy, once this presence of actual beings is acquired (and they are always present due to the everlastingness of the species), the process of *genesis* of the puppy will be entirely explicable in "Presocratic" terms. For Aristotle, therefore, and this differentiates him very strongly from Plato, a physics of nonliving objects, of the Galilean type, which appeals only to material and efficient causes, is possible, and one may add that it would be, in Aristotle's own eyes, a real physics. But doesn't this point go against certain of his statements like that of *Parts of Animals* book 1, that say that there are two kinds of causation, material and formal, and failing to take account of both is "to say nothing about nature" (642a16)? No, if we make clear that this physics will be a physics of *generations* that is precisely a Galilean physics in which nothing is considered a cause except the efficient cause and its effect on matter. When a physics like that studies animals, it therefore misses the essential part, in every sense of the word "essential," that is, what is primary and what considers essences. But the fact that material-efficient causality concerns all natural realities constitutes a second factor in the unity of Aristotelian nature.

For Georges Canguilhem, who is one of those who have pushed reflection on this question the furthest, a true biological thought can be defined negatively. It cannot be mechanist, or, in Aristotelian terms, it cannot be Democritean, because it would then ignore the specificity of its object, life or the living. Nor can it adhere to a "vitalism of exception," which would, in Aristotelian terms, amount to ignoring the definitive achievement of Democriteanism, and fall back into mythic thinking. According to Canguilhem, biological thought has to grasp individual realities, and as such carriers of values, without tearing them away from the causality of the laws of nature, that is, that the fact that life is essentially *organization* does not prevent it from also being a *mechanism*.[61] This position is directly translatable into the terms of Aristotelian etiology, and it can be found in the second schema presented earlier. For Canguilhem, in fact, the unifying factor that brings together the nonliving and living worlds is the fact that physiochemical laws belong to both, and that can be said just as much about Aristotle. But what brings it about that the well-tempered vitalism of Canguilhem cannot be attributed to Aristotle comes from the first theoretical schema, namely the explanatory function of the "superior" state for the "inferior," which constitutes a factor of unification of the physical universe that is very different from that which we find in other philosophers and scholars, ancient as well as modern. Because, we repeat, there is a great deal of difference between saying that the production of iron and the generation of an animal are both processes subsumable under the same physiochemical laws, and arguing that the causality put in

61. Cf. the article "Vie" (Life) that Canguilhem published in 1973 in the *Encyclopaedia Universalis.*

play in the production of iron is a truncated and analogical form of the causality that operates in the reproduction of animals. We say it again, in the first approach, which is that of vitalists like Canguilhem—the living is "nonliving plus"—in the second, that of Aristotle, the nonliving is "living minus." It is often asked whether or not Aristotle was a vitalist. The answer is that he is a vitalist, except that his vitalism is so different from that of others that he could hardly coexist with them within the same category.

Chapter 4

Diversity

Aristotelian causality provides both a principle of unification of the nature to which it is applied, and a principle of ordering. Natural beings are *fundamentally* the same because they are composed of the same elements that are subject to the same physiochemical "laws." But natural beings are also affected by what one might call a coefficient of causality that brings it about that some of them are subject to a more complete causal explanation than others: not only do they appeal to more elements of the Aristotelian causal system, but those elements seem to be more separated from each other. Thus, sexual reproduction appeals to a final causality that is absent from the production of minerals, but it also reveals a less confused causality than that which governs the spontaneous generation of shellfish, in which the role played by the soul in sexual reproduction is not held by an individual of the same distinct species. This chapter will try to show the limits of that unification of nature by demonstrating that none of the procedures for ordering animal diversity that the commentators think that they have discovered in Aristotle's texts manage to reduce that diversity. Aristotle's zoology is more *diverse* than it is *one*.

In the program that he sets out at the beginning of the *History of Animals*, Aristotle proposes to show the *differences* between animals, concerning "their ways of life, their activities, their characters, and their parts" (κατά τε τοὺς βίους καὶ τὰς πράξεις καὶ τὰ ἤθη καὶ τὰ μόρια, 1.1, 487a10). Then he goes on, right to the end of the first chapter, to enumerate cases of the three first sorts of differences, without distinguishing them clearly; then, four lines after the passage cited, he undertakes to present them together, not without changing their order of appearance, which becomes: "the ways of life, the character, the activities." Then he distinguishes aquatic animals from terrestrial animals; then among the first, those that cannot live out of water and those that look for their food in the water but breathe air and give birth on dry land, like the crocodile; then those that have paws, or wings like aquatic birds, or are footless; and finally those that live in the sea, in rivers, or in swamps. Furthermore, some breathe and others, like insects, do not, some live at first in water before changing form to live out of the water, the examples given by the various manuscripts being unfortunately faulty (ἀσπίδων, corrected by Karsch to ἀσκαρίδων, followed by Dittmeyer).[1] Then there is a passage about modes of local movement (487b6–32), and an announcement of a presentation of differences (with the definite article: αἱ τοιαίδε διαφοραί) "according to way of life and activities" (487b33), with the following differences: gregarious/solitary, with or without a leader, carnivore/fruit-eaters/omnivores, nesting/non-nesting, nocturnal/not nocturnal, tame/wild, plain-dwelling/mountain-dwelling/living with people, seeking sexual pleasure/not, living on the high seas/on the coast/in rocks, fight back/hide. Then at 488b11 to the end of the chapter Aristotle enumerates differences of characters: sweet, irascible, obstinate, aggressive, timid, mean, treacherous, and so on, to end with this sentence: "We will speak more precisely about the characters and ways of life of each family later" (488b28).

In other words, not only do the first three criteria of difference function together, but they also combine with the fourth, the parts, as we have seen for the modes of local movement. Also, a criterion like that of "wild" or "tame" is ambiguous: Are we talking about a character trait, which could then be included in the categories "ways of life" and "activities," or from a human perspective, that is, pragmatic and not biological? Finally, we need to notice a vast gray zone that, in books 5 to 7, mixes the study of reproductive organs with that of modes of intercourse, times of gestation, sexual maturity, seasons of intercourse, and so forth.

It goes quite otherwise in the study of the parts, which does not present that kind of indistinctness and disorder, even in what I have called the "gray zone" of the study of reproduction. Aristotle's fundamental

---

1. 1.1, 487B5. Dittmeyer's correction translates to "worms" (Aristotle, *Aristotelis, De Animalibus Historia*). D'Arcy Thompson, in his English translation, follows him and gives "river worms" and continues "out of these the gadfly develops" (Aristotle, *Historia Animalium*). Dittmeyer excises the last clause. Peck translates "bloodworms" and translates the οἶστρος of the text as "gnat" (Aristotle, *Aristotle Historia Animalium*).

position is, in fact, that the study of the diversity of the parts allows a rational approach to the diversity of animal forms, otherwise impossible to grasp. That assumes that every animal is an organism that makes a certain number of parts work together, that these parts are of several kinds, that they are defined by their function, that they function together in being among each other hierarchical in the same way as the functions that they perform. The parts of animals are studied in the first four books of the *History of Animals* and in the *Parts of Animals*, as well as books 5 to 7 of the *History of Animals* for the parts serving reproduction.

The most original aspect of Aristotle's theory concerning the understanding of animal diversity by way of the study of the diversity of their parts is doubtless his analysis of the identity of parts with each other. In conformity with the Aristotelian doctrine of unity, animal parts can be called "one" according to form, class, and analogy. Here is one of the canonical texts:

> There are parts with the same form,[2] for example the nose and the eye of one man in relation to the nose and eye of another man, flesh in relation to flesh and bone in relation to bone.[3] It is also the same in the case of the horse and other animals, all those that we say are identical with each other in form. For there is the same relation between one whole and another whole, and between each of its parts and the parts of the other. Those,[4] on the other hand, that are the same, but differ by more and less, are all of the same family. I call "family," for example, "bird" or "fish." Each of these two, in fact, belong to a different family, and there are several forms of fish and

2. "Form" and "family" are my usual translations of *eidos* and *genos*, often translated by others as "species" and "genus."

3. Camus understands that it is still about the flesh and bone of a man. He is doubtless right, because there are differences between fleshes (cf. 486b9 and 3.17, 520a5, where Aristotle talks about the greasy flesh of animals that have a small stomach). But, given that the term "form" does not designate a fixed level of generality (see later), it is possible, and from an Aristotelian perspective, orthodox, to understand that the text attributes the same form to the nose of two men, and the flesh of a man and the flesh of a dog.

4. Or "these": Aristotle is either talking about parts or about whole animals. The examples suggest that he is talking of animals.

> birds. In a general way parts differ within the same family[5] by contrary qualities, like color, form, the fact that some are more or less subject to the same property, and also larger or smaller in number, large and small, and in a general way, excess and defect. In fact, some animals have soft flesh, others hard flesh, some have a long beak, others short, some have many feathers, others few. But of course, that is not always the case, because even in these animals[6] different parts are found in different animals, for example some have spurs, others not, some have crests, others not. But one may say that most of the parts of which most of the animal is made up are either the same, or differ by contraries, and excess and defect. In fact, one may consider the more and the less as excess and defect. Some animals do not have their parts identical either according to form or according to excess and defect, but according to analogy, for example, bone and fishbone, fingernail and hoof, hand and claw,[7] feather and scale. For that which is a feather of a bird is a scale of a fish. (*History of Animals* 1.1, 486a16)

At least two relevant points from our previous analyses. First, I have tried to establish in the first chapter that by way of research that finds a profound identity beneath apparent diversity, Aristotle constructs a true *biology*, a project that remained without descendants until the age of Cuvier, whom Aristotle profoundly resembles. Cuvier, we repeat, takes up again an Aristotelian project by considering each living thing as an integrated whole, with hierarchically arranged functions, in which the function is definitory of each organ. But the parallelism between Aristotle and Cuvier goes further in that both of them, even though denying the existence of a plan common to all animals, nevertheless posit a common structure for all. The study of the different forms of organisms found in nature results in Cuvier in a comparative anatomy, largely already outlined by Aristotle. From that point of view, the first volume of Cuvier's *Leçons*

5. Reading αὐτοῖς with the manuscripts and not αὐτοῖς at 486b5, which would mean "most of the parts are distinguished from each other." Aristotle, *Les Parties de animaux* (Louis).

6. Animals in the same family. This difference according to presence and absence is found again in animals of the same form but different sex.

7. Or "talon."

d'anatomie comparée, especially the first two articles, is a truly Aristotelian treatise on animal life. We need to start by saying a few words about this common structure before trying to see how it extends into more or less large classes of animals.

Secondly, if we look at things more closely, we notice that there is a division of labor between the treatises of the biological corpus, and that the *History of Animals*, in abstaining from consideration of the function of the parts that it surveys, puts the accent on diversity, while the *Parts of Animals* is rather concerned with the functional unity behind the diversity of the parts. Whence a flourishing diversity encompassed by the *History of Animals* alone, notably with correlations empirically established, which, as we have seen, is not without echo in Cuvier.

What the Word "Animal" Names

In fact, it is undeniable that we find in Aristotle a common foundation on which the diversity of living things branches out: a monkey and a plane tree are, in a sense, the same thing because they are both alive, and a dog and an oyster are, in a sense, the same thing because both are animals. We need to devote a few remarks about this approach in which one may say that unity fires its parting shots at diversity. As we saw in the previous chapter, a living thing is a natural being that has a nutritive soul, at least if we restrict ourselves to examples of sublunary organic life, and an animal is a living thing that has in addition a sensitive soul. The division of the soul into "parts," something that Aristotle inherited from the Platonic tradition, is not what we are concerned with here, especially since it has been already dealt with in a profound and interesting way,[8] and we have discussed it earlier. Anyway, Aristotle prefers to talk about "faculties" or "powers" (δυνάμεις) rather than "parts" of the soul. But Aristotle has himself somewhat obscured the question of the relationships between these powers. This question is dealt with, notably, in *De Anima* 2.3. In the previous chapter Aristotle had presented a list of faculties of the soul, saying that it "is defined by the nutritive, sensitive, and thinking faculties, and by movement" (2.2, 413b11), and he makes two important revisions to this list. In the first place, he excludes the thinking faculty

8. Victor Caston has devoted several studies to this topic, which makes him the best current specialist on the question. See, for example, Caston, "Aristotle's Psychology."

from the list because "it is not yet clear, but it seems that this may be another kind of soul, and that it alone can be separated as the eternal is from the perishable" (413b25). The second suggests, before it is confirmed by the rest of the treatise, that some faculties lead to others, "for if there is sensation, there are imagination and desire" (413b22), which needs to be understood not as an affirmation that the presence of sensation necessarily leads to the presence of imagination and desire, but as an indication that imagination and desire presuppose the presence of sensation. The next sentence, on the other hand, that says "where there is sensation there is displeasure and pleasure" (413b23), says exactly what it seems to say.

That puts us in a better position to comment on the list given in chapter 3: "We say that these are the faculties, nutritive, desiring, sensitive, local movement, thinking" (414a31). If we add to the passages cited the one to which allusion was made in the preceding chapter, comparing the unity of the soul to the unity of a figure, a passage also found in *De Anima* 2.3, we can make the following remarks on the relationships between the parts or powers of the soul. Some powers cannot exist without others, while the second can exist without the first. Nothing alive has a sensitive soul without first having a nutritive soul, because it wouldn't be alive, while plants have a nutritive soul without a sensitive soul; nothing alive has a motive soul without a sensitive soul, although some animals have a sensitive soul only, that is, there are animals that do not move. It must be remembered that the growth of plants does not depend on a motive soul, but on their nutritive soul. But there is another kind of derivation of faculties from each other, not going up but going down, because there is a difference between saying that the sensitive soul as it were presupposes the nutritive soul, and saying that the imaginative soul presupposes the sensitive soul. Let's look a bit at these two forms of relationship.

In the first case, a power cannot exist unless there is another one already present. It's in relation to that, that Aristotle introduces the comparison with a geometric figure, a comparison that must be taken with several considerations. Let us quote the text:

> It is therefore evident that a similar sort of unity will characterize the notion of the soul and that of a figure, for there is no plane figure without a triangle, and its implications, nor is there soul without what we have been talking about. And if in the case of figures, one were to produce a common notion that would harmonize with all, it would not be peculiar to any one

> figure. But it's the same in the case of the souls we were talking about. It is thus absurd to search in either case for a common notion that would not be peculiar to any reality, rather than a formula that would conform to a proper and indivisible species, setting aside a definition of this kind. But if the case of souls is somewhat similar to that of figures, it's in fact the case that the consequent always implies an antecedent power, both in the case of the figures as in that of animated beings: thus, in the quadrilateral there is a triangle and similarly the sensitive in the nutritive. So, it is necessary, for each animated being, to look what sort of soul it has (which is that of a plant, that of a human being, or an animal), and to examine the cause for which they appear as in a series. For without the nutritive, the sensitive does not exist, but the nutritive is apart from the sensitive in plants. (*De Anima* 2.3, 414b19, a passage already partially cited)

The mathematical construction to which Aristotle here alludes rests on the idea that the triangle is the basic figure composing all other figures, which seem here to be reduced to the set of polygons, more or less in the way that Plato composes all his figures that constitute the universe in the *Timaeus* out of triangles. The analogy between geometrical figures and faculties of the soul is thus not a strictly Aristotelian analogy; that would put four terms arranged two by two into the same relationship (A is to B as C is to D; bone is for terrestrial animals what fishbone is for fish), because the triangle is not in the same relationship to the rectangle as the nutritive soul is to the sensitive soul. All the same, one may find in this passage a weak and marginal analogy in which the sensitive soul presupposes the nutritive soul as the rectangle presupposes the triangle. Anyway, Aristotle does not talk about analogy, but about situations that are "somewhat similar" (παραπλησίως, 414b28). Furthermore, the idea of power used here is not the technical sense of power that Aristotle uses elsewhere, for divisibility into triangles, or composed of triangles, is not the same as having triangularity potentially in itself. Anyway, the sensitive soul is not made up of several nutritive souls. Finally, to say that the sensitive soul is potentially in the nutritive soul is imprecise if one takes "potential" in its technical sense, for it is difficult to think that in the case of animals and in any case false in that of plants. The only thing that this comparison clarifies is that the two series, that of figures and that of souls, are sequences that cannot be followed except in the direction that is also a sequence of meanings, as Aristotle says just afterward (415a3): an animal cannot have vision, hearing, and so forth, without having touch, but there are animals that have only touch. The theoretical advantage of this comparison ultimately seems rather weak. Perhaps it is addressed first of all to people whom he needs to convince, notably Platonists.

To summarize, the sensitive faculty presupposes the nutritive faculty, but does not proceed from it in any sense of "proceed." These two faculties are furthermore not species of the same genus. They are two faculties that one may call "basic." Things are completely different when we come to the relationships between the sensitive faculty and the faculties of imagination and mind. In fact, these are all different instances of the same process, even if they are not exercised in the same way on the same realities, bodily in one case, incorporeal in the other, nor in the same circumstances. Perceiving and thinking are two different ways of discriminating (κρίνειν). Different, because Aristotle does not adopt the opinion of the "ancients" according to which "thinking (τὸ φρονεῖν) and sensing (τὸ αἰσθάνεσθαι) are one and the same thing" (*De Anima* 3.3, 427a21), but two processes, of sensible perception and of thought, produced when an object, of sense or thought, impacts the faculty corresponding to it, bringing it about that the faculty identifies with the object: vision becomes the visible as thought becomes the thinkable.

This is not the place for a precise analysis of, notably, *De Anima* 3.2–4, which would show in these passages a perfect example of the way that Aristotle uses previous philosophers. According to him, it is not surprising that the ancients, or many of them, thought that perceiving and thinking were the same thing, and that intellection would be a bodily process. The major reason for this widely shared error is that in fact the preceptive and intellectual processes are homothetic:

> If, then, thinking is like (ὥσπερ) sensing, either it will consist of undergoing something under the action of the intelligible, or something else of the same kind. Thus, it is necessary that there be an unmoved principle, but capable of receiving the form, a principle that is potentially the same as the form, but is not the form. And the relation of the sensitive to sensibles must be like that of the intellect to intelligibles. So, since it grasps everything, the intellect must be "unmixed," as Anaxag-

> oras[9] says, in order to "command" its object, that is, in a way to be able to know it. . . . Thus, what one calls the intellect in the soul (I mean that which allows the soul to think and to grasp concepts) is not in actuality any of the realities before it thinks them. For it would then have a quality, becoming hot or cold, and it would use some organ, as the sensitive faculty does, but in fact there is none. Thus, it is for good reason that it is also said that the soul is the place of forms, although this is not true of the soul as a whole, but of the intellective soul, and the forms are not in it in actuality, but potentially. (3.4, 429a13–29)

But the sensitive faculty has a privileged relationship not only with the noetic faculty. In fact, the *De Anima* gives a lot of importance to φαντασία, traditionally translated "imagination," but perhaps sometimes better rendered "representation." I'm not going to provide even a summary analysis of this difficult notion, the object of many controversies, but we can at least recall its relationships with the sensible and the intelligible, between which it plays a kind of intermediary role. It's enough to cite a few passages. The definition of imagination is as follows: "If imagination is that by which we say that some representation (φάντασμά τι) is produced for us, and we are not speaking metaphorically" (3.3, 428a1). But imagination works on sensible perceptions: when Aristotle writes that "imagination is the movement that is produced under the effect of actual sensation" (3.3, 429a1), a formula repeated in *On Dreams* 1, 459a17, he means to signify that the production of images by imagination, which happens even in the absence of the object, since these images "appear to us even when our eyes are closed" (3.3, 428a16), relies on the remains of movements caused in us by sensible perception. Since imagination is a movement (which argues, by the way, in favor of the translation of φαντασία by "imagination" rather than by "representation," at least to take this latter term in an active sense), Aristotle assimilates it to the movement that accompanies sensation, from the angle of a distinction, frequent in him and one that we have already met, between, for two entities, to be identical, but different in their being:

9. Fr. A 100 DK.

> Since we have discussed imagination in the *De Anima*, and the imaginative faculty is the same thing as the sensitive faculty, although the being of the imaginative faculty and the being of the sensitive faculty are different, since imagination is a movement by actual sensation, and a dream seems to be a kind of image . . . , it is clear that the dreaming belongs to the sensitive faculty, but in that it is imaginative. (*On Dreams* 1, 459a15–22)

By speaking of the sensitive faculty "in that it is imaginative" (ᾗ φανταστικόν), this passage makes the sensitive and imaginative faculties very near each other. Nevertheless, it remains that, as is explained in a passage in the *De Anima*, which editors usually set apart in its own paragraph (3.3, 428a5–16), imagination is not sensation, if only because sensation belongs to all animals, while imagination is absent in animals like the ant, the bee, and the worm.

There is, anyway, a well-established doctrine in Aristotle that "the soul never thinks without an image" (ἄνευ φαντάσματος) (3.7, 431a16),[10] so that by way of the faculty of producing images, the activity of the noetic soul is connected to the activity of the sensitive soul. Sensation, imagination, and intellection are different versions of the same activity, which allows Aristotle to say that they are simultaneously the same and different in their being, and this unity is that of the power of discrimination. As he says in the *Movement of Animals* 6, 700b19: "Imagination and sensation occupy the same space (χώρα) as the intellect, because both discriminate (κριτικά)." From all that one may conclude that faculties like imagination or intellection, which Aristotle does not fail to designate sometimes as "souls" (cf. the imaginative τὸ φανταστικόν, faculty, or soul, *De Anima* 3.9, 432a31), share the same space as the sensitive soul, which has in relation to them the position of a principle, for imagination and intellection rest on sensation, and not sensation on them. No relation of this kind, which might be called derivation, exists between the nutritive and sensitive faculties.

In the lists of various faculties, or "souls," offered by Aristotle, there are two that often appear, the moving and desiring souls, for example, at *De Anima* 2.3, 414a31: "We want to talk about the nutritive, sensitive, desiring, and locally moving, faculties" (κινητικὸν κατὰ τόπον, passage

10. Cf. *On Memory* 1, 449b30: "since we have dealt with imagination in the treatise on the soul, and there is no thought without an image."

already cited). There are animals that have only the nutritive and sensitive faculties, like shellfish that live attached to rocks, and sponges. Aristotle calls them "imperfect" animals (*De Anima* 3.11, 433b31). Other animals need to move from place to place to look for food, escape from predators, find occasion to reproduce, and so on. The local motive faculty is another basic faculty in that it is not derived in any way from other faculties; Aristotle shows this with great precision in *De Anima* 3.9–11. Thus, there are three basic faculties of the soul, nutritive, sensitive, and motive,[11] arranged in an order such that the following cannot exist without the preceding ones, but they do not proceed from each other, and even less are species of the same genus. A basic faculty can, however, branch out into synonymous faculties, in the Aristotelian sense of the word "synonymous," like the fact that it is from the sensitive faculty that the imaginative faculty originates.

This image is blurred by the appearance of two additional terms, the reasoning faculty and the desiring faculty. In a passage in the *De Anima* that gives a list of faculties arranged from the widest to the narrowest, we read:

> Without the nutritive, the sensitive does not exist, but the nutritive exists apart from the sensitive in plants. In its turn, without touch, none of the other senses exists, while touch exists without the others, since many animals have neither vision, hearing, nor smell. And furthermore, some animals endowed with sensation also have the faculty of local movement, others not. Finally, a very small number also possess reasoning and thought. Those mortal animals that are endowed with reasoning are, in fact, endowed with all the others, while those that have the other faculties do not all have reasoning: on the contrary, some do not even have imagination, while others have it only to live.[12] As for theoretical intelligence, that's another question. (2.3, 415a1, passage follows another cited earlier)

11. Cf. *Parts of Animals* 1.1, 641b5: "However, it is not the case that the whole soul is the origin of movement, nor all of its parts, but the origin of growth is found even in plants, that of alteration is the sensitive part, that of local movement is another part, and not the rational. Local movement is in fact present in animals other than human beings, while thought is in none of them," which amounts to saying that the nutritive, sensitive, and motive "souls" are basic faculties for every animal.

12. One should understand that to live they have only imagination, without thought.

According to this passage, animals endowed with the reasoning faculty necessarily also have the faculty of local movement, while the reciprocal is not true. But we have seen that the reasoning faculty is in a way an avatar of the sensitive faculty, in that it ought to be possible for it to belong to an animal that does not have the faculty of local movement. In fact, when he talks about animals that have "reasoning and thought" (λογισμὸν καὶ διάνοιαν), recognizing that they are a "very small number," Aristotle is thinking of human beings, because, for him, although other animals may be intelligent, none possesses reason. Two opposed logics are thus at work here: on the one side, an animal ought to be able to possess every discriminating faculty without having the moving faculty; the First Mover is, in any case, a living nonmover, and it is perhaps of the First Mover that Aristotle is thinking when he limits his analysis to "mortal" animals (415a9). But, on the other hand, among mortal animals, it is difficult not to attribute mobility to an animal as perfect as the human being, or any other animal that would have reasoning and thought.

Animal movement at issue here does not reduce to local movement, even if, like every movement, it involves local movement.[13] To be convinced of that, read the passages dedicated to movement in the *De Anima* and *Movement of Animals*, with, in addition, the *Progression of Animals*, which considers more precisely the various modes of progression of different animals—walking, crawling, flying. One sees there that Aristotle has his sights on a grandiose general theory of movement that tries to organize under common principles the creeping of worms and human initiative rationally determined by the individual pursuing some goal. From that point of view, the *Movement of Animals* is remarkable: after having clarified the conditions of movement in general, or at least some of them, including the movement of the universe dependent on the First Unmoved Mover, it concentrates on the movement of sublunary animals. Aristotle's theory presents two important components. The first is anatomical-physiological: the source of all animal movement is located in the heart, which is also the center of sensation: "Besides, it is reasonable that it goes thus. We say in fact, that the sensible faculty is also found in that place, so that when the area around the source is altered and has been changed due to sensation, the contiguous parts change at the same time, dilating and contracting, in such a way that necessarily, as a consequence, movement

13. Cf. Berti, "La suprématie du mouvement local selon Aristote."

comes about in animals" (*Movement of Animals* 9, 702b20). There's a purely physical explanation, one might even say mechanistic, of the fact that sensible perception sets in motion the moving faculty. Because living things without a sensitive soul do not have a motive soul.

In addition, it is convenient to make a distinction between different kinds of movement, more implied than explicit in Aristotle, when it's a matter of the faculties of the soul. Because the nutritive soul is also a source of movement in generating a being and making beings that have this power of soul grow. The most fundamental transformation that a living thing undergoes is the transformation of food into blood, in blooded animals, and into the analogue of blood in the nonblooded animals, and that is obviously a group of movements. But these movements do not depend on the moving soul, as Aristotle explains thus in the *De Anima*:

> That the moving principle is not the nutritive soul is clear. In fact, it is always for the sake of something that this movement [i.e., local] is done, and it is accompanied by imagination and desire, because the animal, if it is not desiring or fleeing, is moved by constraint alone. Furthermore, even plants would be endowed with movement and would have an organic part dedicated for assuring [local] movement. (3.9, 432b14)

But, in Aristotle, "by constraint" (βίᾳ) is one of the kinds of necessity, the one that makes the movement in question happen just by the physical properties of the being in which it happens to be. Aristotle here indicates one of the characteristics of a living thing that is well explained by Canguilhem, that of choosing or fleeing from *stimuli* coming from the environment, as *values*, represented in this case by pleasure and pain. The necessary movements that lead to growth and digestion do not depend on the moving faculty, nor on a special faculty distinct from the nutritive faculty.

But the explanation of animal movement also depends on a second component, developed in the *Movement of Animals* and *De Anima*, a "psycho-practical" component. In an analysis whose boldness has been overlooked for a long time, Aristotle shows that

> we see that what moves the animal is thought, sensation, imagination, decision, wish, impulse, and appetite. In fact, all of this comes down to intellect and desire; actually, imagination

> and sensation occupy the same space as intellect, because they all serve to discriminate, though they are distinguished from each other by differences presented elsewhere. Wish, impulse, and appetite all come from desire, while considered choice is common to thought and desire, in that the desirable and that which is the object of thought are the first to initiate movement; not, however, every object of thought, but the end of activities. (*Movement of Animals* 6, 700b17, already partially cited)

But *De Anima* 3.9 proposes a supplementary step that is, to tell the truth, a giant step. After having, after 432b7, disqualified the nutritive and sensitive souls as sources of movement, Aristotle writes:

> But in addition, it is not the intellectual faculty nor that which one calls the mind that is that which causes movement, because the theoretical part does not theorize at all about that which is practical, and furthermore says nothing about what one ought to flee or pursue, but local movement always concerns what should be fled or sought. But even when the intellect theorizes about an object of this kind, it never commands flight or pursuit; for example, it often thinks about something frightening or pleasant, without commanding flight: it is the heart that is set in motion, and if it is a pleasant object, it is another part of the body. Also, even when the intellect orders and thought says to flee or pursue something, there is no movement, but the animal acts according to its appetite, as the weak-willed person does. (432b26)

In other words, knowing what is just, good, or beautiful does not set an individual in motion: one has to *desire* the thing that is just, good, or beautiful. Cognitive analysis of situations can certainly clarify the decision, but it does not bring it about. Aristotle here opens a pathway that, in breaking with the Socratic tradition, ceaselessly reborn, and its incapacity to think weakness of will (which makes one choose the devastating evil when one knows the healing good), at least until Freud. As the *De Anima* sums it up, "The first mover is unique: it is the desiring faculty" (3.10, 433a21).

But the desiring faculty poses at least two difficult problems. First, it depends on the motive faculty, which presupposes that in the animal that has it, there are both nutritive and sensitive faculties. But the relationship of the desiring faculty to these two faculties is not the same, and one gets the impression that Aristotle establishes a privileged relation between the sensitive and desiring faculties, to the point of making it a necessary connection. Thus, at the beginning of his study of the soul and its faculties, Aristotle notes: "If there is sensation, there is also imagination and desire, for where there is sensation, there is displeasure and pleasure, and where those are present, necessarily there is appetite" (*De Anima* 2.2, 414b22, already cited). But it is at the end of the treatise that Aristotle faces this problem, and proposes a solution . . . and it is problematic:

> We must look again at what is the case in imperfect animals, in which, in animals that have only the sense of touch, what is the mover? Is it possible that they have imagination, or not? How about appetite? Because it does seem that they have pain and pleasure, and if they have those, they must also have appetite. But how would they have imagination? Wouldn't it be the case that, since they are moved in an indeterminate way, they have them in an indeterminate way? (*De Anima* 3.11, 433b31)

Following ancient commentators, modern interpreters understand the second "in an indeterminate way" (ἀορίστως) as meaning "poorly differentiated." It would then be necessary to understand that an oyster, from the fact that it perceives, possesses an embryonic form of imagination, or is on the way to possessing it. If one thinks, from the angle of what has been said earlier, that imagination is a kind of by-product of sensation, or sensation continued by other means, it's not that surprising that animals that have rudimentary sensation, specifically having only the sense of touch, would also have embryonic imagination and even embryonic thought, if one thinks that the lowest degree of thought is the simple power of discrimination. As for desire, Aristotle reveals that it is in tension between two faculties: although it depends officially on the motive faculty, because it is the "first mover," nevertheless it remains that it is set in motion by the sensations of pleasure and displeasure that in turn depend on the sensitive faculty. What is, in fact, most difficult to understand in the last two lines of this passage is what is meant by "moved in an indeterminate way" (κινεῖται ἀορίστως, 3.11, 434a4). If Aristotle thinks that oysters, attached to their rock, and completely immobile, are nevertheless animated by a kind of movement, so that every animal would be capable of moving in one way or another, that would reinforce the privileged place of the sensitive soul in animals. One might think that the movement of retraction of oysters, animals otherwise immobile, has something to do with the motive soul (perhaps here too one should speak of indeterminacy), but I am hesitant, for reasons too long to set out here, to think that it's the same for movements implied by perception itself, in which images strike the senses.

Finally, the difference between the three basic faculties is not of the same intensity. Animals share with plants the nutritive faculty, and add to that the sensitive faculty; then a *perfect* animal adds also the motive faculty. But this notion of perfection is decisive. We will have occasion to return to the notion of perfection in Aristotle's biology, particularly to distinguish a notion of relative perfection (which we have already spoken about), from the absolute perfection that includes all living things. But one cannot say that a plant is an incomplete animal or a sketch of an animal, because it's a matter of two schemas, one might say two logics of life, that are entirely different. But one can, on the other hand, think of the oyster as a kind of imperfect version of an animal that has a motive soul that is not "indeterminate." The connections between sensitive and motive faculties are many and easier to apply because both faculties concern the animal kingdom.

The second problem posed by the motive/desiring faculty is the confusion of levels that it seems to imply. One of the major epistemological features of Aristotelianism is to have given up on a science of everything in favor of dividing up the field of knowledge into different sciences not only according to the object to which they refer, and that in several levels (physics is different from mathematics, but also geometry is different from arithmetic), but also in the kinds of knowledge (theoretical sciences are different from practical and technical sciences). But because it tries to construct an explanatory schema that takes account of all movements, from local movement of the tiniest animals to the ethically determined action of a human being, the analysis of the motive faculty of the soul, which because of its epistemological status belongs to physics, a theoretical science, also touches *practice* in the Aristotelian sense of the term. This is not the only example of this kind.[14] In principle, the analysis of processes that rely simultaneously on physical causality, more precisely physiological, and ethical causality, does not raise an insurmountable problem, since a person who robs the purse of his neighbor has made a decision that is ethical in nature, judging that this conduct would be

14. This problem is examined in my *Endangered Excellence*, in the chapter "A Biological Politics?"

beneficial to him, that the advantages outweigh the disadvantages, but the actualization of the project presupposes a set of movements, which put in action a physiological causality that implies a great number of parameters. In our example, the dishonesty of the thief will be actualized via material realities and states of affairs: thus, the hotter or colder, thicker or thinner, blood of the thief, these characteristics themselves flowing from several factors, including the climate of the thief's place of origin, which would give him more or less courage or cowardice. Aristotle here formalizes theoretically something that had been a largely dominant position throughout ancient medicine, in which not only pathological states but also the conduct of people in good health, are largely determined by physiological givens, notably by the humoral mixture. Similarly, in a well-known passage at the beginning of the *De Anima*, Aristotle explains that anger can be described both as boiling of the blood around the heart, and as a response to an insult (cf. *De Anima* 1.1, 403a29).

An animal is thus a living thing endowed with sensation. As a living thing, it is able to maintain itself by assimilating nonliving elements from outside (food) and reproduce itself, in one way or another, studied earlier. As an animal it has, among other things, a sensitive faculty that becomes complex and ramified into several other faculties, which Aristotle sometimes describes as so many "souls." Then perfect, or rather not imperfect, animals have a motive faculty that should also be counted as a basic faculty because it does not derive from another faculty. That's why in a passage of the *Politics* that has attracted attention because of its theoretical boldness, and about which I have often changed my opinion, Aristotle *defines* the animal by these three basic functions. In that passage, he is looking for a method for thinking about the plurality of constitutions. This method, developed throughout *Politics* book 4, winds up at the end of the book in a combinatory form: by defining each constitution as a combination of three aspects, which in modern terms one may call legislative, executive, and judicial, and varying instances according to all these factors, one gets a kind of political Mendeleev periodic table. To accomplish this, Aristotle introduces a zoological comparison. We cite here this passage on the biological significance to which we will return at length later, especially for what it says about the Aristotelian way of grasping animal diversity:

> It's as if we were to decide to grasp animal forms; we would determine first what every animal must have (certain sensory organs, and the part that works on the food, and that which received the food, namely the mouth and stomach and, besides,

> the organs by which each of them moves itself), and if the number of these necessary parts are indeed only these, but they present among them various differences (I mean, for example, that there are several sorts of mouth and stomach, and also of sensory and locomotive organs), the number of their combinations would necessarily give a plurality of animal families (because the same animal cannot have several kinds of mouth or, in the same way, of ears); so that when one has put together all the possible combinations, that would result in the species of animal, as many as the combinations of necessary parts. It is the same for the constitutions we have been talking about. (*Politics* 4.4, 1290b25)

This passage says clearly that "what (ἅπερ) every animal must have" (1290b25) are the sensitive, nutritive, and motive faculties. Thus, it's a matter of a perfect animal. This text, which is presented for the sake of comparison, is not in a biological work, but political. If we take it literally, it would seem to indicate that only the different forms of necessary organs, that is, those that fulfill the functions of the three basic faculties, need to be taken into account for *defining* the animal, with the simple addition of a principle of nonredundancy: no animal can have "several sorts of mouth" (1290b33). One would still need to reconsider that limitation, since ruminants indeed have "several sorts" of stomach. This passage makes no reference to the two fundamental aspects of Aristotle's zoology noted earlier, that which since Cuvier has been called the law of organic correlation: the presence of organ A (multiple stomachs) implies that of organ B (split hooves), but prevents that of organ C (solid hoof) to be present, as well as the law of subordination of characters, which subjects secondary characters and functions to primary characters and functions. So let us be content then, for the moment, to think that this passage confirms for us that an animal is a set of organs put in operation by the nutritive, sensitive, and motive parts of the soul. We will study in more detail what it says about animal diversity.

The motive faculty cannot function without the sensitive faculty any more than that can function without the nutritive faculty, including, all the same, a kind of imperialism of the sensitive soul, upstream because the way in which food is acquired, which determines the functioning of the nutritive faculty, is very different in sentient beings than it is in plants, but downstream especially not only because the motive faculty cannot function without the use of the sensitive faculty, but also because it is the sensitive faculty that creates the need for the motive faculty: desire, which is a by-product of sensation, is the "unique first mover." The nutritive faculty, on the other hand, does not create a need for the sensitive faculty. So an animal is above all a living system provided with a capacity of discrimination coming from the sensitive faculty and the faculties derived from it, the imaginative and intellectual faculties, which arouse the desire that provokes movement. Hence the conjunction of naturalistic and practical approaches to movement, in the Aristotelian sense of the word "practical," which makes the sensitive and motive faculties work together, and which rests ultimately on the pair pleasure/displeasure, for by being capable of sensations, the animal rouses in itself an attractive or repulsive desire. In fact, Aristotle repeats several times, as if driven by the evidence, that whatever possesses sensation experiences pleasure and pain. In fact, the very idea of "animal pleasure" is far from being simple to understand in Aristotle, and we will come back to it in the next chapter.

One may grasp this prevalence of the sensitive soul in animals, which occurs to the detriment of the other "souls," notably the nutritive soul, which is however the most basic and thus, apparently, the most necessary of all, if we return to what was said about wind eggs in the previous chapter, and more particularly on the fact that the nutritive soul belongs *potentially* to wind eggs. The nutritive soul is not present potentially in actual animals, but absolutely actually: that soul regulates the assimilation of food, and reproduction. But wind eggs are not animals and the nutritive soul is not in the same ontological situation for a wind egg and for the bird that laid it. But there is a case, sufficiently studied by Aristotle, although he unfortunately skipped over a good many details, in which he affirms that the nutritive soul exists potentially, and that is in embryos. At *Generation of Animals* 2.3, 736b8, we read:

> As far as the nutritive soul is concerned, it is clear that we must posit that seeds and embryos that are separable[15] have it potentially, but not actually until they go looking for food, as do

15. Keeping the χωριστὰ of the manuscripts. Drossaart Lulofs prints τὰ μήπω χωριστά (not yet separated), which he found in the margin of a Florentine manuscript and in the translation by Gaza; P. Louis follows Bussemaker and chooses τὰ ἀχώριστα (the not separated). I think that we can understand χωριστά as meaning "separable but not separated." Anyway, the sense is clear: it's a matter of the embryo that is still in the womb.

> embryos once they are separated.[16] At first, in fact, all embryos of this kind [not separated] seem to live the life of a plant.

A text poorly established and difficult.[17] Thus from line 736b10 ("until they go . . .") Aristotle forgets about the seeds, dealing with the embryos only. But his argument leaves no doubt: in embryos, until they are born, the nutritive soul, which ought to be present since it's a question of living beings, is present *potentially*, and will be present actually when the offspring is born and can look for its food itself. In other words, the meaning of the fact that the nutritive soul is potentially in embryos, and also in wind eggs, is that it is not performing its function as it is in an autonomous living animal: we saw that earlier, but this passage says it explicitly.

But we know from elsewhere that for Aristotle the unborn embryo is able, at least at a certain stage of its development, to perceive sensations. In *Generation of Animals* 5.1, Aristotle, without dealing explicitly with this question, indicates that for him, animals sleep in the uterus of their mother, and thus that they are sometimes awake. He supports this assertion with a curious "empirical" proof: "It seems that they are awake even in the uterus, as is evident in dissections and in ovipara" (5.1, 779a7), which is certainly an allusion to the fact, noted elsewhere and obvious to everyone, that the fetuses move before their birth, and that Aristotle had to notice in his precise observation that we know that he carried out on chicken eggs.[18] But the waking state is accompanied by sensation:

> Whereas we have previously elsewhere defined what are called the parts of the soul, and since the nutritive part is separate from the others in bodies that have life, even though none of the other parts can exist without it, it is clear that all the living things that share only in growth and destruction know neither sleep nor waking, as is the case in plants, for they do not possess the sensitive part, whether it is separate or not.

16. Embryos once they are separated, that is, offspring after their birth.

17. Fortunately, largely clarified by Lefebvre, "Aristote sur le sommeil de l'embryon et du nouveau-né."

18. It is difficult to decide which dissection he is talking about here. D. Lefebvre proposes the hypothetical example that dead human embryos are observed with their eyes open ("Aristote sur le sommeil de l'embryon et du nouveau-né," 177, n. 41). Another possibility: if a pregnant quadruped is slaughtered for ritual sacrifice or food, any late-stage embryo would likely be moving (A.P. suggestion).

(*On Sleep* 1, 454a11)

Whence this remark, made in passage: little babies "sleep a lot, because even in the womb, *while they are acquiring sensation*, they sleep all the time" (*Generation of Animals* 5.1, 778b21). Which leads us to strongly correct the comparison, several times invoked by Aristotle, between embryos and plants, because plants do not sleep, because they are never awake.[19]

Thus, we have this situation: embryos, as also doubtless wind eggs, possess a nutritive soul potentially, because it is not performing its normal functions, while they do possess sensitive soul that is perceiving and thus is actual. This paradoxical situation (the conjunction of a nutritive soul potentially and a sensitive soul actually) is true of embryos only, for after birth, the animal acquires an actual nutritive soul. The requirement that nutritive soul is a precondition for the existence of the sensitive soul is thus formally respected in all cases, since before they acquire a sensitive soul, embryos have a nutritive soul, even if only potentially. But there are cases, at least in embryos, in which the nutritive soul cannot acquire its *functions*, that is, exist actually, until *after* the sensitive soul has the ability of exercising its functions. All of this also confirms the proper space that Aristotle attributes to animality: as was previously said, animals are not plants to which something has been added, and vegetable life is not "basic" life. It would have been necessary to make more room for plants in our parallel study of spontaneous generation and the production of nonliving natural bodies, but then it would have been necessary to rely on the botanical treatises of Theophrastus and to decide going in whether they would share Aristotle's viewpoint on the relationships between sexual generation, spontaneous generation, and the production of nonliving bodies.

An animal, as we have seen is above all a living system able to discriminate, thanks to faculties that, by provoking desire and repulsion in it, bring about movements. This structure, common to all animals, which arranges their different "souls," is distributed into an unbelievable diversity of forms; Aristotle is alone in antiquity, and beyond, to be aware of that diversity. This study of animal diversity, which is one of the two aspects of the inquiry undertaken by the *History of Animals*, is another route that leads to the question of the perfection of the animal world and, consequently, that of the entire cosmos. Nevertheless, it remains that for

19. "There is no sleep without waking, but it is true of plants that they have the analogue of sleep without waking" (*Generation of Animals* 5.1, 779a2).

Aristotle animality is a field of study that is coherent and distinct from the others, that there is a frontier that not only separates animality from nonliving nature, but can also be traced within living nature.

Continuity and Diversity, Perfection and Harmony

The problems of the diversity of animal forms and their possible continuity within a *scala naturae* are closely tied to each other, and some of the texts that present these problems have already been considered. If Aristotle thinks, as some seem to believe, that the various species of animals are continuous with each other, even if, as we have seen, it is in a weak sense of "continuity," that is, if one adopts a position of the Lamarckian variety, minus the chronological and evolutionary aspect, obviously, then this fullness of Nature would be an additional proof of its perfection. That would integrate, or reintegrate, Aristotle into the great Greek naturalist symphony from which he distances himself on other crucial points. But we have shown that the two passages that are usually credited to support continuity, one from the *History of Animals*, the other from the *Parts of Animals*, are far from supporting such a doctrine. We have, in fact, remarked that these passages concern themselves in fact with the two extremes of the vegetable world, the place where minerals seem to "become" plants, and where plants seem to "become" animals. Furthermore, these limits between nonliving, vegetable, animal are not at all obscure for Aristotle, but can be for inexperienced observers, since the criteria that permit going from one of these "kingdoms" to another are precise, as we will see in more detail: it's a matter of either having or not having a particular psychic power, the nutritive power in the first passage, and the sensitive power in the second.

So, it is time to return to the text of *Politics* 4.4, previously cited, since it is the one of the two that seems to go further, and perhaps the one that goes furthest of all, in grasping the diversity of animals; some, including me, have even attributed to it the project of an a priori reconstruction of the animal world via the combination of the characteristics that belong to animals. If one combines this passage with certain others, notably the selections from the *De Anima* examined earlier, it seems at first glance that one may find in this Aristotelian procedure a resemblance with the "deep" classification of Cuvier already mentioned: for Aristotle, the digestive-reproductive system is the most fundamental and definitory of life, animals adding to it a sensitive system and, for the more perfected ones among them, a motive system. The other systems are subordinated to these and depend on each other: in animals in which digestion produces blood, there is a system of distribution of this blood in the body and a system of cooling subordinated to this vascular system. Thus, we have to construct a positive interpretation on the basis of this famous passage in the *Politics*. Today, my position is the reverse of that which I have adopted in the past, along with other interpreters: not only is it the case that this text does not anticipate an a priori reconstruction of all animal species, nor summarize a *scala naturae*, but it is equally false to say that it is not echoed in Aristotle's zoological works.

It's a matter of a text that appears very "tough" in that it seems to present a combinatory procedure bringing together exhaustiveness and necessity. We start with the *necessary* parts ("those that it is necessary that every animal have," ἅπερ ἀναγκαῖον πᾶν ἔχειν ζῷων, 1290a26), "namely" (οἷον) certain organs of sensation, those that work on and contain the food, "namely" the mouth and stomach (unless the second οἷον means "for example," for not all animals have these parts; but we can keep "namely" if we remember that some animals have a mouth by analogy and then give the two occurrences of οἷον one right after the other the same sense) and finally the motive parts. Then comes a short declaration by Aristotle, very precious to us: "and if indeed these necessary parts are only so many" (εἰ δὴ τοσαῦτα εἴη μονον, 1290b29), which obviously is equivalent to an affirmation: it really is a matter of the *only* necessary parts for an animal. We note that this conforms with the analysis proposed earlier according to which the basic and nonderivable functions of animals are the nutritive-generative, sensitive, and motive functions, and that the imaginative, intellectual, or desiring functions, for example, are derived from these basic functions. If we rely on two principles, namely that each necessary part has several variations and redundancy is impossible (to which one doubtless must add a principle of compatibility, not mentioned by Aristotle, because no animal would have the ears of an elephant and the stomach of a mouse), then "the number of combinations of these necessary parts would necessarily yield several kinds of animal" (1290b32), so that one would have as many forms of animal (εἴδη ζῴου) as the possible combinations of necessary parts, a conclusion that Aristotle repeats twice in slightly different forms.

In order to correct my mistaken reading of this text, we must begin by insisting, in a more complete manner than those that have been adopted until now, on the purpose, and thus the context, of this passage. It is a matter, as was recalled earlier, in setting up a comparison between the animal world and the world of cities, of putting together a method of thinking about the diversity of constitutions, which is an important reality of the world of cities, and which ought to be, for this reason, understood in an adequate way by the Aristotelian legislator. But the Aristotelian approach to this question is not linear. In the last three chapters of *Politics* 4, Aristotle in fact presents a grandiose method, which has not yet received the attention it deserves.[20] He starts by distinguishing the three aspects that every constitution is supposed to have, a deliberative part, a part concerning magistratures, and a judiciary part, a tripartition that, as we have said, is reflected, without corresponding exactly, to our distinction between the executive, legislative, and judiciary powers. These instances would parallel the parts "that every animal must have." Examining each of these political parts one after the other, Aristotle shows that one may reconstitute all the forms by varying a small number of parameters. Thus, concerning the deliberative part, "all deciding everything" is a characteristic of popular regimes, while "some deciding all" is characteristic of oligarchic regimes. All can decide on some things, and some decide others, and so on, but that does not give us an immense list of possibilities. If one means to consider all the modalities of "all deciding everything," one doesn't get a large number of possibilities either (for example: the power belongs to all, but there is no assembly except for important questions, and magistrates are elected for the rest, etc.). Similarly, for the nomination to magistratures, it depends on three factors, Aristotle says: a) who picks the magistrates (all or some[21]), b) from which group one takes them, and c) according to which method, election or by lot, and some by election, others by lot. Each of these modalities is politically oriented: drawing lots, for example, is a democratic procedure.

Thus, the last three chapters of *Politics* 4 are concerned with presenting a method of (re)constructing a priori all the possible forms of constitution; this is part of what I have called the "theoretical tools" of the

20. Cf. Pellegrin, *Endangered Excellence*, 315–323.

21. "But each figure can take four forms. In fact, if all nominate, it can do so either by election, or by lot (and 'among all' can be understood: either taking turns in the position, each tribe, deme, or phratry, one after the other, until all the citizens have had a turn, or in each case from the entire citizenry, or for certain magistracies the first way, and for others the second) [or among some by election, or among certain by lot]" (4.15, 1300a22). Thus, one has a combination of three times four, equals twelve terms.

Aristotelian legislator.[22] Because what is at stake for Aristotle is to think of an existing reality in order to reform it, if necessary. Faced with a city whose constitution does not work well, the Aristotelian legislator could, for example, propose tempering the democratic character by introducing into its legislation and/or its judiciary institutions aristocratic or oligarchic elements. But in order to do that, one must first determine precisely where the constitution in question is located in the order of constitutional forms. So, what would be more natural, then, than to think that the zoological branch of this comparison also derives from the same a priori schema of reconstruction? But that is not the case.

This a priori method of reconstruction is not present in 4.4, in which the zoological comparison occurs: in book 4, there is a progression that must be taken into account in order to correctly appreciate the theoretical goal of the zoological comparison. Aristotle's goal in chapter 4 is to show that not only is there a generic diversity of constitutions, which leads him to distinguish, like his predecessors, kingship, tyranny, oligarchy, aristocracy, democracy, and another regime he calls *politeia*, but there are also several kinds (species) of oligarchies, democracies, and so on. Hence the richness and stupefying subtlety of his analyses of the different species of democracy and oligarchy (the two dominant constitutions) in terms of the sociological composition of the classes that vie for power in them. From *Politics* 4.14 on, however, the perspective changes and Aristotle clearly marks that in the first sentence of the chapter: "Again, concerning the constitution in general as well as each of them individually, let us speak of them in order, taking as our starting point what is most convenient for them" (*Politics* 4.14, 1297b35).

It is then, and only then, that Aristotle introduces his combinatorial method of reconstructing constitutions.

The two sides of book 4 have a lot in common: it's a matter of dealing with the *diversity* of constitutions in referring it to the diversity of the *parts* of the city. I have characterized the change in perspective that happens in *Politics* 4.14 as a change of purpose: before chapter 14, Aristotle is interested in the diversity of the *parts of cities*; from chapter 14 on, he considers the *parts of constitutions*.[23] But the zoological passage belongs to the first part of the text, not only because of its position well before chapter 14, but also by its very perspective, such as it appears in the two

22. Pellegrin, *Endangered Excellence*, chapter 7.

23. Cf. Pellegrin, "Parties de la cité, parties de la constitution."

sentences that frame the passage. The first one says: "We have said that there are several constitutions, and their causes. But we need to say why they are more numerous than those we have mentioned, what they are, and why, in taking as starting point what was indicated earlier; we in fact agree that there is not one single part, but several parts, that every city has" (4.4, 1290b21).

So, before the start of chapter 14, Aristotle bases the diversity of constitutions on the diversity of the parts *of cities*: "One reason that there are several constitutions is that every city has several parts" (*Politics* 4.3, 1289b27). These parts are, from a certain point of view, the communities that compose the city; Aristotle mentions here only families, and from another point of view the social groups that also make up the city. Every city, in fact, is composed of a rich class, a poor class, and a middle class, but also different trade groups, distinctions to which one needs to add others, by birth or by virtue. But there is no question, at this stage of Aristotle's reflection, of reconstructing a priori the possible forms of constitution: it's a matter of presenting the differences that exist between, for example, a democracy in which there is a formal equality between the common people and the notables (either because each individual is politically equal to every other person, or because the assembly of the people has equal weight with that of the assembly of the notables, which yields different species), all that under the authority of the laws, and a democracy in which the masses are above the law, and which has thus darkened into a demagogy. These are distinctions made according to political criteria, but Aristotle also adds to the analysis sociological characteristics, like the prevalence of the farming class, or the wealth of various classes. That's also the logic of the zoological passage, as the second framing sentence says clearly: "It's the same for the constitutions we have been talking about. Because cities also are not composed of a single part, but of several, as we have often said" (4.4, 1290b37). And Aristotle goes on to enumerate the necessary parts of every city: "one, the group concerned with food," "two, the artisan class," "three, the commercial group," "four, laborers," "five, the military class" (1290b39–1291a7).

To think, in contrast, that the foundation of the diversity of constitutions needs to be sought at the level of the parts of the constitution and not at that of the parts of the city, places us on a purely functional level: every part is defined by its functions independently of any sociohistorical content. That constitutes not only great theoretical progress, but also an important impact on the practical level, where it is as decisive in the case of the practical science that politics is. In fact, the legislator, in his reforming activity, can instigate a variation in each part and/or review the compatibility of the parts between each other, in a way that rebalances a constitution while keeping its form (by making an oligarchy more acceptable by the popular classes, for example) or by changing the form (transforming an oligarchy into a democracy, for example). It's when it gets to this level that Aristotle can put to work a *combinatorial* method, taking account of the parts of every constitution. But, we repeat, Aristotle is not yet there when he brings in the zoological passage, and even if it works well, as in the case of constitutions, to grasp the diversity of animals via the diversity of their parts, it's a good idea to look again at the theoretical ambitions of this passage in the *Politics*, from a deflationary perspective.

So, we must reread this zoological passage having abandoned the idea that it presents a methodological convergence with the last three chapters of book 4. Jules Tricot, in a note to our zoological passage in his French translation of the *Politics*, could put us on the road to a correct interpretation, in spite of himself. He figures that this passage runs counter to the method employed in the biological treatises because in those his method is inductive, while here it is deductive. The zoological passage in effect results, according to Tricot, "in the constitution of different species, thus appealing it seems, to a principle that can be phrased: 'when the individuals composing a group have all their parts similar, the group constitutes a species,'" this last sentence found in William Ogle's translation of the *Parts of Animals*, published in 1882.[24]

It is absolutely impossible that the method anticipated in the passage of the *Politics* consists in getting different sorts of animal (i) by varying the necessary parts according to the differences that they can offer; (ii) by combining the results of this operation with, as sole constraint, the principle of nonredundancy (and possibly an assumed condition of compatibility), while that is possible for constitutions. It's in this latter case, as we have said, the differences are very restricted in number, while in the case of animals, a procedure of that kind would yield an infinite number of parts and thus an infinite number of species. But then how are we to understand the words "that will yield species of animal, as many as the combinations of the necessary parts"? If Aristotle is not outlining an a priori procedure for a (re)construction of the animal world, it must be an a posteriori procedure.

24. Tricot found this citation in Aristotle, *The Politics of Aristotle*, vol. 4, 163 (Newman).

Thus, I have now revised my previous reading on two crucial points. First, the part of the sentence "the number of combinations of their [necessary parts]" does not refer to an a priori procedure of construction of the possible combinations of such parts, but the combination of parts taken from a set list. It's about the kinds of stomachs, sensory organs, and so forth, which are duly listed in works such as the *History of Animals*, and not kinds of stomachs obtained by variations of the characteristics that a stomach may have (larger or smaller, shorter or longer, more or less fleshy, situated in such and such a part of the body, etc.). Furthermore, not only is it the case that Aristotle cannot want to say that an animal can combine the stomach of a dog with the feet of a chameleon, but even combining parts relatively close (for example, the stomachs and feet of a dog and those of a fox) would lead to the construction of fantastic species and thus leave zoology behind. Second, one must "go up" from animal species to the combinations of parts, which brings us back to Tricot's commentary, but displaced. I think that this passage means that all species of animals can be understood as the combination of different nonredundant, doubtless compatible, necessary parts. So we must posit a thesis that is the reciprocal of Ogle's and say that two animals are different if they combine in each different necessary parts. This difference can be more or less profound.

One might almost paraphrase the lines 1290b34–37 thus: "For every viable combination of necessary parts there corresponds a kind of animal, and there is no kind of animal that is not a combination of necessary parts." Similarly, Aristotle continues, every particular city, whose constitution differs from the constitutions of other cities, corresponds to a viable and nonredundant combination of the form of the agricultural, artisanal, commercial, laboring, and military classes, this difference being greater or smaller. We can recall here Aristotle's analyses about the different kinds of military force according to their political leanings, notably in *Politics* 6.7: aristocratic regimes tend to have an army in which the cavalry predominates, or is at least important, and so on. It remains to be determined whether these five classes are necessary parts or if cities could do without any of them. And, even as there is more difference between a terrestrial mammal and a fish than between two fish, so there is more difference between a warlike aristocracy and a mercantile oligarchy than between two mercantile oligarchies.

In fact, in an unexpected way, the passage points toward the Aristotelian relationship, one that we have often invoked, between identity and difference. Aristotle in fact thinks that different animal species can have such and such a necessary part (and doubtless that goes for the nonnecessary parts as well) of the same form:

> The human stomach resembles that of a dog. In fact, it is not much bigger than the intestine, but looks like a kind of intestine that is a little wider. Then comes the simple intestine with folds, then a large intestine.[25] The lower intestine is like that of a pig, because it is large and the section that goes toward the buttocks is thick and short. (*History of Animals* 1.16, 495b24)[26]

And that is not the case for neighboring species like the wolf and jackal, cited in the preceding note:

> Much the same can be said for the stomach and intestines of oviparous quadrupeds, like the land tortoise, the lizard, the two varieties of crocodiles [aquatic and terrestrial, the latter doubtless a kind of monitor lizard] and in a general way all the animals of this kind. In fact, they have a simple and single stomach, comparable either to that of a pig, or to that of a dog. (*History of Animals* 2.17, 508a3)

Although drawn from the *History of Animals*, a passage like this doubtless should be read from the functionalist perspective of the *Parts of Animals*, rather than that of the *History of Animals*, which we have characterized,

25. This is the text of some manuscripts followed by Schneider. Most manuscripts read: "folded, large enough." But it is probable that Aristotle here distinguishes between the small and large intestines.

26. It's worth quoting a passage that attributes a kind of paradigmatic function opposing the stomach of a dog and the stomach of a pig: "Animals with two ranks of teeth have a single stomach, for example human beings, pigs, lions, wolves. As for the *thōs* [jackal?], it has all its internal parts like that of the wolf. Thus, all have a single stomach and following it the intestine. But some have a larger stomach, for example the pig and the bear (and the stomach of a sow has a small number of smooth folds), others have it much smaller, hardly bigger than the intestine, for example the lion, the dog, and the human being. For the others, the forms vary according to the stomachs of the animals mentioned. In fact, some have a stomach like that of a pig, others like that of a dog, and it goes similarly for larger and smaller animals" (*History of Animals* 2.17, 507b16).

in the previous chapter, as a treatise about animal diversity. The stomach and intestine of the crocodile are certainly not identical to those of the pig (or the dog), but from the point of view of their general appearance and of the way that they fulfill their function (with, in this case, a difference between the relative sizes of these two organs), one may think that they are of the same kind (the same "difference").

This passage in the *Politics* thus leads us to think about the idea of difference in zoology, which, like the ideas of *genos* and *eidos*, is also relative. There are animals that vary according to their necessary parts—a deer and a lobster, for example. Then there are some that differ only in terms of some aspects of their necessary parts, or even a single part: "The family of snakes is identical and almost all its characteristics are comparable to those of lizards among the oviparous quadrupeds, provided one think of them elongated and footless" (*History of Animals* 2.17, 508a9). The organs for local movement are in fact among the "necessary parts." The same goes for lobsters and spiny lobsters,[27] although the difference is a little fuzzier than the previous example, since the lobsters use their claws also for local movement (Aristotle says that this is an "unnatural" use of the claws), while spiny lobsters use them only for grasping (*Parts of Animals* 4.8, 684b1). But Aristotle would doubtless consider that as belonging to the same sort of comparison as of two birds differing only by color.

There remains a final source of error in the reading of our zoological passage in the *Politics* that we must avoid: a retrospective illusion can cause us to think that the expression εἴδη τοῦ ζῴου, occurring twice in line 1290b36, designates species of animals. One of the results of my *Aristotle's Classification of Animals* that I have not abandoned and has, in

27. Translator's note: *Homard et langouste*: The French *homard* is the standard English-language lobster, with large claws; it belongs to the family Nephropidae. The French *langouste* is, in American, known as "spiny lobster" or "rock lobster," although not closely related to the true lobsters; it has no or minimal claws. It is in the family Palinuridae. *Parts of Animals* 4.8 distinguishes four families of crustacea: transliterating: Caraboi, Astakoi, Karides, Karkinoi. Liddle and Scott's *Greek-English Lexicon* gives: crayfish, lobster, shrimp, crab (see https://en.wikipedia.org/wiki/A_Greek%E2%80%93English_Lexicon). Astakoi are *homard* or lobster. What are the Karides at *Parts of Animals* 4.8? They "have a tail" and "do not have a claw" (684a15). Pellegrin thinks *langouste*, or spiny lobster (rather than shrimp?), plausible for the comparison he makes in the text, except that spiny lobsters don't have (noticeable) claws. "Crayfish" seems odd for Caraboi, since they are rather similar to lobsters, except that they live in fresh water.

fact, been generally accepted, is that, when Aristotle uses the terms *genos* and *eidos*, he is not designating a particular level of generality. Here, in particular, εἴδη τοῦ ζῴου are not more restricted than the γένη ζῴων of 1290b33. We find that it is not at the level of species that this method that makes a correspondence between a combination of necessary parts to an animal family can find the most relevance for zoology, but in reference to larger classes. And when one liberates oneself from the false idea that Aristotle tries, in this passage in the *Politics*, to elaborate a method for reconstructing animal species, one can see that a method of the kind developed in this passage in the *Politics* is one that is also employed, at least sometimes, in the biological treatises. If one takes the example of mollusks, one may see it by citing the passages in the *Parts of Animals* that deal with the three necessary parts that they have:

> Mollusks have around what one calls their mouth two teeth, and in the mouth, instead of a tongue, a fleshy part by which they can distinguish what is pleasant in what they eat. . . . After the mouth, in mollusks there is a long throat, after which there is a crop, as in birds, then, going on, a stomach, followed by a simple intestine going to the anus. (4.5, 678b7)

> For the parts concerned with nutrition, which all animals must have, it goes the way that has been said, but obviously there must be a part analogous to that which exists in blooded animals for governing perception, for that must belong to all animals. So, in mollusks there is a liquid found in a membrane through which the throat extends in the direction of the stomach; this part is attached rather to the back, and some call it *mytis*. (4.5, 681b13)

> We spoke earlier about the internal parts of mollusks, as of those of other animals. On the outside they also have the trunk that contains their body, with no definite shape, and the feet in front of this trunk, around the head, between the eyes and around the mouth and teeth. Other animals that have feet, then have them, some in front and back, for others along the sides, like the nonblooded polypods. But the family of mollusks has characteristics that are peculiar to them in comparison with those [the bloodless polypods], for they have their feet on the

> side that one may call the front. The cause of that is that their rear is joined to their front, like in the spiral-shelled shellfish. (4.9, 684b6)

Thus, the family of mollusks, one of the great families of nonblooded animals, is fairly well characterized. For their nutritive parts, they have a fleshy organ that is the analogue of a tongue, a long throat and crop, and a simple intestine; for central sensory organ they have a liquid in a membrane; as organs of local movement, feet arranged appropriate for their structure. But, on the one hand, that does not allow us to place all the mollusk species in a chart, and on the other hand, a construction like that is not always possible. Thus, in insects, another of the four great classes of nonblooded animals: "The parts concerning nutrition are not the same in all [insects], but there are many differences" (διαφορὰν ἔχει πολλήν) (*Parts of Animals* 4.5, 682a9). For insects, it would then be necessary to apply the "descending" method of the *Politics* toward narrower classes, in the direction of species, a procedure that is not completely impossible, but much more difficult: one would have to distinguish wingless insects from those that have more or less numerous wings, and the same for parts devoted to nutrition and sensation.

Because the method anticipated by the zoological passage in the *Politics* can also be applied to species. Thus, if we consider certain varieties (γένη) of eagles enumerated in *History of Animals* 9.32, the eagle called "hare-killer" and the eagle called "duck-killer" have, on the background of dominant identity (which allows them all to be called eagles), differences that affect all their parts and thus all, and especially, their "necessary" parts. But in the unchanging world of Aristotle, in which living species cannot improve themselves, from the fact that Nature always brings about the best, one finds that the "hare-killer" is a "perfect" animal, in the sense we have seen, which means that the stomach of a "hare-killer" must only be combined with the wings and eyes of a "hare-killer" because it is in this precise combination that Nature "brings about the best." In fact, that's probably what our text says, at least between the lines: to affirm that "the same animal cannot have several varieties of mouth (πλείους στόματος διαφοράς) nor, similarly, ears," surely means that a dog cannot have at the same time several kinds of mouth. But we have seen that that is not always true, since ruminants have several kinds of stomach. But the sentence in question can also mean that an animal cannot have any one kind of stomach and the variety "hare-killer" can only have the "hare-killer" kind of stomach. Cuvier takes up that sort of program and pushes it much further than Aristotle when he declares that "whoever has rational control of the laws of organic economy can remake every animal," starting from one organ or one bone.[28] Cuvier's method is, for sure, strongly inductive.

Thus one may apply to each class of animal, no matter the degree of generality, Ogle's analysis: when two birds exhibit the necessary parts proper to the "hare-killer" eagle, these two birds belong to the *eidos* "hare-killer eagle," which is found to be what we call a species, just as when two animals exhibit the same digestive, sensorial, and local movement organs characteristic of mollusks, they belong to the *eidos* "mollusks," which we consider a phylum. In contrast, that would be impossible for insects, but it is possible for the *eidos* Hymenoptera or that of ants. All these *eidē* are thus constituted in a way that combines nutritive, sensory, and local movement parts of a certain kind. The Aristotle who wrote the passage in the *Politics* is thus the same Aristotle who was the author of the zoological corpus. To finish with the passage from the *Politics*, we must make two more remarks. First, if it were a matter of an a priori reconstruction of animal species, one might imagine that one could construct a species of bird that would have the wings of a hare-killer and the stomach of a duck-killer, and it would be viable. But there's no way that Aristotle would call that one of the εἴδη τοῦ ζῴου, because in Aristotle's animal world no class of virtual animals exists. Secondly, something that is surely missing from the zoological passage is the idea of a continuous variation that would allow us to declare that the set of animal species itself forms a continuous whole.

Can we attribute to Aristotle the thesis that the living world, and thus the animal world that is part of it, are complete wholes such that nothing can be added to it, and nothing taken away? We can certainly credit Plato with this idea, since the *Timaeus* answers clearly that the largest classes of living things are four (celestial gods, winged, terrestrial walkers, aquatic) because these four species are the ones in which the divine intellect "discerned the presence in the what it is to be an animal" (*Timaeus* 39e), which means that one may be sure that the animal world is indeed complete because it reproduces a model. A. O. Lovejoy, in his very celebrated 1936 work, already cited, *The Great Chain of Being*, reason-

28. Cuvier, *Recherches sur les ossements fossiles de quadrupeds*, 100. "In a word, the form of the tooth entails the form of the bone joint; that of the shoulder blades that of the fingernails, exactly as the equation of a curve entails all its properties" (100).

ably attributes this doctrine to Plato, and refuses it, with no less reason, to Aristotle, whom he credits, in contrast, with the idea of a continuity between living things. Of course, Lovejoy had read enough of Aristotle to realize that the Aristotelian divisions in the animal world rest on many kinds of differences, which means that a given being can be "superior" in one respect, but "inferior" in another, to another being. But, according to Lovejoy, "any division of creatures with reference to some one determinate attribute manifestly gave rise to a linear series of classes."[29] Lovejoy presents the historical development of this doctrine of the great chain of being from antiquity to the nineteenth century; he thinks that Aristotelian continuity in some way replaced Platonic completeness. And, in fact, there are in the Aristotelian zoological corpus some series that do indeed seem to roll out in a continuous way. David Lefebvre has analyzed[30] a particularly spectacular example found in the first chapter of *Generation of Animals* book 2, which distinguishes five kinds of animal generation, going from that of vivipara to that of larvipara, kinds that are "well arranged consecutively" (εὖ καὶ ἐφεξῆς, *Generation of Animals* 2.1, 733a33). So, we need to trace completeness and continuity in Aristotle's biology.

But I intend to rely on two other passages to see whether one may attribute a continuist position to Aristotle, all the more because one of them has, one may say, completeness accents. Of course, David Lefebvre has mentioned these two passages, analyzing one of them closely. The first is from the *Generation of Animals*, and has already been partially cited; the other is in *On Respiration*. This order is purely for convenience, and could be reversed.

> [Sea water] is much more corporeal than fresh water; it is warm by nature, and shares in all the parts, humidity, breath, and earth, so that it shares with all animals born in different places a tie to each of the elements. Because one may assign plants to the earth, aquatic animals to the water, terrestrial animals to the air. On the other hand, the more and the less, the nearer and the farther, produce large and surprising differences. As for the fourth class, we shouldn't seek it in these places. Nevertheless, we should demand one that occupies the place of fire, for that's what we count as the fourth of the

29. Lovejoy, *The Great Chain of Being*, 56.

30. In his contribution to a 2019 colloquium on "Biology and Psychology in Aristotle."

> [simple] bodies. But fire always appears in a form that is not its own, but is found in another body, because it's air, smoke, or earth that appears to be enflamed. But one should look on the Moon for the class in question, for it does seem that the Moon shares in the fourth element distantly. But that would be a topic for a different work. (*Generation of Animals* 3.11, 761b9)

> The human being is the only one whose top of the body is directed toward the height of the universe, because it is such that it has this part [the lung] (διὰ τὸ τοιοῦτον ἔχειν τοῦτο τὸ μόριον). So, one must count this part, for the human being as for the other animals, as a cause of their essence, on the same basis as any of the other parts. That is the reason for the existence of the lung. As for its necessary cause and efficient cause, one ought to think that animals of this kind are constituted in the same way as those not of this class. Some, in fact, are made with more earth, for example the family of plants, others of water, for example aquatic animals; for flying animals and terrestrial animals, some are made with more air, others with more fire, and each finds its level in the places that are appropriate for it. (*On Respiration* 19, 477a21)

These two passages ought to be considered together because they both offer an approach to the diversity of living things, animals and plants, considered in their relations with the four elements that constitute the sublunary world. One might then suspect that these two passages might be able to furnish a basis for the application to the world of living things the principle of completeness. But their approaches are different.

The passage taken from the *Generation of Animals* considers two different questions, one after the other. Up to 761b15, Aristotle notices that sea water is richer in the elements than other environments,[31] to say next that the sea "shares (μετέχειν) with all animals born in different places" (761b11) because of the presence of the different elements in these animals, which means that the sea has an elementary communion with all animals wherever they live. This assertion has often been misunderstood. It does not mean that the sea is one of the components of all animals, nor

31. Aristotle speaks of "the humid, breath, and earth." Perhaps we should take breath as a mixture of air and fire. See later discussion.

that it includes in itself all the forms of all animals,[32] and even less that it is the origin of all living things, an Anaximandrian idea that Aristotle is not interested in sharing, as we have seen. That can be seen in the fact that (cf. the γὰρ at 761b13, which explains that which follows rather than what precedes, a usage well known to Greek grammarians) "we may assign plants to the earth, aquatic animals to the water, terrestrial animals to the air." "Assign" translates τίθημι with the genitive that we can call "attributive," a usage for which the Liddell and Scott dictionary gives just one example, drawn from Demosthenes (*Olynthian* 1.10): τῆς ἡμετέρας ἀμελείας ἄν τις θείη δικαίως, "one may justly attribute to our negligence" (it's a matter of losses undergone during the war). The sense is doubtless not as decisive here, because it is hard to think that Aristotle is going to hold the earth responsible for plants, and so on, and if it's an attributive genitive, probably it's a feeble attribution: Aristotle does not claim that plants and animals are *produced* by the earth, aquatic animals by water, and terrestrial animals by air, a thesis that, in contrast, is found in the passage in *On Respiration*.[33] One may perhaps rely on a vague relation and translate τὰ φυτὰ θείη τις ἂν γῆς by "plants have a relationship (or something to do) with the earth," but we ought to be able soon to make things more precise. A general consideration: Aristotle includes under the title "terrestrial" (πεζά, footed) animals that walk and fly, which differs from his usual usage of the term.

But why does Aristotle make this remark about the elementary richness of sea water? Because, in the examination that he is leading about spontaneous generation of shellfish and other marine animals, it is important to underline that sea water can play the role of a genetically active milieu in the formation of living things of this kind. Then comes a line

32. "It includes samples of all beings that properly belong to each of the elements" (Louis), a thesis that is all the same hard to accept . . . David Lefebvre translates ὥστε καὶ πάντων μετέχειν τῶν καθ' ἕκαστον γιγνομένων ἐν τοῖς τόποις at 761b11 by "in such a way that it also has its share in all the animals born in different places," which still goes, though less than the Louis translation, in the direction of the idea that the sea contains animals of all possible forms. To me, the passage posits no more than an elementary community.

33. Platt, followed by Peck, is perhaps not wrong to suppress the reference to places, which could have been introduced here to make this passage agree with others like that in *On Respiration*. He translates: "[Sea water] has a share in all the parts of the universe, water and air and earth, so that it also has a share in all living things which are produced in connexion with each of these elements."

and a half (761b14–15) that has prodigiously stimulated the imagination of commentators, some of whom have seen in it a summary by Aristotle of a projected *scala naturae* within which living things would go from one to the other by a kind of continuous deformation: "the more and the less, the nearer and the farther, produce large and surprising differences."[34] If a teaching like that were proven, Lovejoy would be right. But it is too much to ask from this remark. It is very important to understand that these words put in parallel different living species and elementary combinations of the environment in which they are born, but it does seem that this remark principally, and doubtless uniquely, concerns the case of marine animals that reproduce by spontaneous generation. In the case of marine shellfish, what explains that mussels and clams are formed here, besides the vanishing "male" principle that we discussed in the previous chapter, is also the elementary composition of the place where they are born. For the sea has an extraordinary elementary richness, because it contains earth, water, and breath in varying proportions. Due to this fact, the sea is a propitious place for the engendering of animals, and that is why, as we have seen, it "shares" the elementary realities "with all the animals that are born in different places" (761b11). And in fact, the slight variation in the elementary composition of the silt from one place to another leads to remarkable differences in producing in this place oysters, and one may add, these kinds of oysters, in another place mussels of such and such a sort. Aristotle adds that this is also the case, but to a less marked degree, for different sorts of fresh water, rivers, swamps, and so on.

After this misleading sentence, Aristotle re-asks the question about the diversity of living things in relation to the elements of the sublunary world, and reveals his true perspective, indisputably introducing the question of the *completeness* of the living world, while finally letting us understand what it means "to assign plants to the earth, aquatic animals to the water, terrestrial animals to the air." For as there are four elements there needs to be a kind for fire. Living things, especially animals, can contain fire, because, as the second-to-last sentence of our passage says, "fire always is present in a form that is not proper to it" (761b18), that is, not as fire, but mixed with other elements, which makes smoke, hot air, or hot earth. Some animals can even be born and live in a fiery environment, since the *History of Animals* says that there are certain winged bugs that form in molten copper in Cyprus, and that the salamander is

34. Τὸ δὲ μᾶλλον καὶ ἧττον καὶ ἐγγύτερον καὶ πορρώτερον πολλὴν ποιεῖ θαυμαστὴν διαφορᾶν.

among those animals "constituted in a way that it does not burn" (5.19, 552b15), which presupposes that there are others.

But far from saying that these fiery places can accept animals like salamanders, Aristotle is posing the question of the relationship between animals and elements from a cosmological perspective. In effect, what Aristotle seems to want to say here is that all the *regions* of the world should include animals. In effect, to speak of the "fourth separation" (τῆς τετάρτης ἀποστάσεως, 761b22) about fire is a clear reference to the Aristotelian structure of the sublunary world as described in *Meteorologica* 1.2: "The four [elementary bodies] being fire, air, water, and earth, the one among them that occupies the highest position is fire, the lowest position is earth, the two others being between them in similar relationships, for air is closer to fire and water to earth" (1.2, 339a15). An image of the world that would for centuries inspire representational images of the universe in illustrated medieval manuscripts, presenting the world, including the sublunary part, as a nesting of concentric spheres. From which the reference to the Moon. There is completeness, because, even though lunar animals have never been observed, "it is on the Moon that one should seek the kind in question" (761b21). This is not a matter of a natural science argument, which would require that the phenomenon studied first be observed, but a logical procedure (one might say a priori). When, in fact, Aristotle distinguishes, in the *Movement of Animals*, two cases that it is impossible to see, he gives sound, which necessarily cannot be seen, and "things on the Moon" (ἐπὶ τῆς σελήνης, 4, 699b19), which are invisible "for us," he does not want to say that someday the Moon creatures will be visible. To go to the Moon is for Aristotle something that is definitively impossible, so that if we don't see Moon creatures today, we will never see them. So, therefore, what Aristotle here asserts is the necessity that every region of the sublunary world, four in number, can shelter living beings. Thus, there is indeed completeness, but a cosmological completeness. But not all living things can be assigned to the same sublunary region. Terrestrial animals, for example, are assignable to the airy level of the sublunary world. We also should note that the fact of saying that all the regions of the sublunary world should include life does not amount to saying that *everywhere* one should find living beings: for example, there are in the earthy sphere of the sublunary world vast rocky formations that are empty of all life.

This assignment of living things to the Moon considered as the place for the fourth element has both posed serious interpretative problems and encouraged sometimes fantastic hypotheses, ever since antiquity. I need to make two remarks about the Moon as a fiery place. First, in Aristotle's cosmology, the Moon is a celestial body and is not composed of fire, but of ether. If the Moon were fiery, the only place it could be would be in the tangential part of the last sublunary circle, the circle of fire. Therefore, the expression ἐπί τῆς σελήνης (761b21) needs to be understood as signifying "in the region (or the environs) of the Moon," which is grammatically fine, but is going to conflict with the sense that this expression seems to have elsewhere, for example at *Movement of Animals* 4, 699b19, cited earlier, where it indeed seems to mean "on the Moon." Although . . . Next, if Aristotle is making an allusion to hypothetical animals living in the region of the Moon, that region, like every other region where animals live, is not entirely fiery, but just mainly fiery.

That's what a difficult passage in the *Meteorologica* seems to me to suggest: "As for the body on high that extends as far as the Moon, we say that it is different from fire or air, but it is nevertheless true that here one of the parts is purer, there another part is more mixed, and it varies, especially in the area where it leaves room for air and to the world that encircles the Earth" (1.3, 340b6). One wonders whether Aristotle is talking about the space between the sphere of fixed stars and the Moon, or the space that extends from the center of the Earth to the Moon, included. In the second part of the passage, in any case, he designates the region of junction (interface) between the supralunary and sublunary worlds. It's in the most mixed region (the text says, "less unmixed," ἧττον εἰλικρινές, 340b8) where one might find animals. There remains a difficulty: this "most mixed" region actually seems to be a region, or a part, *of the ether*. As, in any case, animals have to be found *somewhere*, because they can't float in space, they have to be on the Moon, but at the common limit of the orbit of the Moon with the fiery region of the sublunary world, so they would have, one may say, their feet on the Moon and their head in the fiery region, something that seems to me confirmed by the sentence at 761b22: "For the Moon seems indeed to be in community with (κοινονεῖν with the genitive) the fourth separation," that is, the sphere of fire.

We must at least mention the speculations that have been generated by this reference by Aristotle to the Moon as a place with life. In a very erudite article, William Lameere,[35] relying on the texts we have just cited and some others, taking as proven that Aristotle's lost works were

35. Lameere, "Au temps où Franz Cumont s'interrogeait sur Aristote."

juvenilia, and adopting a chronological hypothesis à la Jaeger (but relying more on François Nuyens), concludes that the young Aristotle thought that space at the limit of the sublunary world, that is, in the region of the Moon, was peopled by living things who were not like the animals that we know, nor divine, but of a demonic nature. To be "demonic" is to be superhuman but nevertheless not divine.[36] The most solid foundation, but not a remarkable solidity, of W. Lameere's reading is a passage in Michael of Ephesus's commentary (published under the name of John Philoponus) on the *Generation of Animals* 3.11, 761a13, which I cite following his translation:

> Fire is not determinate in its nature, as it occurs in our regions; it is only in the superior regions, those that are close to the sphere of the Moon (here's what I mean by the completely pure element of ether: it is the region contiguous with the sphere of the Moon, and is unified with it, up to a certain height); consequently, the living beings that are born there and have been born, reside in the sphere of the Moon. There exist, in fact, and are born distinct living beings, endowed with reason, ethereal in their bodily nature, who neither eat nor drink, and whose only occupation is visual and contemplative: they have ether and air for their home. And each of them can live more than three thousand years, but they are subject to death. These matters, truly, can be dealt with elsewhere. (CAG XIV.3, 160.11)

Michael of Ephesus attributes this sort of doctrine to Plato, whom he considered the author of the *Epinomis*, but without telling us by name on what authority he says that Aristotle shares this doctrine.

In order to rest his thesis on "authentic" texts of the Aristotelian corpus, W. Lameere cites, among other passages, a famous and surprising text in the *Generation of Animals*:

> Every soul seems indeed to share its power (δύναμις) with a different and more divine body than what we call the elements.

36. Cf. two remarkable passages, one that says that dreams are not sent by the gods "but are demonic, for their nature is demonic" (*Divination in Sleep* 2, 463b12), the other defining as "demonic" those acts by nature or chance without depending on human will (*On Dreams* 1, 453b23).

> Just as souls differ from each other in respect of dignity or indignity, similarly its nature differs from [the elements].[37] For in the seed of all animals there is that which makes all seeds fertile, what one calls the hot. But this is neither fire, nor a power of that kind, but the breath that is enclosed in the foam of the seed as well as the nature that is in the breath, and which is analogous (ἀνάλογον) to the element of which the stars are made. That is why fire does not generate any animal, nor does any seem to be constituted[38] in fiery places, whether humid or dry. In contrast, it is what makes the heat of the Sun and of animals, not only that which is transmitted by the seed, but if there exists also a residue different in nature that too possesses a vital principle. It is thus clear that the heat that is in the animal is neither fire, nor does it have a principle derived from fire. (2.3, 736b29)

According to a passage like that, for all animals (and not only for those animals that may be found on the Moon), that which makes their sperm fertile (and at 737a4, Aristotle says that it's the same for animals that reproduce nonsexually), it is, among several other factors not envisaged here, the heat of the Sun[39] and animal heat that are transmitted in the breath that is the foam of the sperm (or in its analogue). In fact, this passage has been accused of having a doctrine that it does not include because of the allusion to "the element from which the stars are made," that is, ether. But it's a well-known position of Aristotle that the celestial bodies are living immortals, and this passage in the *Generation of Animals* establishes an analogy (the word ἀνάλογον is in fact in the text at 736b37) between these living immortals and the living things that we have around us, an analogy, that is, in Aristotelian terms, a relationship in four terms: that which ether is for the celestial bodies, the hot breath of the sperm (or its equivalent) is for mortal animals, because ether and breath are both principles of animation for the living things. It's the same doctrine that

37. The interpretation of ἡ τοιαύτη διαφέρει φύσις is difficult. One might possibly understand that it is the nature of the soul that differs from the rest, or that the body in question itself or the soul itself admits of differences.

38. The text says, "nothing seems to constitute itself," but we may assume that it's a question of living beings.

39. Cf. *Physics* 2.2, 194b14: "Man engenders man, and also the Sun."

one finds in Cicero's *On the Nature of the Gods* 2.42, typically assigned to Aristotle's lost *On Philosophy*:

> As, then, some animals are generated in the earth, some in the water, and some in the air, Aristotle thinks it ridiculous to imagine that no animal is formed in that part of the universe which is the most capable to produce them. But the stars are situated in the ethereal space; and as this is an element the most subtle, whose motion is continual, and whose force does not decay, it follows, of necessity, that every animated being which is produced in it must be endowed with the quickest sense and the swiftest motion. The stars, therefore, being there generated, it is a natural inference to suppose them endued with such a degree of sense and understanding as places them in the rank of Gods.[40]

There's no suggestion here that possible demonic animals live in the heavens, but celestial bodies, everlasting beings composed of ether.

We come to the passage in *On Respiration*. This text does not lack obscurities; fortunately, we are not obligated to clarify all of them for what we are interested in. Once we have discovered the final cause (which is to cool the organism), it's a matter of figuring out the "mechanical" causes of the nature of the lung. So, we have here Aristotle's usual connections between final causality and material plus efficient causality, and it's not so much an explanation of the generation of the lung as of its composition and conformation (the word συνέστηκεν, "have been constituted," at 477a27 is not about a process, but a state). It seems to me difficult to see, as Jules Tricot does, if the expression "the animals of this kind" refers to animals that have a lung, to relate "those that are not of this kind" to nonliving realities. He would have to translate, "many *beings* that are not of this kind," which seems to correspond less well with the cadence in the Greek sentence (καὶ τὰ τοιαῦτα . . . καὶ μὴ τοιαῦτα, 477a26–27). It seems more natural to think that those "who are not of this kind" are animals having a cooling system other than pulmonary.

Before our passage, Aristotle has explained at the same time the cause of the fact that some living things have a pulmonary respiratory

40. Translator's note: Cicero, *Nature of the Gods* (Yonge).

system, and why having a lung as a means of cooling is a sign of greater "dignity" (cf. "the more dignified animals," τὰ τιμιώτερα τῶν ζῴων, 19, 477a16). The reason is that these animals are hotter, the human being, being the animal with the purest and most abundant blood (doubtless he means "the most abundant in the lung") and whose upper parts are aimed toward the upper parts of the universe. Then comes our passage; we may be surprised that it has not become one of the cardinal references for those who study Aristotelian etiology.

We need to understand the meaning of the expression "in the places that are appropriate" (ἐν τοῖς οἰκείοις τόποις, 19, 477a30). To do that we need to consider what happens immediately after this passage, at the beginning of paragraph 20 of *On Respiration*: Aristotle here criticizes Empedocles, who thought that the animals containing the most fire would live in the water, because the water would temper their natural heat. Aristotle rests his criticism on two arguments. The first, which says that the animals in question would have had to leave terra firma to go live in the water, is historically remarkable in that it proposes, in the form of an absurd hypothesis (for Aristotle, no animal species can change its life environment), a schema the reverse of that proposed by Anaximander, who imagined a marine origin for all animals; the second calls attention to the fact that aquatic animals are not hotter, but less hot, than terrestrial animals. It would be better to think that "those with a wet constitution live in a wet environment, those with a dry constitution live in a dry environment" (20, 477b24).

One may thus conclude that in our passage, Aristotle maintains, as he did in the first text, taken from the *Generation of Animals*, the thesis according to which living things, whether plants or animals, occupy different places characterized by at least the preponderance of one of the four elements that all and only constitute the entire material universe, and that, as the passage from *On Respiration* adds, if the place where these animals are found is "appropriate," it's because—a very strong thesis that Aristotle introduces in *On Respiration* 19 and 20—these animals are constituted of the same material as their environment, that is, they are characterized by the preponderance of some one element. But there is a necessary relationship between, for fish for example, to live in the water, to be constituted of "more water," and to have gills, because lungs would not be convenient for them. Thus, Aristotle explains here the presence of gills, but also the absence of a respiratory system in plants, not because it is "better," but following *necessarily* from the material composition of their life environment.

Before drawing any conclusions from all this, we must remove two hermeneutical doubts, or obstacles, for there are two difficulties in the assertion that "for flying animals and terrestrial animals, some include more air, others more fire." First, most translators understand that flying animals include more air, and terrestrial animals more fire, while it is a good deal more likely that, always following the Aristotelian method of chiasmic exposition, it would be the contrary, the flying animals having more fire, and the terrestrial animals more air,[41] which would suppress a contradiction between our two passages, since the one would say that terrestrial animals are "assigned" to the air, and the other that they are constituted of air, and thus situated in regions mainly airy. In fact, this resolves the contradiction only halfway, because the passage in the *Generation of Animals* calls "terrestrial" both walking and flying animals. Next, one may well ask oneself what Aristotle means when he asserts that aquatic animals are mainly constituted of water, and, more difficult to understand, that terrestrial animals are mainly constituted of air, and winged animals of fire.

Animals with horns, like bovines, about which we have said quite a lot already, are largely made of earthy matter, and all terrestrial or flying animals have flesh made entirely of earth and water. Plants, on their side, are mainly made of water. The interpreters are doubtless wrong to think that Aristotle wanted to say that, among the elements of which they are composed, terrestrial animals include above all air, and plants contain an earthy part more elevated than their share of water or air. In fact, saying that plants are ἐκ γῆς πλείονος συνέστηκεν (19, 477a27) means rather that they have more earth than other living things contain. From that, the fact that terrestrial animals contain "more air," meaning than plants or fish implied, is understandable, if only because their system of cooling, which is respiratory, make them include much more air than other animals. Similarly, perhaps one may suppose that birds, that generally drink little,[42] have an internal heat higher than that of many other animals and, thus, have a fiery nature, more fiery anyway than those of the others, fish and bovines.

But why are fish both relegated to water and composed of more water than other animals? One might at first think that we have here an

41. That's how Michael of Ephesus understands it (CAG XXII.1.137.30).

42. Cf. *History of Animals* 8.3, 593b30: "The family of birds, as a group, drinks little, and raptors indeed drink not at all."

example of Nature always achieving the best. But the treatise *On Respiration* sends us in another direction, one that was suggested earlier: if fish have this relationship with water, it is for the same causes (material and efficient) for which the lung is as it is in animals that breathe, and it was in fact after his examination of the "mechanical" causes of the lung that Aristotle proposed his theory of appropriate places for different classes of living things. This relationship of animals with the elements is thus not explained by a final causality, which means that it is not a demonstration of the excellence of Nature. It's for a "mechanical" reason that animals with horns need air to cool themselves, and thus have more air in their composition than do birds. In the same way, we have previously noted in the case of fish that their possession of gills follows *necessarily* from their material composition and from the place in which they live.

The perspective of this passage is therefore not cosmological, but the distinction that is made is not from the point of view of the places, but of that of the animals. And, lastly, what this passage establishes is an elementary affinity between living things and the places where they normally live. That does not give theoretical means for discovering all possible species, nor even all possible classes of animal, even less to arrange them in a series. Furthermore, given that Aristotle thinks about animals like a real zoologist, and not with an apologetic or metaphysical goal, one must certainly apply a "flexible" reading of a text like this. To say that water is appropriate for aquatic animals because they have more water in their constitution than birds do is true at different levels. As, in fact, the first chapter of the *History of Animals* tells us, among aquatic animals, some breathe air, others give birth on dry land, and so on. To them too, the water is more appropriate than to birds. In the same way, air can be more appropriate to birds when it is warm and thus is in harmony with their fiery nature. Similarly, in the *Generation of Animals* passage, the assignment of different kinds of animals to places characterized by the preponderance of an element is itself of variable force. That passage speaks, in fact, of "all animals that are born in different places in relation with each of these elements" (761b11).[43] But it seems difficult to say that terrestrial animals are born in the air, in the sense of suspended above the earth. A passage that seems to contradict this one might help us. In the *Meteorologica*, Aristotle remarks that "there are no animals except on the earth and in the water, none in the air or in fire, because these

43. Πάντων μετέχειν τῶν καθ' ἕκαστον γιγνομένον ἐν τοῖς τόποις ζῴων.

are the materials of their bodies" (4.4, 382a7). Pierre Louis is doubtless right to understand that Aristotle means to say that no animal lives only in the air or in fire. So, it seems that there are two degrees of belonging for living things in places determined by the elements: plants, with few exceptions, live truly *in* the earth, and fish live *in* the water, while cows live *on* the earth and birds live, on occasion, *in* the air. Perhaps, then, we should not translate the word γιγνομένων at 761b12 as "that are born," but something like "lead their life." In any case, the relationship between animals and places in the sublunary world is far from being rigid.

From this (too) long analysis of Aristotle's texts, then, we do find the idea that the universe, at least the sublunary part, is "full of life," that like all the realities of this world "down here" animals are composed of all the elements that constitute the universe, and this "all" has a distributive sense. These animals tend to live in the places they resemble from the perspective of elementary composition, their assignment to a given place thus having "material" causes, and not deriving from an intention of rational Nature. This sort of position is a stranger to, and even in contradiction with, the idea according to which living nature is complete and missing none of its components. That thesis cannot be attributed to Aristotle, neither in its strong form, which would say that no species of animal can be missing, nor even in its weak sense, which would say that every animal is appropriate, or assigned, to a place and all places necessarily need to be assigned to living things. This assignment is in fact present in a flexible form: it's because air can be hotter than water that there are animals that both are hot (and thus related to fire, which, we remember, never is present in a pure form in living things) and enclose little water that birds are *appropriate* to the air. But borderline cases and exceptions are likely to be rather numerous. So, what about human beings? Aristotle actually believes that they are the hottest animals: should we then say that they are appropriate to fire like birds, or to air like cows, with which they share the general conditions of their existence, normally walking on earth and breathing air? We need to repeat, again, that Aristotelian and Platonic procedures are opposed on this point. Plato determines the number and situation of living things "from above," because animal diversity reflects the aspects of "life itself," while Aristotle proceeds "bottom up": *one finds* that some living things have more elemental fire than others, which makes them more inclined to live in hot regions, unless other, stronger, factors direct them toward colder regions.

In contrast, the *Generation of Animals* passage indeed introduces the idea of completeness, not of the living world, but of the sublunary part of the universe. In all the sublunary regions, which are spheres nested inside each other, one should, as a matter of fact, find living things. Doubtless that contributes to the recognition of the perfection of that part of the world, since, as we have seen, to be alive is a sign of perfection. From all the speculations of the commentators who, from antiquity until today, have wanted to see in the hypothesis of Moon animals divine or semi-divine living things, one may draw the impression that Aristotle had the idea that the presence of life at each sublunary stage imitates the everlasting life that belongs to the celestial bodies composed of ether. We have, in fact, seen that the *Generation of Animals* established an analogy between the supralunary life of celestial bodies and sublunary life (cf. ἀνάλογον at 2.3, 736b37, commented earlier).

So let us summarize. On the question of the number of living species, Aristotle separates himself completely from Plato because in the *Timaeus*, living things are *parts of the world* (cf. *Timaeus* 30c4), and thus if one living thing is missing, the world necessarily is missing one of its parts. In the same way, we have seen that the distinction between the four great kinds of living things (divine, winged, terrestrial, aquatic) *is read* in the very idea of "Life Itself." I have tried, earlier in this chapter, to show that there is nothing like that in Aristotle, and that for him living things are *in the world*, and since the world is composed of four elements, it is not surprising either that living things inhabit different places in the world that are composed mainly of a given element, nor that a place like that spontaneously gives birth to animals that differ according to the elementary composition of the place, nor that living things themselves would be composed of variable mixtures of the four elements. But we don't see anything in Aristotle that could be the basis of a necessary existence of a given species of animal. After all, in Aristotle's universe, it would be possible that the only existing living thing would be the universe itself. Aristotle *observes* the existence and the diversity of animals, he does not *deduce* it. The kind of harmony that would make of living things, particularly animals, a "great body" is not found in Aristotle. In zoology more than anywhere else Aristotle means to stick to the observation of existence: one cannot say *why* there is an octopus with narrow tentacles, nor *why* the triton (axolotl) is so badly made, but neither, in a general way, why the octopus or the triton exists. Perhaps this refusal to ask why there is being rather than nothing is a synchronic version of the refusal to ask about the absolute origin of things, another version of the Aristotelian stop sign. A metaphysical-apologetical approach like that of Plato could allow *deducing* the number of living things, while the approach of *biologists*, like that of Aristotle, allows only observation, given the number of accidents, strange events, and exceptions that a biologist must account for.

We can complete these remarks about animal diversity with a few others about the harmony that Aristotle finds in the animal kingdom, for diversity and harmony are in a relationship with the perfection of the world of living things. Let us start by repeating a few results. Aristotle's general picture of the animal world depends strictly, as we have seen, on the general position that he adopted to surmount the Parmenidean critique. Thence a zoology fundamentally "immobile." Not only do species seem not to evolve, but they don't adapt: they *are adapted*. The camel did not develop a hard palate because it had to eat spiny food, but it has always inhabited places where the food is spiny, and possessed a hard palate, the conjunction of the two being a proof of the excellence of Nature. We have also seen that the texts do not allow us to yield to the temptation of a "full" reading, under one of the three forms that that temptation can take: Nature does not actualize all possible animal forms, as one might be tempted to believe on the basis of a quick reading of *Politics* 4.4, Nature has not distributed animals according to a scale of growing complexity and perfection, Nature has not varied the material composition of animals by a continuous variation of their elementary composition and/or the elementary composition of their environment. There are, to be sure, animal sequences in which one goes from one species to another by way of small differences, as among the crustacea. There are also scales of beings that can be characterized by the fact that certain animals are more "perfect" than others in reference to such and such a point. But that sort of scale, which applies more to families of animals than to species, is *observed*, without obeying a plan, for which, in any case, there wouldn't be an instance to conceive it. Nature has not even explicitly wanted animals to occupy every natural environment and to possess a nature *completely* appropriate to their living environment.

The Revenge of the Special on the General

Aristotle never raises more doubts about the relationship that he has established between everlastingness and perfection of the universe than in the case of animals, a position that he shares with all ancient thinkers: if animal species are everlasting, they must be perfect, but the superlative used here (nature "does the best," τὸ βέλτιστον, *Parts of Animals* 4.10, 687a17) does not have an absolute value. "The best" is that which leads to, or at least does not prevent, the everlasting survival of a species. From that, as we have seen, if there is a harmony in the animal world, it is mainly deployed within species, and only secondarily at the level of the relationship between species. This stress put on specific harmony and not on general harmony seems to be an important characteristic of Aristotelianism, in opposition to other ancient philosophies. This is something that is particularly striking in the domain where universal harmony ought to prevail, that of the universe. In the *De Caelo*, Aristotle certainly asserts the global perfection of the universe and contrasts it with the special perfection of each of the bodies composing this universe: they are all perfect because they have three dimensions, but each of the component bodies "is limited by its contact with the neighboring body" (1.1, 268b7), while the Whole is "totally" (πάντῃ, 268b10) perfect. But we are far, in the *De Caelo*, from the construction of a system like that in the *Timaeus* in which the planetary orbits result from the division of the circles of the Same and Other, dividing "the interior revolution six times, forming seven unequal circles, each corresponding to a double or triple interval" (36d). Certainly, the uniform circular movement from east to west of the sphere of the fixed stars is the first movement, and is the condition of existence of all the other movements of the universe. But it's not the sphere of the fixed stars and its movement that are the authors, not even the guarantors, of the cosmic order. The celestial bodies are not ordered according to a *plan* and, remarkable example, Aristotle does not feel the need to try to justify their number.

As is well known, Aristotle adopted a concept of the movements of celestial bodies based on the system of concentric spheres going back to Eudoxus of Knidos; Plato had asked him to propose a model that, using circular motions alone, would account for astronomical observations (Simplicius referred to this with the famous phrase "save the phenomena"). Callippus of Cyzicus corrected the Eudoxan system by adding a great number of spheres; the Callippan system suited Aristotle, especially if as an ancient tradition reports, he himself participated in perfecting it. In a construction of this kind, which is a system of concentric spheres with different axes of rotation nested inside each other, what determines the movement of a planet depends on the system that causes the movement of the next planet up in the celestial order. Thus, the orbit of Mercury is controlled by the kinetic system of Venus, since Mercury is located after

Venus going toward the center of the Earth, which is also the center of the universe. But Aristotle did not mean to offer a mathematical hypothesis of the type that had previously been developed by Eudoxus and Callippus, but to propose a *physical* model. To do that, Aristotle combined two apparently opposed imperatives: assure the kinematic autonomy of each planetary system while at the same time integrating all the planetary systems into a whole. Hence Aristotle's introduction of "compensatory spheres."

These compensatory spheres are meant to make each astral system kinematically autonomous. Thus, to make Mercury's system autonomous, it is enough to interpose between the systems of Mercury and of Venus compensatory spheres equal in number to those of the spheres of Venus, turning in the opposite direction. But, "Aristotle saw very well that it was not necessary to use as many spheres as that method demanded; instead of maintaining, in fact, between the various systems, an absolute independence, of the kind that each would move as if the others didn't exist, there is no problem supposing that they transmit to each other diurnal motion, since they all must share in that rotation."[44] The compensatory spheres in the Mercury system would thus be equal in number of the number of spheres in the Venus system, minus one. The entire cosmos thus presents a set of systems of movement, each of which is autonomous, but all remain dependent on the uniform circular movement of the sphere of the fixed stars, the last one in the sky. Although these autonomous elements have a point in common, that is not enough to make them a system.

The way Aristotle refutes the Pythagorean theory of the harmony of the spheres reinforces this impression: taking things in an absolutely immediate and naïve way, he disqualifies that theory by saying that if the celestial spheres were emitting a sound during their course, we would hear it . . . But even though Aristotle himself reports the hypothesis that "their speeds, measured by their distances, are in the ratios as musical concordances" (*De Caelo* 2.9, 290b21), which rests on the idea that the planets are arranged in a harmonic order, he does not offer a criticism of this part of the theory. Aristotle simply ignores the grandiose Pythagorean construction that embraces the entire universe. This preference for the juxtaposition of autonomous entities rather than for their integration into harmonious constructions is met also in other domains. For example, it's at work in Aristotle's choice of the city as the framework for human

44. Duhem, *Le Système du monde*, 127. "Diurnal" rotation is that of the sphere of the fixed stars.

ethical accomplishments, just at the moment when great empires were chasing cities from the historical foreground by integrating them into larger systems. Where Aristotle tried to construct a political, or civic, harmony, the Stoics would demand a cosmopolitan harmony. And one might say that a preference of this kind characterizes well enough one of the "heavy" tendencies of Aristotelianism. Thus, the difference in the status of different kinds of sciences, as opposed to a science of everything, the dialectic of the *Republic*, the same accent put on the ontological coherence of the individual substance, even if all substances are governed by the major schemas of Aristotelian metaphysics—hylomorphism, the doctrine of the categories, the distinction between potentiality and actuality. We have indeed in Aristotle a kind of revenge of the specific on the general.

That does not suppress the general and unifying schemas, as we have seen for the animal world. For example, the fact that Aristotle attributes a really *causal* function to material and efficient causes in a way puts all sublunary material beings into the same world. We may be able to push this unification quite far. In a book that is as remarkable as it is short, Andrea Falcon[45] wonders whether it is possible to attribute to Aristotle the project of a science that would consider the physical universe as a whole, that is, to include both sublunary and supralunary regions. That presupposes thinking anew some fundamental Aristotelian ideas and positions. So, if the celestial bodies are alive, which Aristotle claims, they must have a soul, but that soul must be both rational and the source of a rational movement (for the only property of the matter of which celestial bodies are made is to move in a circle, which is not enough to explain the movements of these bodies), and that movement must be the product of a desire itself rational, but which relies neither on sensible perception nor on imagination. That overturns, without absolutely annihilating, the concept of the soul that we have described earlier. Hence the subtitle of Falcon's book: "Unity without Uniformity." It is absolutely Aristotelian to think that a physics can exist at several levels of generality, notably because every science relates to a *genos*, which can have a variable extension. The science of the physical universe as a whole, the science of sublunary mobile realities, the science of living things or of animals, all deserve to be called *natural sciences*. But, obviously, the more one thinks about an extended *genos*, the more it is necessary to introduce expectations and

45. Falcon, *Aristotle and the Science of Nature.*

corrections to make a unique science for a unique class. Thus, the specific is ceaselessly reborn from its ashes.

It is not only the case that living species do not gain meaning by being inserted into a structure that would fix their number and location in a series, but one may say that the species is the strategic level of the living world as Aristotle conceives of it; as we have seen, the substance is the pivot of Aristotelian ontology. Thus, one must say a few words about the relationships between living species, because that is where a possible supra-specific harmony might reveal itself. But, to take up again the decidedly precious example of little fish, we see that rational Nature has provided living things with countermeasures allowing them to defend themselves against the activities of other species. Thus, she has made the little fish prolific to respond to that appetite of cetaceans. But she does not provide, in one species, characteristics *in favor* of other species. One cannot say, for example, that she has given sheep characteristics that make them more easily captured and digested by wolves. Similarly, it is not for the benefit of the little fish that Nature has placed the mouth of sharks toward their belly, but to prevents sharks from succumbing to their gluttony. It's only by one step removed that Nature uses this characteristic favorable to sharks for the sake of little fish. Let's look at the passage: "It does seem that Nature has done this not only for the sake of preserving other animals (because when they turn over, that slows them down, and the other fish can escape; all animals of this kind, in fact, eat living animals), but also so they don't satisfy their gluttony for food" (*Parts of Animals* 4.13, 696b27). It is the gluttony of these animals that makes them dangerous for small fish, because if they take in only moderate quantities of food, the fish can survive. But this gluttony is *especially* dangerous to themselves.[46] We notice that Aristotle extends this system between disadvantages and correction of disadvantages more than he should, because he extends it to dolphins that do not have a ventral mouth (cf. 696b26). For them, the rule applied to the others is as it were suspended, and that on both sides of their relationship with the small fish: are dolphins less gluttonous, which would prevent them from dying from overeating, and

46. We find in this passage in the *Parts of Animals* a beautiful illustration of the progressive character of the word ἀλλά in the expression οὐ μόνον . . . ἀλλὰ καὶ. Cf. Jean Humbert: "There is no doubt that the very frequent expression οὐ μόνον . . . ἀλλὰ καὶ (not only . . . but also) plays a large role in the '*progressive*' use of ἀλλά; the negation of the first term serves as a *springboard* for the second term." Humbert, *Syntaxe grecque*, 376.

prevent the little fish from being decimated? This introduction of the unexpected and exceptional is completely within the style of Aristotelian zoology, contributing to making it a "real" zoology.

For Aristotle, living species are not primarily elements in a larger structure, but autonomous entities that are integrated, *for better or worse*, in a larger structure, so that the perfection of nature is revealed first of all within each species. The perfection of the living world taken as a whole is only secondary, derived, and incomplete. Are we thus faced with a deficit of perfection in comparison with Platonistic nature, and in a general way, with the nature of other Greek philosophers, with few exceptions? Well, no, on the contrary, we have here an additional evidence of perfection on the part of this living Nature that exhibits no global rational structure. One of the results of Aristotle's extremely detailed studies of animals is that each species of animals *does not need* to be integrated into a global rational structure to be perfect, in the sense of perfection that we have previously explicated. "Local" arrangements are enough: there need to be prey for predators and mechanisms of compensation to palliate the losses that each animal undergoes from the fact of its own organization or its relationships with its milieu. For Aristotle, a nature like that of Plato, which derives the perfection of a species from its insertion into a larger structure, because that structure itself manifests a perfection at a higher level, would be *in vain*, since each animal is perfect in itself. Such a dependence of the particular in relation to such a universal would be a sign of imperfection.

We find an interesting additional contradiction between the specific and supra-specific levels in regard to vital resources. It's an unsurpassable given in Aristotle's "everlastingness" biology that each species has sufficient means, particularly in terms of food, to perpetuate itself.[47] But if we look at the relationships between animal species, the image appears of an animal world torn apart by unending turf wars. That goes well beyond predator-prey relations, which are in providentialist systems a kind of necessary evil, strictly circumscribed.

A lot can be gained from the first chapters of *History of Animals* 9, which notices that there are species that are "at war" (πόλεμος) and others

47. The same for human beings. In my *Endangered Excellence*, I remarked that Karl Polanyi had correctly noted that "Aristotle does not accept the principle of scarcity that condemns human beings not only to technical inventions meant to make up for their natural weaknesses and to an endless extension of exploitation of the environment, but also to a permanent confrontation with groups of human beings other than their own" (110).

that are "friends" (φίλοι). Thus the fox and the snake are friends "because both are troglodytes" (ἄμφω γὰρ τρωγλοδύται, 9.1, 610a12), which seems a rather weak basis for friendship. Animals are enemies "that live in the same places and get their life from the same things" (ὅσα τοὺς αὐτοὺς τε κατέχαι τόπους καὶ ἀπὸ τῶν αὐτῶν πιεῖται τὴν ζωήν, 608b20). Obviously, predator and prey are at war, as indeed "all are at war with carnivores" (9.1, 608b26). The examination of these relationships of hostility and friendship are part of the study of the "character" of these animals, one of the great chapters of his zoology that Aristotle enumerates at the beginning of the *History of Animals.*

Let's look at friendship/hostility relationships among several animals, still in *History of Animals* 9.1:

> The snake is at war with the weasel and the pig, with the weasel when they are both in the house, because they live on the same food, and the pig eats snakes. The stone falcon is at war with the fox: it strikes and claws at its fur, and kills its young, because it is a raptor. The raven and the fox are friends, because the crow fights against the stone falcon, so when the stone falcon attacks the fox, the raven comes to its aid. (609b28)

We have here interlocking relationships. The fox and the snake are friends because they are both troglodytes, but when living in the same place leads to competition for food, as in the case of weasels and snakes, which people keep in their houses to get rid of rodents, these animals become enemies. The case of the raven and the fox illustrates the saying, "the enemies of our enemies are our friends," as Aesop could have put it. The general rule regulating relationships between animals seems to be a state of war. Violence seems, in certain cases, to be inevitable, because carnivores must kill their prey. But, in fact, this propensity for war that one reads in texts like this is not always a naturally characteristic part of these animals, but is imposed on them mainly by the scarcity of resources. We can cite two examples. Certain fish are friends "whose food is found in the same place or neighboring areas, if the food is abundant" (*History of Animals* 9.2, 610b12). In the same way,

> It's possible that if food were not lacking, animals that are now feared by human beings and savage would live harmoniously with them and would act in the same way with other animals. What shows that is the way that in Egypt animals are cared for,

> because they have access to an abundance of food, and they live together, even the most savage. In fact, from the care that they take of these animals, they are peaceful; for example, in certain places, crocodiles are at peace with priests because the priests keep them well fed. One may observe the same thing in other countries and various territories. (608b30)[48]

The case of the snake and the weasel is interesting from another point of view, because their hostility is not natural, but "cultural," because it is due to their both inhabiting a *human* place, the household, that they are in competition.

From all that one may conclude that Nature has to introduce aggressiveness and disharmony between animals in order to assure their survival. But shortages that cause warfare come from a situation in the natural environment, itself natural, because dependent on natural processes, like the climate and the production of plants. There is, however, a limit to that, because, as we have seen, all animals are at war with the carnivores. If crocodiles and lions have an abundance of food, it has to be human beings who are feeding them by taking care of the killing of the prey. Rational nature thus introduces war between otherwise peaceful animals as a consequence of a lack of harmony between them and their environment. War thus appears to be a corrective to a disadvantageous situation, like the fecundity of small fish that corrects for the gluttony of cetaceans. The absence of natural harmony brings about war: in fact, we are far from Bernardin de Saint-Pierre and very close to Darwin. It is important to note that Aristotle nowhere points out a secondary benefit of this war introduced by the fact of poverty, as Darwinian biologists would do, saying, for example, that the aggressiveness of hungry crocodiles regulates the reproduction of the animals that constitute their habitual prey, although he does propose elsewhere an analysis of this kind, in the case of the little fish, which one might assume would become excessively numerous if the cetaceans weren't eating them. But would they become *too* numerous? Aristotle doesn't say. Thus there is lacking in Aristotle a harmony between the mineral, vegetable, and animal "kingdoms," something that is proposed by nearly every finalist account of the universe.

48. See also the case of the lion, which from the "most ferocious" becomes "the tamest" of animals (*History of Animals* 9.44, 629a8).

This prevalence of diversity in the living, especially animal, world, can be found in many domains and at many levels. I would like to come back to an example already noted several times: a general law of the living world, both physiological and "ethical," brings it about that animals with hot blood are more courageous than animals with cold blood. Within the same species, then, the male, with hotter blood than the female, will be more courageous than her, and that is what is observed, or rather what his sexist prejudices make Aristotle observe. But it is also found that female bears and panthers are more courageous than the respective males. Similarly, as we have seen, in the explanation of the absence of both antlers and two rows of teeth in does. There too the individual interest of this particular animal, the doe, with its particular characteristics overrules the general law of distribution of earthy matter, for the necessity of having antlers to use the earthy matter needs to yield to a more imperious necessity, that of the survival of the doe. But the teeth? In this case, Aristotle, as the good metaphysical thinker that he is, "recovers" things by invoking a "law" more general than that which regulates the distribution of earthy matter, namely that reproduction necessarily (or almost) takes place between animals of the same species, and the doe belongs, like the stag, to the species "animal with antlers" even if she doesn't have horns, which she couldn't deal with. But animals with horns do not have two complete ranks of teeth. But there is nothing like that for the bears and panthers.

This situation, in which general rules constantly need reconstruction or rectification because threatened by an exception that observation has rudely discovered, seems to me one more characteristic of biological thought.

To all these considerations, we may add, as a kind of appendix, a well-known Aristotelian doctrine that may serve as a remarkable proof *a contrario* of the prevalence of the particular, and thus of the secondary character of the general. It is a position shared by many ancient thinkers that the world around us depends on superior existences, whether celestial, divine, or the gods, and apparently Aristotle agrees with that determination, because in relating the superlunary to the sublunary he offers an image of the universe as a rather Platonistic structure. Of course, as we have seen, it is no longer that of participation of the inferior in the superior, but the imitation by the lower of the higher, which places the superlunary, a region where there are only regular circular movements of bodies composed of subtle matter, in the role of paradigm, of which the sublunary is a degraded copy. If that which has been said several times is true, namely that the superior is in the position of *explanans* and not of *explanandum*, then the zoologist too must go by way of the heavens to understand the world of living things in our sublunary world. And, as a matter of fact, Aristotle makes biological entities depend on the course of the stars, something that we find in a surprising passage in the *Generation of Animals* that aligns what we may call biological rhythms with the movement of the celestial bodies:

> It is reasonable that the lengths of gestations, of generations, and of lives, tend to be measured by the natural periods. I mean by "period" day and night, months, years, and the times that they measure, but also the periods of the Moon. Lunar periods are full Moon, new Moon, and the intermediary quarters. In fact, the durations of these lunar periods are in a certain relationship with the Sun, because a month is a period common to both. The Moon is a principle due to the fact of its community with the Sun, that is, that it participates in its light, for it is like a smaller Sun. That is why it contributes to all generations and all completions, for to a certain proportion (μέχρι συμμετρίας τινὸς) heat and cold produce generations and subsequent corruptions. But the limits of these transformations, their beginning and their end, depend on the movement of these heavenly bodies. (4.10, 777b16)

The passage continues by saying that in the same way bodies of water, the sea and others, are moved by winds that depend on the lunar and solar periods, with this remarkable assertion "there exists a kind of life of the wind, in that it is born and dies" (βίος γάρ τις καὶ πνεύματος ἐστι καὶ γένεσις καὶ φθίσις, 778a2).

This passage should be read in parallel with one in the *Generation and Corruption* that asserts that "we see that when the Sun draws near there is generation and when it moves away there is diminishment" (2.10, 336b18), the two periods being equal, and several other passages give the Sun, but not the Moon,[49] a causal role in the generation of living things.[50]

49. But as the Moon is presented as drawing its powers (light and heat) from the Sun, the doctrine is not different.

50. Note the famous "man generates man and also the Sun" (*Physics* 2.2, 194b13).

Although this could have led to cosmobiology, that development is quickly terminated, but in a way that is interesting to point out. Let's go back to what follows the passage in the *Generation of Animals* just cited: "Nature tends to count generations and deaths by these numbers [i.e., those of the periods of the Moon and Sun] but cannot do so exactly due to the indeterminacy of matter and the intervention of numerous principles that often obstruct natural generation and corruption, and are causes of unnatural events" (778a4). The schema that Aristotle proposes here is thus indeed a kind of directive come from on high, against which the sublunary *resists*, for two reasons. First, since natural beings are material, their material resists the imposition of a numerical structure come from the celestial bodies. This strongly recalls the Aristotelian doctrine according to which, in fertilization, the female matter resists the movements contained in the male sperm. It is a little more difficult to decide which "principles" obstruct the natural processes of generation and corruption. We must note that Aristotle does not say that these principles are themselves contrary to nature, but that they have the effect of impeding generations and corruptions from conforming to the celestial movement, which introduces into generations and corruptions an antinatural factor. Thus, it would be completely conformable to nature that the pregnancy of women would always be of the same duration, but, as the *Generation of Animals* explains (4.4, 772b6), the length of gestation in the human species is variable. As Aristotle makes clear, this irregularity is not found in other animals, and is attributable to the human ambivalence between uniparity and multiparity, an ambivalence due to combination of these two facts, equally natural, of the warm and humid nature of the human body, and the size of human beings. These facts are indeed principles in that they produce results and supervene on the regulation of reproductive periods by celestial movements.

Second, astral causality intervenes in the animal world only partially and with exceptions. That can be explained particularly by the kind of causality that is brought to bear, explained by a passage in the *Metaphysics*: at Lambda 5, 1071a15, the Sun and its ecliptic (i.e., the Sun moving on the ecliptic) are said to be "neither material, nor form, nor privation, nor of the same form, but moving causes" of the human being. In other words, the superlunary entities are active neither on the being nor on the coming-into-being of animals, but on the processes that bring those about. The control of superior entities of the living world is thus reduced to a minimum. The divide between superlunary and sublunary, one of the main obstacles that modern physics has had to surmount, hardly compromises Aristotle's zoology at all.

Chapter 5

Animal Nature and Human Nature

In an often-cited passage of his *Introductory Lectures on Psychoanalysis*, as completed in 1917, Sigmund Freud declared that humanity has undergone, in the course of history, three great *Kränkungen*, a term that connotes both wounding and humiliation; these events have damaged human "naïve self-love" by exiling human beings from the central and royal places they had taken for themselves. The first *Kränkung*, the Copernican revolution, showed that, far from being in the center of the universe, the Earth was nothing but a tiny part of it, verified in a "hallucinating" manner by modern cosmology; the second, the Darwinian revolution, denies to human beings a special place in the animal world; while the third, the Freudian revolution, showed that "the self is not master even in his own home."[1] These three revolutions probably, the first two surely, can be described as refutations of the spontaneous anthropocentrism of human beings. But the universe that the first of these revolutions upended was an Aristotelian universe. This despite the absence of a posterity of Aristotelian cosmology in antiquity and the high Middle Ages, and because of the central place of Aristotle among the Arab thinkers and the Christianization of Aristotle imposed by the Dominican theologians of the thirteenth century. It is necessary to recognize once more that the Aristotelian cosmos, which is alive, top to bottom, right to left, front to back, is radically anthropomorphic. What about the second *Kränkung*? After having tried to show that Aristotle's animal world does not offer

1. This passage can be found at the end of the eighteenth lecture. Freud, *Introductory Lectures*.

the harmony and completeness often attributed to it, I would like, in the first section of this last chapter, to ask myself whether and to what degree Aristotelian zoology is anthropocentric.

It can be anthropocentric in two ways. The first, already considered and refuted earlier, is that of David Sedley, who thinks that animals, and even all of nature, has been arranged as it is, in the interest of the human species. The second, much more interesting for one who has studied Aristotle's biology, leads to asking oneself whether, for him, the animal world is constructed following a plan that takes the human being as a model. The answer to that question cannot be entirely negative, because the human being, or rather *man*—obligatory phallocratism—is surely an animal of reference for Aristotle. It is incontestable that according to him, man is *globally superior* to the animals, even if some of them outclass him in some domains, since, as we will have occasion to see again in detail, a dog perceives more odors than we do. But does this superiority make the human being a *model* for other animals? Given Aristotle's fixist positions, it cannot be the case that man would be for some animals a model in the sense in which these animals would try to achieve a human biological project. The only way that man could be a model in the Aristotelian universe would be by furnishing the zoologist the means of making more intelligible the organs and properties that are found in animals. Such a relationship would conform to the important requirement of Aristotelianism that it's the more perfect and more developed that explains that which is inferior to it.

To examine this question, we need to reconsider several well-known Aristotelian passages concerning the relation between human and animal. After that examination, this relationship will seem a good deal more complex and variegated than is generally thought, because some of these passages are at times read too quickly. We'll start with a text to which I have already alluded. In *History of Animals* 1, Aristotle declares that one must begin the study of the parts of animals with those of the human being, because "the human being is necessarily (ἐξ ἀνάγκης) the animal best-known to us" (1.6, 491a19). Since no interpreter, to my knowledge, has credited Aristotle with human dissection (that had to wait until the Alexandrine physicians in the century after Aristotle),[2] one may well ask

2. Did Aristotle try to evade this taboo by observing cadavers exposed for various reasons? We remember the passage in Plato's *Republic* 4 in which Leontius allows his eyes to gorge themselves on looking at cadavers. Pierre Ménétrier thinks that Aristotle

oneself what that "necessarily" means. But it's wrong to use this passage to establish any paradigmatic status of the human over the animal world, because what is best known is what is directly accessible to perception—head, face, eyes, ears, etc. In chapter 16 of book 1, Aristotle in fact writes: "The parts visible from outside are arranged in this way and as has been said: these are the ones that have especially been named and are well known. This is not the case with the internal parts. In fact, the internal parts of human beings are especially unknown, so that to think about them we need to refer to the parts of other animals that have a nature that resembles his" (494b19). It seems from this that some animals have been dissected, perhaps with a research goal, but also perhaps simply on the chopping blocks of butchers and fishmongers. The word "necessarily" thus has a sense that one may call trivial, just indicating what is obvious. If the social obstacle of the prohibition of human dissection had been lifted, it is not certain that that would have made human beings the model of intelligibility, because there are functions that are better understood when one looks at them in certain animals, if only because the organs that perform those functions are larger. The multifunctionality of human organs, very developed in the hand, but also true of the lips, for example, that serve both to protect the teeth, and for language (*Parts of Animals* 2.16, 659b32), also risk making the functions more obscure.[3]

This reversal between the external and internal parts in terms of the knowledge we have of them is remarkably illustrated in *History of Animals* 3, chapters 2 and 3, where we find a very meaningful counterexample to this rule. Aristotle there explains that "the nature of the blood and blood vessels" (3.2, 511b11), which doubtless should be understood as a sort of hendiadys to mean the way that blood acts in the blood vessels, that cannot be observed precisely, because in dead animals the major blood vessels collapse because the blood has left them, "as for living animals, it is impossible to examine how the parts work; in fact, their nature is internal. So those who observe dead animals that have been opened have not observed the most important principles of the blood vessels, while those who observe very emaciated human beings have found the principles insofar as they can be traced from outside" (3.2, 511b18).

claimed that the human liver is trilobed from having observed cadavers of infants: cf. Pierre Ménétrier, "Comment Aristote."

3. Cf. the passage cited later at *History of Animals* 8.1, 588a25: "because some of these traits are more manifest in human beings, certain others more in other animals."

This is an illustration of the difference between observation of internal parts and observation of external parts, since in emaciated human beings, the vascular system counts as external observation, but permitting a more exact observation: then one grasps that blood and the vessels that contain it have the status of a *principle*, something that looking at flattened blood vessels, empty of blood, will not reveal. But immediately afterward, Aristotle presents in detail the conceptions that have been presented of the human vascular system[4] by Syennesis of Cyprus, Diogenes of Apollonia, and Polybus of Cos, the nephew of Hippocrates. The first and second of these systems are both reported, in a somewhat more developed form, in the pseudo-Hippocratic treatise *On the Nature of Bone*, a compilation later than Aristotle, and the last in *The Nature of Man*, a treatise attributed to Polybus. Then Aristotle presents his own account of the blood vessels.

What does this obvious violation by Aristotle of the principle that one ought to examine internal parts in animals and not in humans demonstrate? There is first the precise case under study that is in question: it's a matter of seeing how blood acts in the blood vessels, but that cannot be done on living animals, or on dead human beings. The passage does at least tell us that Aristotle did not practice animal vivisection. And, in fact, it is easier to see the blood vessels from outside in a living thing that does not have thick fur, thus better in a human being than in a dog. But that proves especially that Aristotle's remarks have a very restricted methodological application: resorting to observation on a human being and observation on an animal in turn are purely circumstantial conveniences. But we also find, in the present instance, that Aristotle has at his disposal in the medical literature several accounts of blood vessels. The presentation of vascular systems by others is doubtless very precious in that it permits *correcting* their observations, all the more so, since their systems, especially those of Diogenes and Polybus, are "exact" enough, according to Aristotle's point of view of course, to be useful. And that's what Aristotle does, but with an important innovation: when, after having presented the systems of others, he goes on to his own, he launches into a comparative study of blood vessels, distinguishing what is typical of all blooded animals, for example the distinction between the large vessel and the aorta, the one located forward and to the right, the other smaller, toward the back and to the left, as that which is normal in certain families of blooded animals. But different arrangements are mentioned as such. Thus, in the case of the vessel that, when it leaves

4. Cf. "in man" (ἐν τῷ ἀνθρώπῳ), 511b31.

the heart, "divides, going toward two places. In fact, these divisions tend toward the ribs and clavicles, and then go across the armpits in humans, to the arms, in quadrupeds toward the front legs, in birds to the wings, and in fish, to the ventral fins" (*History of Animals* 3.3, 513b34). And even for the common characteristics, Aristotle indicates that the large blood vessel is situated "more (μᾶλλον, 513a19) on the right," and the small vessel "more on the left," which leaves, if one may say so, some play in the exposition, which may thus be adapted to different species. Thus, Aristotle is doing comparative anatomy rather than medicine. But the route that the vessels take toward the armpits in human beings does not make the path of these same vessels in quadrupeds any more intelligible.

In none of these texts that one can call practical methodology does one find, then, an affirmation that the human being is a model of intelligibility for the animal world. Aristotle says only that blooded animals, and he is doubtless thinking of viviparous blooded animals, have the same arrangement of blood vessels with a few variations: there is never a suggestion that the arrangement in the human being shows something better than that of another species.

Nevertheless, Aristotle claims that human beings are superior to other animals according to certain criteria. We will consider several of those that seem to belong to the biological domain, before saying a few words about "politicalness" and language. First about standing erect. We can begin with a remark made in the *Progression of Animals* (4, 706a19), according to which man is "that one of the animals that is most (μάλιστα) in conformity with nature (κατὰ φύσιν)," this thesis being established on the basis of the fact that it's human beings in whom right and left are most distinct, the right being better than the left, "but by nature the right is better than the left and most distinct from it" (706a20), hence the conclusion that in human beings the right is "more better" than in other living things,[5] therefore the human is more in conformity with nature. This conclusion depends on a general principle applying to the cosmos as a whole. The same for other passages, like this one, already cited, from *On Respiration* (19, 477a21), that says that man "is the most erect of animals, the only one to have the top of his body that goes in the direction of the top of

---

5. Aristotle marks this in a remarkable way: "The right is more right in humans" (διὸ καὶ τὰ δεξιὰ ἐν τοῖς ἀνθρώποις μάλισστα δεξία ἐστι) 706a21. P.-M. Morel cleverly translates "plus adroite"; "in human beings, the right is righter."

the universe." All that rests on Aristotle's basic thesis that the universe has absolute dimensions, up and down, right and left, front and back.[6] This human preeminence relies therefore here on the greater or lesser degree of affinity with the spatial structure of the universe itself, which presupposes a sort of difference in degrees of distance of various animals in relation to the model; this is not, of course, enough to introduce a total ordering of the animal world, and thus found a *scala naturae*, but it does play a role in the theoretical framework of the study of the living world as a whole. One must not forget that the Aristotelian cosmos is alive. From the point of view of standing erect, the real model is therefore not so much man as the universe, in such a way that the human situation is, one may say, bordered on both sides, since the human being is in the position of a copy in relation to the universe, and a model for the animals.

Same for up and front. On the occasion of a passage in the *History of Animals* that says that man has in "front" the parts that quadrupeds have "down," Andrea Carbone notes: "This generalization establishes a principle of comparison permitting the *translation* of the polarity front/back in man to the analogous polarity in quadrupeds, below/above."[7] Provided that one says that that translation goes only one way, because the human schema is the original version of the schema that is found in quadrupeds.

So, we can end up favorable to a restricted and weak interpretation of Aristotle's assertion that "man is the most in conformity with nature," though this affirmation is incomprehensible to people today. This thesis is advanced as part of the construction of a general theory of animal movement, at a moment when Aristotle is explaining that this movement occurs within a framework of three dimensions, defined by the three pairs, up/down, front/back, right/left. We know from elsewhere that the world itself is framed by these three dimensions, and that it, like all living things, in an axiomatically determinate way, up, front, and right, being better than their opposites. To say that the human being is the animal in whom right and left are most distinct one from the other and the right better than the left, to a higher degree *because* (διὰ τὸ, *History of Animals* 8.29, 607a19) he is the animal most in conformity with nature, seems here to rest on the thesis that these two conditions, distinction between right and left, and superiority of the right, are actualized in the living thing

6. Cf. *De Caelo* 2.2.

7. Carbone, *Aristote illustré*, 140.

that is the world. So, it seems that here the being "most conforming to nature" means the one that conforms most to the nature of the Whole. In the same passage of the *History of Animals*, "it is mainly in the case of man, compared with other animals, that the up and down are defined by relation to natural places" (1.15, 494a26). The "natural places" are here the *absolutely* natural places, that is, the up, down, and so on, of the universe. In this sense and on this precise point, man is the most natural of animals.

Animal relationship to space, and thus kinds of movement, can in fact be ranked according to a scale going from the most to the least perfect, but that is only in relationship to the greatest classes: bipeds (bipedal motion exists in two ways: human and bird), quadrupeds, animals that swim, and those that crawl (this in two ways: like serpents and like worms). And there is, in fact, a sliding scale when one goes through this series. Thus, in *Progression of Animals* 4, Aristotle shows how the neatness of a principle fades away as one descends this scale. This principle, that by nature movement starts on the right, is notably true for the movement of celestial bodies:

> The part of the body whence local movement naturally begins in each sort of animal is the right; the opposite, which naturally follows, is the left. But this division is more marked in some than in others. In those, in fact, that use for movement organic parts (I mean feet, wings, or some other part of this kind) the distinction in question is better marked. . . . The others have the parts in question but not as clearly discernible. (4, 705b18)

Snakes, caterpillars, and earthworms (the examples given in the passage skipped) obey the same kinetic laws as man, but in a degraded fashion.

Finally, there are cases in which the degradation is even stronger, in that the movement is brought about in an entirely unnatural way (here, with unnatural parts). Thus, this prodigious passage, which manages singlehandedly to sum up the panorama of Aristotle's biology:

> Spiny lobsters and crabs all have a right claw larger and stronger, because all these animals naturally carry out their actions more with their right parts, and nature always gives to each being, exclusively or more than others, the tools that it is able to use, for example tusks, teeth, horns, spurs, and all parts of this kind, all destined for defense and fighting.

> Only lobsters have one claw larger than the other at random, and that is true both for females and males. They have claws because they belong to a class of animals with claws, but they have this part irregularly because they are mutilated and they do not use them in the natural way, but for local movement. (*Parts of Animals* 4.8, 684a26)

Lobsters have their claws "in an irregular way" (ἀτάκτως) because their right claw is not always stronger and the natural way of having them would lead to their using them for grasping and defense (that would be using them "in a natural way"), while they also use them for local movement, as we noted in the lines analyzed earlier following this passage. One may then conclude from all these cases here mentioned, and many others, that the agreement of man with the universe has indeed the effect of making more intelligible the properties of animals. Both for those whose movement starts from the right, but not very clearly, and for those in which the right and left do not seem to differ from the point of view of movement, the human situation furnishes an explanation.

But erect posture has an explanation in addition to the spatial disposition,[8] which also provides a relative image of human superiority:

> Having little heat, the ovipara are sufficiently cooled for long periods by the simple movement of the lung, since it is airy and empty. And that goes with the fact that most of these animals are in general rather small. Heat actually encourages growth, and the abundance of blood is a sign of heat. Furthermore, warmer animals are more erect, which is why man is the most erect of all, and among quadrupeds the vivipara are more than the others. In fact, none of the vivipara, whether footless or footed, live in burrows as [ovipara] do. (*Parts of Animals* 3.6, 669a35)

One can, of course, recall that for Aristotle heat is a sign of perfection—the male, for example is more perfect and hotter than the female—but here erect stature is *comparatively* clearer in human beings (cf. hotter animals

8. The spatial disposition has an explanatory function that can be added to the explanatory functions of the four causes, as Andrea Carbone has shown in the remarkable work just mentioned.

have their body *more* erect, ὀρθοῖ τὰ σώματα μᾶλλον, 669b5). In the same vein, one finds the famous passage in the *Parts of Animals* saying that animals that have weight too great to carry "have become quadrupeds" (τετράποδα ἐγένετο, 4.10, 686b1), which obviously is not meant to be taken diachronically. Erect stature is thus produced by the heat and lightness of the human body, and does not reflect any absolute superiority, heat and lightness being eminently subject to the more and less.

But there is in Aristotle yet another approach to the erect stature of human beings. Look at this passage in the *Parts of Animals*:

> The human being has arms and what we call hands instead of front legs and paws. Because he is the only animal that stands erect, because his nature and substance are divine. But the most divine function is knowing and thinking. And that is not easy if a large part of the body is weighing down the upside, because the weight makes thinking and common sense hard to move. That's why, when the weight of the corporeal elements becomes excessive, the body necessarily bends toward the earth, so that, for their stability, nature has provided front legs instead of arms, and paws instead of hands. (4.10, 686a25)

Aristotle's reasoning here deduces erect stature from the fact that man is divine, and that erect stature is the only one that permits exercising this divine function par excellence that is thought. Nature thus gives erect stature to human beings because they are the only ones able to profit from it, just as she gives them hands because they are the only ones with the intelligence that allows the use of hands. The human being has hands because he is more intelligent than the other animals; it's not because he has hands that he is the most intelligent, as Anaxagoras said.[9] It would doubtless be Anaxagorean to say that man participates in the divine because he stands erect, but that is not Aristotelian. But erect stature does not offer any vital advantage to human beings as, for example, webbed feet offer to ducks in relation to aquatic birds that have their toes separated, but it does permit human beings to lead a fully human life, one that includes thought. The difference between human and animal seems indeed, in texts of this kind, to be an absolute difference: even if Aristotle recognizes the existence of animal intelligence, only human beings think (it's a well-known

9. Cf. *Parts of Animals* 4.10, 687a6.

Aristotelian thesis that only human beings have *logos*) and that is why Nature has given them erect stature. Right away, the human being cannot be, from that perspective, in the position of a model for other animals, because his superiority rests on a characteristic that animals do not have, but here too, the human condition makes intelligible for us the condition of certain animals: by understanding why in cows, for example, the "ideal" situation of having the head up high cannot be realized (because of the excessive weight of their body), one understands why Nature had to work with necessity and make these animals quadrupeds.

Aristotle's explanation of the erect stature of human beings is thus in conformity with Aristotle's regular practice of explaining a phenomenon by applying several causal dimensions to it. After an explanation that could be called formal, which aligns human beings with the universe, and an explanation by material and efficient causes, according to which erect stature flows from the abundance of hot blood, and lastly a final cause explanation is formulated in the usual terms of the teleology at work in the animal world: Nature gives erect stature to men because they are the only ones to be able to get from it the intellectual advantage that they do. The first two approaches establish the superiority of human beings over animals, and thus can give man the function of a model in relation to animals, while the third approach seems to establish an impenetrable barrier between human and animal. But all these approaches contribute to intelligibility.

But the barrier between man and beast is only apparently impenetrable, because the human privilege consists in having a "divine" nature and substance (*Parts of Animals* 4.10, 686a27) that itself can be relativized. Thus, in a remarkable passage in the *De Anima*:

> The most natural of functions (φυσικώτατον γὰρ τῶν ἔργων) of living things that are complete and not defective, and which do not reproduce by spontaneous generation, is to produce another like itself, animal for an animal, plant for a plant, in order to participate in divine eternity (ἵνα τοῦ ἀεὶ καὶ τοῦ θείου μετέχειν) to the degree possible, for all have a tendency to this, and it is for the sake of this that they do everything that they accomplish by nature. (2.4, 414a26, already partially cited)[10]

10. Cf. *Generation of Animals* 2.1, 735a17; b24; 1.23, 731a24–b9; 3.10, 760a35. We note that this passage asserts that in spontaneous generation there is not, properly speaking, *reproduction* of an animal.

This passage is worth a bit of discussion. Read literally, it says that Nature, which must be rational Nature, has given living things a tendency to reproduce "in order to participate in divine eternity" (taking the expression τοῦ ἀεὶ καὶ τοῦ θείου, eternity and the divine, as a hendiadys). But Nature has done that above all so that species survive, as was said, among others, by the passages cited in the note. There is no doubt that this permanence of species is a sign of the perfection of the universe. But Aristotle takes an important step when he assigns to reproduction "and all actions that living things accomplish by nature" the goal of sharing in the divine. The insistence of the text ("in order to participate," "it is for the sake of this") shows that it's not simply a casual expression: although it is not without echoes in the Aristotelian corpus, as we will see later, particularly when we examine a passage in the *Nicomachean Ethics*, 7.13, 1153b25, we have here one of Aristotle's rather Platonizing texts, if not the most, in the corpus, in that it puts a kind of desire for the divine as an internal tendency in all living things, making the divine a factor in the unification of the living world. In fact, one may think that nonliving realities do not share in a desire of this kind, or share it very little, or share it only metaphorically. Animals are less divine than human beings, but they are divine all the same. It is difficult to construct a *scala naturae* out of that, but nevertheless it puts human beings in the position of a model, even if the consequences of that position for our understanding of animals seem rather minimal.

The superiority of the human corporeal arrangement is sometimes demonstrated in relation to a particular function. Thus, for the disposition of organs contributing toward the process of nutrition, the human arrangement is seen to deteriorate as it is reproduced in other animals, to the point where it is reversed in bloodless animals. Thus, "the animals called mollusks have exterior parts as follows: first come what are called the feet, second, behind them, the head, third, the cavity that encloses the internal parts and what some call the head, which is not correct, and then there are also fins arranged in a circle around the cavity. In all mollusks the head is to be found between the feet and the belly" (*History of Animals* 4.1, 523b21). This passage shows at least two things. First, there is a unique arrangement that goes from human beings to mollusks by way of animals that are closer to human from the point of view of the disposition of the parts discussed here, and not two (or more) distinct arrangements; but second, and especially, it's the mollusks that present a degraded organization of the human arrangement, and not the reverse, and that the mollusks, from the point of view of the disposition of these parts, do not exhibit an organizational plan that is their own. If we were to consider mollusks in themselves, one would be right to call "head" the cavity that contains the internal parts, while functionally it is their belly. The "true" head is found situated between the feet and the belly. Here is a passage in the *Parts of Animals*, which introduces an account with letters, comparable to accounts of this kind that Aristotle uses in the *Physics* or in the logical treatises, gives a more complete picture, and has the advantage of confirming some of the analyses proposed in the previous chapters, among other things:

> In a general way, in fact, animals with shells from a certain point of view look much like crustacea, and from another, the same as mollusks. . . . This is true in a way for all of them, but especially for those with spiral shells. In both cases[11] their nature[12] is presented as if one were to think of it aligned in a straight line, as is the case in quadrupeds and human beings: first, a kind of mouth at point A of this line, then the gullet B, stomach C, and then the intestine to the exit point of excrement D. That's how things look in blooded animals, and around this straight line there is the head and what one calls the chest. . . . At the same time, the arrangement in a straight line of the internal parts seems to be presented in the same way [as in blooded animals],[13] while in the service that they offer to their external parts they differ from blooded animals. . . . Mollusks and spiral-shell animals are similar to each other, but opposite of the previous ones. In them, in fact, the final extremity has been curved toward the beginning, as if, in curving a straight line, called E,[14] by leading D toward A. (*Parts of Animals* 4.9, 684b17)

---

11. That of shelled animals and mollusks, and not mollusks and spiral shelled, as Düring says, whose correction of at 684b24 I accept (ἐπὶ ἄκρῳ τὸ ἄνω στόμα τι τῆς εὐθείας instead of the manuscripts' τῷ ἄκρῳ τῷ ἄνω στόμα τι τῆς εὐθείας).

12. One could understand "the nature of shelled animals and mollusks," but it is likely that it's a question of the nature of their internal parts, as the continuation shows.

13. Other translators understand that the disposition is the same in crustacea and insects.

14. The line ABCD is now called E, something that, contrary to what Düring says, occurs elsewhere in Aristotle: at *Physics* 6.7, 238b6, for example, he labels a finished size with a single letter. One can also think that after having drawn the line ABCD, Aristotle draws another one that he calls E.

Passages like this lead us to clarify a fundamental point that was raised several times in our first chapter: like Cuvier, who distinguished four branches that are not subsumable under a unique plan, we have noted that Aristotle posited without more justification a great distinction between animals with blood and animals without blood, and within them distinguished four heterogenous groups, shelled animals, mollusks, crustacea, and insects, and these are precisely the large classes that appear in the passage that we have just now cited. Cuvier writes that within each branch the different classes of animals "are only modifications founded on the development or addition of some parts but do not change anything essential,"[15] and a position like that can be attributed to Aristotle. But this very general division seems here to be subverted by Aristotle, because this fundamental function, nutrition, is exercised in all animals via modifications of a unique arrangement. It goes the same way, in fact, for Cuvier, up to the point where he shows himself, as we will see, more Aristotelian than Aristotle. We have seen that Cuvier's branches differed from one another on the basis of irreducible differences in their nervous systems, these differences manifesting themselves by the functions of sensation and movement (it's not only a matter of local movement, but also every movement due to the irritation of animal tissue). But for Aristotle too, animality is not defined by nutritive capacities (which are definitory of life, both vegetable and animal), but by sensation. So that in this respect, Cuvier, in relying on sensation more than on digestive function, is more faithful to Aristotle's teachings than Aristotle himself in the passage from the *Parts of Animals* that we just quoted. But in fact, this passage, in speaking of mouth, gullet, stomach, and intestine, indeed considers the *animal* ways of exercising the nutritive function.

We have already seen that for Cuvier the nervous system of vertebrates was placed above the digestive canal, while it is placed underneath in mollusks; this does not seem to introduce a hierarchy between the two arrangements. But in the description that he gives of the four branches, Cuvier actually adopts a reference plan that is that of vertebrates, which have a brain, spinal marrow, and a system of nerves organized around a large sympathetic nerve. Mollusks and arthropods do not have spinal marrow; "asteroids or zoophytes" don't even have a brain, and present only "traces" of a nervous system. This reference to a more perfect organization, that of vertebrates, nevertheless does not allow including all these animals

15. Cuvier, *Leçons d'anatomie comparée*, 1, 69. [A.P.: My translation.]

in one unique plan. Furthermore, if, among the branches the vertebrates represent the superior degree of organization and the zoophyte the lowest degree, Cuvier seems to hesitate about the rank to attribute to the two other branches. He seems there again to adopt an Aristotelian position: if you take the point of view of perception, arthropods (especially insects) are above mollusks, which in turn are ahead from the point of view of other functions.[16]

To return to the passage quoted from the *Parts of Animals*, one sees in it Aristotle flirting with the idea of a unique plan for the entire animal world, but in respect of the digestive system. This unity that, in a way, would be superimposed on the profound diversity of animals, does not however establish a *global* scale going from the most to the least perfect, but scales according to the functions considered. Thus, we see introduced in the passage the point of view of locomotive organs, illustrating that it is the combination of identical and different "necessary parts" according to *Politics* 4.4, that was analyzed in the previous chapter. But this passage from the *Parts of Animals* is remarkable for another reason: it can contribute to an illustration of human superiority that it cannot establish on its own. In terms of the nutritive function, the model is represented by quadrupeds and human beings that have the digestive organs in a straight line, and one may presume that among animals that have these organs in a straight line, the quadrupeds (Aristotle surely is thinking here of viviparous quadrupeds) and human beings have an exemplary system, in that it puts in a straight line in the "good" order better organs than those of other animal families. But to base human superiority on a text like this, one must get out of the framework of the nutritive function. If, as was said, among other passages, by *Parts of Animals* 4.10, 686a2, cited earlier, quadrupeds have become such by a sort of degeneration in relation to the human model, Aristotle offers, of course just as a suggestion, a rationale something like this: for the nutritive organs, quadrupeds and human beings have a model arrangement, but four-footedness is a kind of degeneration in relation to standing erect, so the human arrangement is the real model. It's also a matter of a model in the sense that the better arrangement conveys intelligibility: without reference to the model, it would be difficult to identify the head and belly of mollusks.

16. Cf. Cuvier, *Leçons d'anatomie comparée*, 1, 38–39, which puts arthropods ahead of mollusks. It seems however that *globally* mollusks win out over arthropods, cf. 1, 69.

If we consider things from an anatomical-physiological perspective, it's in relation to bodily proportions[17] that Aristotle approaches most closely a global superiority of the human being over animals. If the human frame is superior to that of animals, it's because the human body is better proportioned than theirs. It's that proportion that shows through in the amazing remark that "all animals are like dwarves in comparison to man" (*Parts of Animals* 4.10, 686b4), because in man the upper part, "which extends from the head to the anus" (686b6), is not too large in relation to the inferior part, as it is in dwarves, babies, and animals, but is "proportionate" (σύμμετρον, 686b7). Furthermore, the proportion is not the same over time, since in human beings the upper part of the body is at first larger than the lower until things get their definitive look, while in certain animals it is the opposite, and others keep the same proportions their whole life, like the dog (*History of Animals* 2.1, 501a4).

How does Aristotle demonstrate this thesis, which seems to us to be aberrant, for why would it be better for a deer to have human proportions? It seems that it goes like this: dwarves, which are human, have proportions that are different from "normal" human beings, and it is the same for babies, who are also human, but undeveloped. But it is the adult man who is not afflicted with dwarfism who is the norm. Since Aristotle thought that he could assert that animals are nearer to dwarves than to "normal" people, proportionately, he deduces that animals are badly proportioned. But why not say that man is a badly proportioned deer? Because that would lead to claiming that the "normal" man is a dwarf, or a baby, consequently badly proportioned, which is impossible, precisely because a normal man is normal and thus natural, since the natural is that which occurs always or mostly.

We have in this case an interesting figure of human exemplarity in Aristotle's zoology. In his fixist system, as we have said, there is no question that other animals tend toward or try to achieve human proportions; furthermore, to give the deer human proportions, far from being a proof of the excellence of Nature, would be disadvantageous for it. So, the human model is thus not a model to be imitated. And, in contrast to the disposition of organs of nutrition, human proportions do not make the proportions of other animals more intelligible. That does

17. For the following, read in Andrea Carbone's book already mentioned the chapters "Schematization Put to Work: The Human Body as a Model" and "Schematization Put to Work: Animals Other than Man," 105–175.

not prevent a text like this from reaffirming the superiority of man, even if this human norm fails to furnish us with a criterion of intelligibility for various animals: the zoologist—and Aristotle would say the "natural scientist"—does not understand the deer better when he knows that it is a dwarf in comparison to man.

Other points of view in addition to the spatial situation and the anatomical structure can be considered as establishing a superiority of humans within the animal world. The first can depend on a famous passage at the beginning of *History of Animals* book 8:

> In most other animals, traces (ἴχνη) of dispositions of the soul are found, which in human beings are more clearly differentiated. Sociability and savagery, sweetness and crossness, courage and cowardice, fear and confidence, spiritedness and deceit, and resemblances [with man] in regard to intellectual understanding (τῆς περὶ τὴν διάνοιαν συνέσεως) are met with in many of them, as we have said for the parts. In fact, some differ from man by the more and less, the same for man in relation to many animals (for some of these traits are clearer in man, some others more in other animals), while some others differ by analogy. For example, what are art, wisdom, and understanding (σύνεσις) in man correspond in certain animals with another natural capacity of this kind (τις ἑτέρα τοιαύτη φυσικὴ δύναμις). (8.1, 588a18)

This passage is more difficult to interpret than it looks, because it seems to hold two contradictory theses one after the other: on the one hand, these "traces" that we find in animals are less clear than that in human beings of which they are the traces; on the other hand, for some of the qualities in question, human beings can be inferior to certain animals. It seems that Aristotle does not grant the same status to traces, which are traces of "dispositions of the soul" (τῶν περὶ τὴν ψυχὴν τρόπων) and concern what may be called the "ethical soul," and to "resemblances" (ὁμοιότητες, 588a24), which are a matter rather of the functions of the intellectual soul, all derived as they are (as we have seen) from the functions of the sensitive soul, are no less proper to human beings (and eventually to demonic entities). The case of animal superiority concerns only the traces of ethical states—a lion, for example, can be more courageous than a man (whether we should understand "any man" or "a particular man"), while for intellectual performances, the relationships between human and animal can only be analogical; the words "as we have said for the parts," definitely apply to the traces, but even more so to these resemblances. Thus, for σύνεσις, a word that designates the quickness of mind in rapidly grasping ideas: to say that in animals it is "another natural capacity of this kind" is to give it an analogical status.

The beginning of *History of Animals* book 9 introduces an additional parameter, the division between male and female, which provides an exposé that is ridiculous in our eyes, in that it reveals a naïve and self-satisfied sexism. If we have the daily experience that the woman is "more impudent, more dishonest, more deceitful, more resentful" (9.1, 608b11) than the man, if she cries more easily and sleeps less, it's because of characteristics that belong more generally to females of several species: "In all the classes where there are male and female, nature has established approximately the same relationship between male and female characters. This is especially obvious in human beings, in large animals, and in viviparous quadrupeds. The character of females is in fact sweeter, more easily tamed, admits more easily of caressing, and learns more easily" (9.1, 608a21). We find again, as in the case of digestive organs, human superiority submerged in a much larger group. But, after a long account of psychology compared according to gender, Aristotle again takes up his argument at the beginning of book 8 in these terms: "Traces of these character traits can be found so to speak in all animals, but they are clearer in those that have a more developed character (μᾶλλον ἦθος) and to the highest point in man, for man has the most developed nature, so that in him these states are clearer" (9.1, 608b4).[18]

18. Arnaud Zucker, in his 2017 article "Sur un prétendu anthropomorphisme aristotélicien en zoologie," criticizes translators for having rendered ὥστε καὶ ταύτας τὰς ἕξεις εἶναι φανερωτέρας ἐν αὐτοῖς by "so that in him (human being) these states are clearer" (my translation), while he thinks that ἐν αὐτοῖς ought to be understood as "in them (animals with more developed characters) these states are clearer." That's to ignore Aristotle's grammatical complexity, which translators know well, and which often leads to prioritizing sense over grammar: in animals that have more character, the states are clearer, but that is true in the highest degree in human beings, so that in him they are the clearest. The words ὥστε καὶ would be incomprehensible if the αὐτοῖς did not send us, contrary to strict grammatical logic, to human beings. So Balme is right to translate "are more evident in humans" (71). In the same article, Zucker claims that the beginning of the passage from book 8, cited earlier, does not attribute traces to animals and dispositions of the soul to humans, but traces to all, to which I must make the same comment: if there is a trace, there is something of

All translators, including me, translate the word ἀποτετελεσμένην at 608b8 with the comparative "more developed," but in fact the text says simply that the human being has a developed nature. For another example of a positive having the value of a comparative, refer to *Generation of Animals* 2.4, 737b25: "One must begin with those that are first: the first are developed animals, but such are the vivipara, and among them, the human being is first." This "developed" in fact means "more developed."

If we put this passage together with the beginning of *History of Animals* book 8, cited earlier, we may well wonder what exactly in animals, among the traces and resemblances, can be found in a more developed form than in human beings. Because the human has a "more developed" character (Aristotle says "more character," μᾶλλον ἦθος, 608b6) and that "to the highest point." And, in fact, the example of the courage in the line cited earlier is deceptive. In fact, we have already seen that true courage is that of a virtuous citizen. Thus, we may think that the beginning of *History of Animals* book 9 corrects the beginning of book 8 by adding to all levels of comparison between human and animal a kind of coefficient of analogy. In fact, man has the most developed character, that is, he alone *really* has character. As for any superiority of animals over human beings, we need to fall back on purely physical characteristics (we will see, when we discuss pleasure, that animals have sensory faculties superior to those of human beings), or innate know-how, which does not require deliberation, like the spider weaving its web, or the sophisticated migratory practices of certain birds and fish.

We will return to the traces and resemblances between humans and animals, because that will be useful to us in our examination of two other characteristics that reveal the relationships between human beings and animals, politicalness and language. I need to summarize some of the results in my book on Aristotle's political philosophy.[19] When Aristotle says that man is naturally a political animal, he seems to be expressing himself as a zoologist, since he calls "political" animals as different from each other as "the bee, the wasp, the ant, and the crane" (*History of Animals* 1.1, 488a10). In fact, as I have tried to show, the biological character of human politicalness, which is a property relevant to the theoretical natural

which the trace is a trace, and that is something that belongs only to humans, even if it is possible that some humans have only traces. Thus, my translation is correct.

19. Pellegrin, *L'Excellence menacée*, 79–108; English translation, *Endangered Excellence*, 67–93.

science, does not prevent it from being concerned with politics qua the dominant (and I think unique) *practical* science. Because if, for an animal to be political, is to live in a group with animals of the same species that participate in "a unique function common to all" (488a8), human beings share with other animals this "biological" politicalness. But people *carry out* their political character by means of *actions* that are both individual and collective, whose end point is the establishment of the most political of realities, the *polis*, or city. There is no animal city.[20]

Let us return, then, to the questions raised by the last two books of the *History of Animals*, considered earlier. The matter of the "traces" in animals of human intellectual faculties has been treated in a masterful way by Jean-Louis Labarrière.[21] He has a balanced approach to the question because, even though he separates himself from those who attribute to animals only metaphorical prudence, he does not confuse human and animal prudence.[22] Prudence (φρόνησις), whether one takes it to be the primary ethical virtue, or one of the intellectual virtues, is taken to calculate the means to an end that is proposed by something else, for one must assuredly agree with the interpreters who, during a famous dispute, argued that prudence is applied to the means and not to the end. Thus, prudence presupposes an analysis of a situation, foresight, deliberation, and decision. Here is a remarkable (and remarked) example of animal prudence offered by the *History of Animals*:

> The cuckoo seems to act with prudence in the disposal of its progeny; the fact is, the mother cuckoo is quite conscious of her own cowardice and of the fact that she could never help her young in an emergency; and so, for the security of the young, she makes of him a suppositious child in an alien nest. The truth is, this bird is preeminent in the way of cowardice; it allows itself to be pecked at by little birds, and flies away from their attacks. (9.29, 618a25)

20. "It is not only in order to live, but more for the sake of a happy life that people come together in a city, because otherwise there would be a city of slaves and a city of animals, but those do not exist because they do not participate in happiness nor in a life guided by reflective choice" (*Politics* 3.9, 1280a30).

21. In addition to his 1990 contribution, "De la phronèsis animale," see his collected essays, *La Condition animale*.

22. Here I recall some analyses from *Endangered Excellence* (86ff).

This prudence of the cuckoo is, to be sure, a sign of the excellence of rational Nature, since it contributes to the survival of the cuckoo species. But this character trait of the cuckoo contributes to the analyses presented earlier, in the second chapter. Just as Nature did not deliberate before giving horns to cattle to defend themselves, or make little fish prolific so they might survive, the cuckoo did not deliberate, taking account of the parameters of the situation, most importantly, her own cowardice and inability to care for her offspring, to *prudently* adopt the solution of having them raised by birds of other species. That's what a human being would have done. As in the case of other end-directed properties of living things, Nature contributes a kind of timeless deliberation and decision, which leads to providing the cuckoo with prudence of this kind.

Even if we were to agree with Labarrière that the prudence of the cuckoo is real prudence (and that in fact the cuckoo acts after evaluating a situation by arranging the means to realize an end, which is the survival of her brood), it is difficult to decide, once we compare this prudence to human prudence, whether we are dealing with two different kinds of prudence or the same kind with different intensities. In the first case, the very idea of a human superiority or exemplariness doesn't make sense, since an animal cannot make use of a sort of prudence other than the kind that is useful for it. It would thus be necessary, in order to talk about human superiority, to think that animal prudence is a "trace" of human prudence. Cuckoo prudence would then be modeled on human prudence, but it's lucky for the cuckoo that it's not the same. It seems, on the other hand, that other human intellectual activities that Aristotle also attributes to animals cannot be considered as anything but analogues of human faculties. That's how it is for the intellectual sagacity mentioned earlier (περὶ τὴν διάνοιαν σύνεσις, *History of Animals* 8.1, 588a23), calculation (λογισμός), and even *logos*, and cleverness, since in the *History of Animals* one kind of spider is called "the most clever" (σοφώτατον, 9.39, 623a8).

What Labarrière shows is that animal that exercise of all these intellectual faculties, no matter how close to their human analogues, cannot resort to deliberation and judgment; animal prudence "does not preside over any *praxis*."[23] That confirms what was said earlier: man alone has character completely, that is, to speak Greek, to be the only ethical animal. In fact, the city is the product of a history and has a content of ethical ends, two domains that do not derive from theoretical thought and thus not from biology. We will look again at the specific example

23. Labarrière, "De la phronèsis animale," 415.

of sheep in the last section of this chapter. In *Endangered Excellence*, I wanted to show that there is no Aristotelian sociobiology. That can also be proven by an argument that I did not use in that book: if sociobiology is a method that tries to clarify human attitudes, values, and behaviors by the analysis of their animal equivalents (one must not forget that Edward Wilson, the founder of sociobiology, was an entomologist and especially a myrmecologist), then a procedure like that is properly speaking impossible for Aristotle, because it explains the more by the less. A sociobiologist would be, in contrast, more compatible with the Presocratic mechanistic philosophies. Human beings, and especially citizens, find themselves at the intersection of two "thoughts,"[24] theoretical, because they are animals, and practical, because they are ethical beings, without that blurring the difference between theoretical and practical. But isn't there a trace in ants of human politicalness, which is historical and thus cannot be a possible model for animal politicalness. Could animal politicalness be a model for its human version?

One may answer that question both positively and negatively. That an anthill may be a model for the city could be the case for Plato, as he is described in Aristotle's *Politics* book 2: if the happiness of the city, which sanctions its harmonious and efficient functioning, can do without the happiness of the citizens, then the anthill is a model for human cities. But, writes Aristotle, "it is impossible that the city be happy as a whole if most, if not all, or even some of its parts are not happy" (*Politics* 2.6, 1264b17). Aristotle distances himself thus from one of the common characteristics of all totalitarianisms. But we must also recognize that when Aristotle writes "man is naturally a political animal," he really means to say that this tendency to live together while accomplishing a common purpose[25] is given to him by nature, like the tendency to leave behind a being like oneself. Man then possesses a political tendency like that of certain animals, which Nature uses for his benefit alone, since he is the only one with the ability to form cities.

The interrelation of human and animal is even clearer concerning language, because not only do certain animals have phonological systems of communication, but also there are some that have, among other things, the material

24. Cf. *Metaphysics* Epsilon.1, 1025b24: "every thought is either practical, poetic, or theoretical" (πᾶσα διάνοια ἢ πρακτικὴ ἢ ποιητικὴ ἢ θεωρητική).

25. The tendency to live together defines "gregarious" animals, of which political animals are a subdivision. Human beings are thus also naturally gregarious.

means, that is, organic, for producing language, and especially because Aristotle applies to human language one of the major conceptual tools from his biological analyses, and that more completely than he does for politicalness, for example. The Aristotelian analysis of language puts to work at least three concepts: language properly so-called (λόγος); voice (φωνή), which conveys significance; and speech (διάλεκτος), which is articulated voice. Speech, according to the *History of Animals* 4.9, 535a31, "is the articulation of voice by the tongue." Aristotle reserves language for human beings, and seems hesitant about speech, since at ten lines of distance, he affirms that speech is "proper" to human beings (536b1), before finding in certain animals, for example the parrot, "articulate sounds that one may consider as speech" (536b11). As for voice, it serves animals by signifying and communicating the pleasure and pain that they feel, and from that point of view, human beings are animals just like the others. Furthermore—another relationship between language and biology—Nature has given human beings the physical and physiological means to speak, and Aristotle explains forcefully and in detail how the human tongue and lips[26] are conceived for the facilitation of speech and the respiratory system to make it possible.

Not only does Nature include in the biological toolbox that she has provided human beings a larynx, teeth, tongue, lips, which make it possible for them to speak, but she also applies to language the same explanatory schema as for the hands: she gives language to human beings alone, because they alone can use it, which means that other animals cannot, and their cries are sufficient for them. On the other hand, Nature can make use of the parts useful for speech for other uses:

> [In human beings] the lips are for the protection of the teeth, as in other animals, but even more for the good. Because they are also used for speech. Just as, in fact, Nature has not made the human tongue like that of other animals, having made it for two uses, as we say she has done in many cases; she has made the tongue for the perception of tastes and for speech, and the lips have two uses, speech and the protection of the teeth. In fact, language expressed by voice is composed of phonemes, but if the tongue were not as it is and if the lips

26. Tongue and lips do not have the same status, because although they seem necessary for articulating sounds, the lips are neither necessary nor sufficient: the parrot articulates without lips, the monkey has lips but does not articulate.

> were not moist, most phonemes could not be pronounced. (*Parts of Animals* 2.16, 659b32, already cited)

Here there is a small bit of Aristotelian exegesis to be clarified. Elsewhere, in fact, Aristotle asserts that the goodness of Nature is demonstrated in a principle of specialization:

> It is better, when possible, to not have the same organ for different functions, and that the organ of defense be sharper, while that which functions as a tongue be spongy and handle food. In fact, where it is possible to use two organs for two functions without their getting in the way of each other, Nature does not usually make, like ironsmiths, spit-and-lampstand combinations, to economize. But where it is impossible, she uses the same organ for several functions. (*Parts of Animals* 4.6, 683a20) [27]

If one wants to harmonize these two positions, while avoiding slicing up the *Parts of Animals* into chronologically distinct strata, it is necessary to think that specialization (one organ for each function) is the best arrangement, but it is sometimes impossible or at least counterproductive, and that consequently plurifunctionality is often a necessary resort. In fact, it would have been difficult to place two independent systems on the front of the human head, both requiring the use of breath, the one for respiration, the other for speech. Furthermore, this plurifunctionality is a proof of Nature's talent in finding and thus obeying a teleological principle. Thus, the fact that, in view of the impossibility of freeing their front feet from the sole function of supporting their imposing corporeal mass to give them a prehensile function, Nature has assigned the latter function to the elephant's trunk, which is otherwise an organ of respiration; this doubtless requires more intelligence than opting for a juxtaposition of organs that would have been in any case difficult (cf. *Parts of Animals* 2.16, 659a16ff.). Nature is economical when she can, and when she cannot, she becomes clever.

27. Cf. *Politics* 1.2, 1252b1, concerning the fact that it is better that the functions of a wife and of a slave should be performed by different people: "for Nature does not scrimp like the blacksmiths that make Delphic knives, but she makes one thing for one use: in that way, each tool accomplishes its task better."

But what is important for our argument is to notice that the double function of the elephant's trunk and that of the human tongue does not have the same status, since the first obeys both the law of organic correlation and that of subordination of characters, which the second does not do. In fact, it is necessary for the elephant to have an organ of respiration and an organ for grasping, the two need to be able to function together, the first being more fundamental. When we say that, we do not leave the organism of the elephant, while language is a function come from elsewhere: there is no biological sense in saying that the nutritive and linguistic functions of the tongue are correlated, and that one is more fundamental than the other. From that, we may thus draw the conclusion that the possession of language cannot be invoked as an element that indicates the *biological* superiority of humans over other animals. The olfactory acuity of dogs, in contrast, confers on them a biological superiority over human beings, for example.

To return to language, despite the rough sketches of language that some animals possess, it is proper to human beings and thus ought to be connected to a characteristic or capacity that is also proper to them, like the hand, which is proper to human beings, having been allotted to man by Nature on the basis of another characteristic properly human, an intelligence superior to that of beasts. More precisely, as is said at the beginning of the *Metaphysics*, humans are the only animals that attain the level of art (τέχνη) that presupposes the use of reasoning.[28] In other words, human beings "deserve" the hand, because they are rational animals. It's the same for language, which has to do with that which is proper to humanity. But, as Aristotle says with a remarkable redundancy, "That which is *proper* to human beings in relation to other animals, is that they *alone* have perception of good and bad, just and unjust" (*Politics* 1.2, 1253a15), and language is the tool that they use to communicate these values, which animals do not do. Here Nature applies to a *practical* trait (for the values in question are ethical) that she usually uses for biological properties. Good and bad, just and unjust, are properly human values, in contrast to pleasure and pain, which means that they are tied to this specifically human quality, the possession of reason, and that all the more in that the word *logos* designates both language and reason. We can see

28. Cf. *Metaphysics* Alpha.1, 980b26: various animals "share somewhat in experience (*empeiria*), but human beings use both art and reasoning" (καὶ τέχνη καὶ λογισμοῖς).

at the same time that Aristotle, in contrast to later empiricists,[29] avoids making language a howl continued by other means, to tie it to a structure proper to the human spirit. Aristotle is definitely closer to Chomsky than to Condillac. That does not bring into question the divide between theoretical and practical, but we cannot say that this human specificity entails a *biological* superiority.

It seems, then, that men and beasts can be placed in a series of comparisons according to the more and the less, with, in the background, an absolute difference that the possession of *logos* makes, something that we will look at more precisely in the next section of this chapter. The relationship between humans and animals according to the more and the less is subject to three limits. First, this classification, far from being global (which would allow saying, for example, that a monkey is closer to the human model than is a sparrow), concerns only specific properties: the type of reproduction, or the organs of perception, courage, or cleverness. Second, if not for psychic dispositions like tameness or courage, at least for functions like perception, there is no single scale, since, for certain qualities, certain animals win out over humans, but, as we will also see, not from the point of view of every sense. Third, it is not even sure that man wins out in every case of intellectual activity, as we have seen concerning spiders and migrating animals.

In fact, we have seen concerning spontaneous generation, and we could bring up many other examples, it is the organs and processes that are the most perfect that help us understand the less perfect. To be more precise, the class of viviparous blooded animals gives us the key to the organs and processes in less perfect animals, mainly thanks to analogy: it's a matter of seeing what plays the role of sperm in oysters, and so forth. In his paradigmatic function, man is often submerged in a far larger group.

Nevertheless, the degradation of the human corporeal schema to which may be added the greater or lesser degrees of dispositions and performances more or less great of certain functions (but which do not always reveal inferiority of human to animal), do not have a sufficient explanatory force to establish man as a model of intelligibility pure and simple for animals. Thus, one may say that anatomical-physiological knowledge of human beings is not a decisive trump card for the Aristotelian zoologist.

29. And many contemporary scholars, notably paleontologists, seriously infected by empiricism, like most scientists.

But a third reality forces us to reexamine the question of a *model*, namely, the baby. We can cite again, but more completely, a passage considered earlier:

> For example, what are art, wisdom, and understanding in man correspond in certain animals with another natural capacity of this kind. This is particularly clear when we look at the period of infancy: in babies it is possible to see as it were traces and seeds (ἴχνη καὶ σπέρματα) of the states to develop later, but their soul does not differ much from that of animals at this time, so that it would not be illogical that some traits would be the same as in other animals, whether that be close relatives or analogues (τὰ μὲν ταὐτὰ τὰ δὲ παραπλήσια τὰ δ' ἀνάλογον ὑπάρχει τοῖς ἄλλοις ζῴοις). (*History of Animals* 8.1, 588a29)

This passage thus supports that idea that some traits can be the same, related, or analogous, in animals and infants, but implicitly it conveys the idea that this is not the case between animals and adult human beings. We have already seen that concerning the proportions of the body. But infants are potentially adult human beings. In the case of infants, we would thus have examples of animals that have the adult man as a model in the sense of a form that they will themselves actualize. From that point of view, animals, at least some of them, would be incomplete human beings that remain permanently incomplete; in a sense, permanently babies. Some adult human beings are also closer to animals, and infants, than to complete adults; that's the case for natural slaves, and we will see that concerning pleasure and pain. For on the crucial question of pleasure too, human beings are at an interface between biology and practice.

Animal Pleasure, Human Pleasure

Pleasure is a particularly interesting idea for us, because, like language and politicalness, it constitutes what I have just now called an interface between the biological and practical domains, but also because pleasure and pain are per se properties for every animal, since they are in a necessary relationship with sensitivity, which is itself definitory of animals.[30] Thus,

30. This section relies on one of my published articles, "Le plaisir animal selon Aristote."

we must, in order to study pleasure, appeal to ethical and political texts, as well as to texts in natural science. At first glance, things seem simple enough, and can be summarized in the form of two theses. (i) There are two kinds of pleasure, bodily pleasures (σωματικαί), and psychic (ψυχικαί) (*Nicomachean Ethics* 3.10, 1117b28). Whence a first division: animals have access to the first, humans to both. (ii) There are two possible uses of bodily pleasures for human beings according to whether they are temperate or intemperate,[31] whence a second division between temperate men on the one hand, and animals and intemperate humans on the other. What follows will add complexity to this simplicity.

Psychic pleasures are in fact of two kinds. First, there are those that are felt "without the body being affected" (οὐδὲν πάσχοντος τοῦ σώματος, 1117b30), for example the pleasure of learning. The study of these pleasures is an important part of Aristotelian ethical research, but doubtless the pleasures that come from the senses are more crucial for Aristotle, as is true for all ancient moral theory. He confirms that when he writes, beginning his discussion of pleasure in *Nicomachean Ethics* 7, "The study of pleasure and pain is the province of the political philosopher" (11, 1152b1). As I tried to show in my *Endangered Excellence*, ethics and politics are in fact one and the same science considered from two points of view, rather than two different and subordinate sciences.[32] Whence, in the *Nicomachean Ethics*, there are two long passages, in two different books (7 and 10), dedicated to pleasure, of which the first is repeated in the *Eudemian Ethics*. The major aspects of this ancient approach to pleasure, which takes a more developed and subtle form in Aristotle than in many others, are well known, especially since they may be opposed to the stance taken concerning pleasure by our religious or post-religious societies, which consider it as at the least suspect. We can summarize without going into too much detail, because that does not concern us directly here.

Certainly, there are for Aristotle, as for other ancient thinkers, bad pleasures, but the general rule is that pleasure is the sign of the success of the activity that it accompanies:

> Activities that differ in kind are completed by things that are different in kind. So, the activities of thought are different in

31. The word ἀκρατής will be translated either "incontinent," "intemperate," or "uncontrolled."

32. Pellegrin, "De la tradition aristotélicienne," 37–48.

> kind from those of sensation, and these all differ in kind among themselves, and consequently the pleasures that complete them differ in kind among them. . . . An activity is intensified by the pleasure that is appropriate to it, and in each domain, those who judge and discern better are those who take pleasure in performing these activities; thus, in geometry those who take pleasure in doing geometry are those who grasp each thing better; the same for lovers of music or architecture. (*Nicomachean Ethics* 10.5, 1175a25)

Nicomachean Ethics 10.4 presented this point very completely. Pleasure is thus a sign that the action that it accompanies is virtuous, which allows making a distinction between the virtuous act and the action of the "continent" (ἐνκρατής) person, who does apparently virtuous actions, but for whom it is painful to abstain from vice: the continent person is courageous, abstains from lying, and so on, but the virtuous person does these same things *with pleasure*. Thus, "it indeed seems that the goal of courage would be pleasant" (*Nicomachean Ethics* 3.9, 1117b1). Aristotle also gives, among others, examples of psychic pleasures that are not directly concerned with ethical virtue. Thus, at *Nicomachean Ethics* 3.10, 1117b29, "love of learning" (φιλομάθεια). This last class of pleasures is particularly significant for us, because Aristotle, when cataloguing psychic pleasures, asks which soul is involved. Obviously not the nutritive soul, or the sensitive soul per se. Aristotle relates this kind of pleasure to discursive thought (διάνοια), which makes them pleasures that we can call "intellectual." Right away we understand why animals cannot experience these in the proper sense of the word, since they have nothing better than "traces" of thought. We will speak again a little later of pleasure as a sign of the *success* of an action.

Psychic pleasures of the second sort engage the senses. "Where there is sensation, there are pleasure and pain" (*De Anima* 2.2, 413b23), which means that even an oyster can experience pleasure and pain. Furthermore, the *Nicomachean Ethics* tells us that "for each sense there is a corresponding pleasure" (10.4, 1174b20), but perhaps this passage applies only to human beings. These psychic pleasures also involve the body, but they are not, properly speaking, pleasures of the body. We will see now that animals, that are obviously concerned primarily with pleasures of the body, can also experience certain psychic pleasures.

As far as animals are concerned, the distinction between bodily pleasures and psychic pleasures, made in the ethical treatises, is also found in the zoological treatises (no implication of a chronological relationship is intended). Animals have at least one sense, that of touch, and up to four more; most likely, then they have different kinds of pleasures associated with each sense, because we don't know why animals would be excluded from "for each sense there is a corresponding pleasure," mentioned earlier. We may presume, although there are few texts, that for the sense of touch, of which the organ is flesh for animals that have flesh,[33] or its analogue for the others, the pleasure would come from the stimulation of what modern physiology would recognize as the nerve endings present in the flesh. There is in Aristotle at least a tendency to figure that the *normal* activity of all the senses is a source of pleasure, and by normal we mean proportional, excess leading first to pain and then to the destruction of the sense concerned:

> If voice is a certain concord and if voice is, from a certain point of view, the same thing as hearing (but from another point of view they are not one and the same thing),[34] and if concord is a proportion (λόγος), it is necessary that hearing also be a certain proportion. And for that reason, each excess, whether high-pitched or low-pitched, destroys hearing. The same goes for the sense of taste with savors, and for vision with colors, if the object is too bright or too dark, and for smell if the odor is strong, whether sweet or bitter, because perception is a certain proportion. That is why sensible qualities produce pleasure—when they are pure and unmixed, they lead to proportion and procure pleasure (ἡδέα μέν, ὅταν εἰλικρινῆ καὶ ἄμικτα ὄντα ἄγιται εἰς τὸν λόγον), for example the spicy, sugary, or salty are then [when they are proportionate] pleasant. In a general way, mixture is a concord rather than the high-pitched or low-pitched, and for touch it is that which can be warmed or cooled. Sensation is a proportion, when excess causes pain[35] or destruction. (*De Anima* 3.2, 426a27)

33. Flesh is only an intermediary, the real organ of touch being internal, in the region of the heart or its analogue.

34. This parenthesis has been suppressed by several editors, but it is found in the manuscripts and does not hurt the sense.

35. The reading λύει (loosens), chosen by Bywater, instead of λυπεῖ (causes pain) is possibly better.

The word *logos* here is not "reason," which Aristotle does not attribute to animals, but its normal sense of "proportion" or "ratio," and thus relates also to animals. So, in the case of fish: "In fact, we can see clearly that they swim away from loud noises, for example the oars of triremes, so that they are easily captured in their holes. Because a slight sound in the open air becomes for those that hear it underwater painful (χαλεπός), strong, and heavy" (*History of Animals* 4.8, 533b5). Within the category of corporeal pleasures, those that are called "corporeal pleasures" per se are those that engage touch and taste, psychic pleasures concerning the three other senses. But animals have a very different relationship with each of the two sorts of pleasure: "None of the other animals, besides man, has pleasure in these senses [vision, hearing, smell] except coincidentally (accidentally). In fact, dogs do not enjoy the odor of hares, they enjoy eating them; odor helped the dogs find them. Similarly, the lion does not delight, except accidentally, in the lowing of the ox that he is going to eat, and whose odor he has sensed" (*Nicomachean Ethics* 3.10, 1118a16). Remember the close relationship that Aristotle makes between touch and taste. Certainly, the sense of taste has as proper sensibles qualities associated with savors, but that does not provide, Aristotle says, more than a bit of pleasure in itself, "in any case not in self-indulgent people" (*Nicomachean Ethics* 3.10, 1118a29), who "look only for the pleasure coming entirely by touch" (1118a29). Then Aristotle tells the anecdote about the gourmand who prayed that his gullet would become as long as that of a crane so that he would take pleasure in the contact, meaning that the pleasure of swallowing outweighed that of savors (3.10, 1118a33). Aristotle even gets to the point of saying that "taste is a form of touch" (*Sense and Sensibles* 4, 441a3, and *De Anima* 2.9, 421a19).

As for animals, they indisputably take pleasure in the exercise of touch and taste: "It is obvious that fish have a sense of taste, because many of them take pleasure (χαίρει) in specific savors, and they often bite on bait made with tuna or fatty fish, as if they had pleasure in tasting (ὡς χαίροντες ἐν τῇ γεύσει) bait like that" (*History of Animals* 4.8, 533a32). "These animals [mollusks, crustacea, and insects] in fact search for different foods and they do not all take pleasure in the same tastes (οὐ τοῖς αὐτοῖς πάντα χαίρει χυμοῖς). The bee, for example, never approaches anything rotten, but just sweet things, and the mosquito, nothing sweet, but acid" (535a2). The agreeable, or not, character of these odors is accidental, because they are agreeable or painful because of something else, namely the fact that they are the odors of foods that the subject that perceives them likes or not, and the fact that it is hungry or satiated. That applies to animals just as much as to human beings, self-indulgent or not: one cannot say that the aroma of a delicious serving is agreeable per se, since it can just as well be neutral or disagreeable, for example for someone who has had too much to eat. But to be hungry or satiated can be the state of a temperate person. To put it simply, Aristotle's position is less bizarre than it might seem at first glance. Things are clearer for vision and hearing than for smell, as we will see again later: a lion takes pleasure in seeing the color and hearing the cry of a gazelle that serves as his habitual food, but that does not make him a lover of paintings or music. We need to make two points more precise. First, animals do indeed have olfactory, auditory, and visual pleasures, because to say that they have them "coincidentally" is not to say that they don't have them at all. Second, a temperate man may also have pleasures and displeasures coincidentally.

And then there are odors that are per se agreeable, like those of flowers: "They do not lead more or less toward food, nor do they arouse any desire, but rather the opposite" (*Sense and Sensibles* 5, 443b28), Aristotle doubtless wanting to say that no one has a tendency to eat something that smells like a flower, for, as the comic poet Strattis says, one must not put perfume on lentils (*Sense and Sensibles* 5, 443b31). In fact, odors that are accidentally agreeable are odors that are related to tastes, while the odor of flowers is not. Aristotle says, anyway, of the first that "they are divided into species according to tastes" (444a6).[36] Aristotle repeats several times that these kinds of pleasant odors per se are "proper (ἴδιον) to the human being" (444a3, 444a28), which means that animals do not perceive them per se, but can only perceive them accidentally. But they cannot perceive them accidentally, because they are not relying on other pleasant perceptions, in this case tastes. Thus, it would be necessary to conclude that not only are they not pleasant for them, they don't perceive them at all.

We need to note that that goes for bad odors as well:

> In the same way, there is no animal other than the human being that is displeased (δυσχεραίνει) by the odor of things that are foul-smelling of themselves, at least if they are not pernicious (φθαρτικόν). The same for human beings who get

36. Cf. earlier 443b1: this type of odor "ranks in the same order as tastes" (τὸ μὲν γάρ ἐστι κατὰ τοὺς χυμούς τεταγμένον αὐτῶν).

> a headache from charcoal smoke, which can often cause death (φθείρονται), in the same way, from the strong fumes of sulfur and bitumen nonhuman animals die (φθείρεται) and they flee because of the sensation they feel. But the bad odor itself does not cause them concern (while many plants smell bad), at least unless it has some effect on the taste of their food. (*Sense and Sensibles* 5, 444b28)

We should understand that human beings sense these foul odors and experience displeasure, while animals do not experience displeasure (they "do not cause concern," οὐδὲν φροντίζουσιν, 445a2), because they are per se, but they are accidentally disagreeable, either because they are dangerous, or because they pollute the food and are thus associated with gustatory displeasure.

Aristotle offers two reasons, very different from each other without being incompatible, for which animals do not take pleasure in the odor of flowers. This odor, in fact, is not simply a luxury available to human beings, tied to a function, intellectual or ethical proper to them, and not available to animals. Odors per se in fact may have a biological function: *Sense and Sensibles* 5 explains that good odors, those of flowers for example, have a warming effect on the brain, a cold organ: "This kind of odor comes to human beings to help them be in good health; it does not have any other function, and it succeeds" (444a14). But the human being is the sole beneficiary of the hygienic virtues of these odors, because the human brain is that one that is proportionately the largest, and also the most cold and moist. All that fits with the general schema of Aristotelian finalism: other animals have a smaller brain, less moist and less cold, so they do not need the heating provided by the odor of flowers. So, if these odors were to warm their brain to the point of giving them pleasure, they would do it *in vain*. Although Aristotle does not say so explicitly, we can think that we are dealing with a liminal effect: as nonhuman animals have a smaller brain, less cold and less moist, the warming action of the emanation of the odor of flowers would be less significant, such that it would be below the threshold at which it could be perceived as pleasant. Because it does indeed seem that it is the warming function of the odor of flowers that gives it its pleasant character. But the odor in question has nevertheless a warming action on the brain of animals, and more on some of them, like quadrupeds, as we will see in a moment.

This position of Aristotle is paradoxical all the same. How can it be that dogs, who have a much finer sense of smell than ours, don't perceive odors, whether good or bad, that are perceived by people? We can understand that a dog does not get any pleasure from smelling a rose, since that odor does not coincide with canine food, but to go from there that the dog smells nothing . . . In fact, we must certainly introduce some adjustments to the thesis that animals *do not at all* perceive odors that are per se pleasant, like that of flowers, or unpleasant per se, like those of things that are per se foul. The passage that says that only the human being perceives odors that are per se good needs to be studied more closely. The text says: "Odors of this kind are proper to human nature because among animals the human being is so to speak the only one to perceive the odor of flowers and other things of this kind, and to get pleasure from it. Their heat and their movement are in fact proportional to the excess of moistness and heat of this place [the brain]" (*Sense and Sensibles* 5, 444a28). In the part of the sentence that says that "the human being is so to speak the only one to perceive the odors of flowers and other things of this kind, and to get pleasure from it" (μόνον ὡς εἰπεῖν αἰσθάνεται τῶν ζῴων ἄνθρωπος καὶ χαίρει ταῖς τῶν ἀνθῶν καὶ τῶν τοιούτων ὀσμαῖς, 444a31), doubtless we must take the αἰσθάνεται . . . καὶ χαίρει as a hendiadys and understand, not that the human being is the only animal capable of sensing this odor *and* to take pleasure in it, which would imply that animals not only would not get any pleasure from it, but they wouldn't perceive it, but that man is the only one to have a pleasant sensation from it. So while the human being will go on smelling flowers *for pleasure*, and that is also useful to him because it warms the brain, as for the animal, it does not get pleasure from smelling flowers, because it doesn't smell the odor for what it is, and because it can't smell it as accidentally pleasant because there is no food with which it can be associated, but it smells it all the same. Not to be perceived as pleasant in itself does not presuppose not being perceived at all.

I believe that we can push the biological analysis of good odors even further by relying on the theoretical toolset forged by Aristotle in his zoological treatises. One must understand that the odor of flowers is "good in itself," because it brings us pleasure independently of any other sensation, but also doubtless because this pleasure is independent of its function (warming the brain), but Nature cleverly uses this quality to accomplish this function. This conforms to the teleological schema that

Aristotle described previously, since the earthy matter, for example, behaves independently of the function that rational Nature gives it, constituting horns or teeth. The difference between a human being (not self-indulgent) and animals is thus that the dog, when it is satiated, does not find the odor of the hare agreeable, and if he perceives it, in any case he doesn't perceive it as a *pleasant* odor, while someone who does not need to have his brain warmed, for example because he has a fever, or because he lives in a tropical climate, will nevertheless find the odor of flowers pleasant, otherwise one could not say that flowers have a good odor per se. Thus, we understand that the odor of flowers is a per se odor, which is useful, by its action on the brain, to the person who senses it, and it is agreeable to all human beings from the very fact of the warming that it produces, even if the person smelling it doesn't have biological need of that warming. As for animals, they perceive it as "neutral."

There is a second reason, very different from the anatomical-physiological reason that invokes the effect of odors on the brain, why only human beings enjoy the odor of flowers. *Nicomachean Ethics* 3.10 presents it best. This reason is in close relation with the division of the senses into touch and taste on the one hand, vision, hearing, and smell on the other, that is, with the division into pleasures of the soul and pleasures of the body (1117b28). But the pleasures of vision, hearing, and smell can certainly be coincidental pleasures for the self-indulgent man, and even, regarding at least olfactory pleasures, for a temperate man (for one may be both temperate and a gourmet), but when they are experienced in themselves, they do not concern the body, even if they are obviously perceived by the intermediacy of the body. These are the psychic pleasures that directly escape intemperance, as Aristotle says at *Nicomachean Ethics* 3.10, 1118a1–12, while he remarks, with his usual perspicacity, that "even in these cases [vision, hearing, smell] there is a way to get pleasure as one ought and, here too, there may be an excess or deficiency" (1118a5). For there is a difference between liking the odor of apples and roses, and liking the odor of perfume and fancy food (1118a10): there is a possibility of intemperance in the second case, when a self-indulgent person enjoys them because these odors remind him of gastronomic or sexual pleasures. But then that's an accidental pleasure.

For these psychic pleasures, *Sense and Sensibles* thus proposes a second approach, to tell the truth, less convincing for smell than for vision and hearing, and so we need to start with them. Aristotle explains that pleasant colors and sounds are those the components of which are in a rational relation with each other. All colors are composed of a mixture of white and black, which doubtless means that white and black can mix together in many proportions (an infinite number, in fact): one part white, one part black; two white and one black, and so forth.

Here's the passage that explains this:

> The product that results [from the mixture of white and black] will not appear to be either white or black, but since it must have some color, and it can't be either of those two, it needs to be a mixed color, a different kind from the first two. . . . One may thus conclude that many of these colors exist according to a proportion (λόγος), white and black in a relation to each other of three to two or three to four, and so on, while there are some that are not based completely on a proportion, but on an incommensurable relation according to excess and defect. In fact, colors involving the most rational numbers, as in the case of musical harmonies, seem to be the most pleasant of colors, for example purple, red, and some few colors of this kind; they are few for the same reason as musical harmonies; the other colors are not based on numbers. Or else all colors are based on numbers, some of them in an orderly way (τεταγμένας), the others disorderly; these, when they are not pure, it's because they are not expressed numerically. (*Sense and Sensibles* 3, 439b22–440a6)

A difficult passage, and one that we won't really comment on here. In *Sense and Sensibles* 3, Aristotle is studying the problem of vision, and thus of colors. From line 439b19 to the end of the chapter, he considers successively three hypotheses for explaining the multiplicity of colors, ultimately rejecting the first two: the various colors are produced neither by the juxtaposition of white and black, nor by the superposition of various colors (as the Sun is white but looks red through fog), but by a real mixture, like that which is analyzed in the treatise *On Generation and Corruption*, to which allusion is probably made a little later ("in the treatise on mixture," ἐν τοῖς περὶ μίξεως, 440b3). But this doctrine of mixture can also be applied to the two other hypotheses,[37] for all of these hypotheses

37. 440b18 says so explicitly: "most of the colors are due to the fact that their component parts can be mixed with each other according to several relationships, some mixtures

consider that the diversity of colors rests on the plurality of relationships between the components, which are white and black.

Concerning unpleasant visual sensations, Aristotle identifies them like those of colors (do not forget that color is the proper sensible of vision), but not composed according to a proportion expressible by numbers, or they are, but "without order," without providing more details, which perhaps shows that the question doesn't interest him very much. It's the same for sounds, which have as elementary components, high and low pitch.

We have here an idea of proportion that is much more demanding than that which was expressed in the passage from the *De Anima* 3.2, 426a27, cited earlier, which was satisfied with calling the relationship in question as "neither too much nor too little." In a remarkable article, Paul Kucharski proposes the hypothesis that this analysis of the diversity of colors has a Pythagorean origin, and more precisely, that it is "based on the theory of music, of Pythagorean inspiration, as it was constructed in the 4th century."[38] But, to the degree that the first *logos*, that of *De Anima* 3.2, is available to many animals, to that extent it seems difficult to imagine that a perceiving subject would be able to grasp mixtures of white and black formed according to numerical proportions if that subject is not able to grasp that sort of relation. A human being, even one with no concept of the proportionate relationships of the notes, is able to perceive that relationship qua rational being, although a connoisseur of music will grasp them more easily, and take more pleasure in hearing the music. Notably from the fact of the theory of the identity of sensed and sensor, at the very basis of the Aristotelian concept of perception.

To say that colors and sounds are pleasant that have rational relations in them goes in P. Kucharski's direction. In *Sense and Sensibles*, Aristotle seems to rely on the common notion that finds musical harmonies pleasant, but there are in the Aristotelian corpus passages that provide the causes of this fact, notably in the Aristotelian *Problems*, but also in *Politics* 8, principally in chapter 7. Let's start with a passage from the *Problems*: "Why does everyone take pleasure in rhythm, melody, and in general in consonances (συμφωνίαι)? Isn't it that we naturally take pleasure in natural movements? The proof is that newborn babies enjoy them" (*Problems* 19.38,

taking place according to numbers, others according to the more and the less; thus, what has been said elsewhere about colors in the hypothesis of juxtaposition and in that of superposition can also be said about mixture."

38. Kucharski, "Sur la théorie des couleurs et des saveurs dans le 'De sensu' aristotélicien," 383.

920b30). Or because, as the *Poetics* says, "imitation as well as harmony and rhythm are natural to us" (4, 1448b20). Since animals are not music lovers, which means in Aristotelian terms that they don't have auditory pleasures per se but only by accident, one may conclude that harmony and rhythm are not natural to them. What the Aristotelian teaching about auditory pleasures per se (but also visual and olfactory) establishes then is nothing less than a separation between human nature and animal nature.

We need to note here a point at which Aristotle's doctrine of psychic pleasures in a way overflows its banks, and that will allow us, as we promised, to find more complexity in the second simplistic thesis we presented at the beginning of this study of animal pleasure. In *Politics* 8.7, Aristotle says that melodies that are little or not at all harmonious may nevertheless provide pleasure:

> And since there are two kinds of spectators, the one free and educated, the other vulgar people, artisans, laborers, and others of this kind, there need to be competitions and exhibitions for their relaxation as well. Just as the souls of those people have deviated from their natural state, melodies are deviations from natural states.[39] What provides pleasure to each person is that which is appropriate to his or her nature (τὸ κατὰ φύσιν οἰκεῖον); that is why one must give permission to those who compete in the competitions during these presentations to resort to music of this kind. (*Politics* 8.7, 1342a18)

Thus, it's a matter here of a pleasure of the soul, but of a soul that has "deviated from its natural state." But as deviant as it may be, the soul of these vulgar people, which one may often consider as "unregulated," remains natural, just as illness is a natural state, and, on the other hand, the pleasure that these people get from obnoxious songs does not seem to be based on sensual pleasures, as the odor of the hare that reminds the dog of its taste. In fact, we are not in the situation to which Aristotle calls attention at *Nicomachean Ethics* 3.10, 1118a12, which speaks of self-indulgent people getting pleasure from the senses relevant to psychic pleasures (vision, hearing, smell), but accidentally because these percep-

39. In which the "color" (the image is conveyed in Greek by the word παρακεχρωσμένα, 1342a24) is made irregular by the presence of small intervals subdividing the genus (chromatic or diatonic) into species called *chroai*. Cf. Bélis, "Les nuances dans le Traité d'harmonique d'Aristoxène de Tarente."

tions remind them of "the objects of their appetites" (1118a13). So what we doubtless have here is the case of psychic pleasures per se, but in self-indulgent people. Because it's not the same thing for self-indulgent people to enjoy a pornographic presentation, which would be an accidental pleasure, recalling their debauchery, and to take pleasure in discordant music that, if not always at least often, will not remind one of any pleasure related to touch. As is often the case, we have here a crucial point dealt with in a fashion that is at the very least laconic. Because if Aristotle elsewhere establishes a split between temperate people on the one side, and self-indulgent people and animals on the other, here the line runs between human beings, both temperate and self-indulgent, and animals.

In fact, self-indulgent people pose a formidable problem that Aristotle does not take the trouble to resolve explicitly. Because as self-indulgent as they are, these people remain no less rational beings, able to appreciate beautiful colors or harmonious sounds. Two situations remain possible, maybe separated from each other by degree. It's possible that a libidinous glutton nevertheless still remains a lover of art and music, even if at times he can accidentally enjoy colors and sounds that remind him of his debauchery. But it is possible that his self-indulgence has reached the point where he cannot enjoy colors and sounds except accidentally. At that point we could talk about "bestiality" (θηριότης).[40] It's still the case, however, that the self-indulgent person can cure his self-indulgence, while beasts cannot cure their bestiality.

In contrast, we must recognize that Aristotle does very little to enlighten us about odors: do odors per se, that is, those that do not coincide with the taste of foods, also have a numerical structure that stems from discursive thought? The sense of smell is surely the "theoretical" sense that most closely approaches the "bodily" senses of touch and taste. Perhaps one may change the terms of the problem, rather than resolving it.

Odors that are pleasant per se (in human beings only), we have seen, have a beneficial effect on those who perceive them. But that does not mean that everything pleasant is beneficial; a passage in the *Sense and Sensibles*, just two lines further than the passage cited earlier, is worth our notice:

40. As contraries are correlated with each other, the *Nicomachean Ethics* tells us that bestiality is so vicious that its contrary is a "heroic and divine virtue" like that of Hector (7.1, 1145a20).

> Food being both dry and moist,[41] although pleasant, is often harmful, but the scent coming from an odor pleasant per se is so to speak always beneficial in whatever circumstances. And it is because this odor comes (γίγνεται) via respiration, not in all animals, but in human beings and some blooded animals such as quadrupeds and all those that participate more in the nature of the air.[42] When, in fact, odors rise toward the brain, because of the lightness of the head that they contain, the parts around it acquire an increase of health, because the odor has a natural capacity to warm. (5, 444a16)

This passage tells us several things. First, that a food can be at the same time pleasant, that is, present a taste that provokes pleasure, and harmful. But taste is one of these sensations, along with tactile sensations to which tastes are closely associated as we have seen, which can provoke in animals a pleasure that is per se because not based on another sensation. We can probably deduce from that that odors associated with this harmful food are accidentally pleasant to them. Odors that are pleasant per se, in contrast, are always beneficial to human beings that sense them, Aristotle being careful to add "so to speak" to leave room for exceptional situations. Next, this passage confirms what was said before: these odors in themselves have *an effect* on animals, at least on some of them, such as blooded quadrupeds, since they warm the brain, but since their brain is proportionately smaller than that of human beings, they do not feel that warming, and either they do not perceive these odors (hard-nosed version), or they perceive them, but don't take pleasure in them (moderate version).

There remains the question that was left hanging: we have seen that pleasant colors and sounds are so because they express a rational structure among their components, and are unpleasant when that harmony

41. Translators generally understand the words καὶ ἡ ξηρὰ καὶ ἡ ὑγρά at 444a16 as meaning "food, whether dry or moist," but George Robert Thompson Ross (cf. Aristotle, *Aristotle, De Sensu and De Memoria*), is certainly right, appealing to Alexander (without citing him), in understanding that the cold associated with the moist is a morbid factor. Jules Tricot is also of this opinion. Cf. 441b25: "neither the dry without the moist, nor the moist without the dry, is a food, for that which nourishes animals is not just one of these two factors, but their mixture."

42. We note that this remark confirms our interpretation of *On Respiration* 19, 477a25, proposed in the previous chapter.

is missing, and that corresponds to the *nature* of those who perceive them—but why are odors per se pleasant or unpleasant? It seems difficult to extend to olfactory pleasures the analyses that apply to pleasure per se due to harmony that supports melodies and well-tempered colors. It seems possible to attribute to Aristotle the thesis that very strong odors are disagreeable, even damaging, but did he think that there are well-tempered odors? Maybe on this question of the conformity of pleasures of smell with other psychic pleasures (those procured by vision and hearing) we need to take a completely different tack. In the *De Anima* we read:

> Smell and its object are much less easy to define than the preceding objects, because what odor is, is not as clear as for sound, light, and color. The reason is that our perception of odors is not as acute, and is less good than that of many animals. The human being, actually, has a poor sense of smell, and does not perceive any odor without either pleasure or pain, which shows that olfaction lacks acuteness. And it is likely that it's the same for the perception of colors by animals with hard eyes: the differences between colors do not appear to them clearly beyond whether they provoke fear or not. (2.9, 421a7)

Ever since Simplicius, commentators have understood that as meaning that, as the human sense of smell is not very easy to impress, only strong odors reach it, but they are accompanied by pleasure or displeasure. If this text means to say that pleasure or displeasure follows sensation beyond a certain threshold, Aristotle's position would be untenable, because the fact that the dog senses odors that we do not perceive in no way prevents the dog from submitting to very strong odors that we would also perceive. So, the dog too should experience pleasure and displeasure coming from odors, and possibly more than us.

This passage does not at all deny that a dog, for example, would experience pleasure, or pain, from smelling strong odors, but it can only be a matter of a coincidental pleasure. I think that we must interpret the assertion that the perception of odors by a human subject does not happen without pleasure or pain by keeping two things in mind: first, it's a matter of odors perceived per se, and that is a privilege of the human species; second, in the case of the sense of smell, there is no neutral sensation, but when it is strong enough to be perceived, an olfactory sensation is either pleasant or unpleasant: the καὶ in the part of the sentence καὶ οὐθενὸς αἰσθάνεται τῶν ὀσφραντῶν ἄνευ τοῦ λυπηροῦ ἢ τοῦ ἡδέος at 421a11 seems to me to have a consecutive value (the human being has a bad sense of smell, *which brings it about* that he does not perceive without pleasure or pain). As far as the other two theoretical senses, it seems, in contrast, probable that one may perceive by way of their mediation without pleasure or pain, for example when we see a white surface or hear an ordinary sound. As for intensity, it cannot be other than painful for these two senses—excessively bright light, or blaring noise. It's through experiencing pleasure or displeasure by way of respiration that a person becomes aware of sensing an odor. Similarly, animals with hard eyes do not perceive colors unless they are very lively; but they make them afraid. So, there is a precondition for an odor being pleasant (and doubtless that should mean odors either agreeable per se or by accident), that they be of moderate intensity. But to explain that an odor is pleasant per se, thus for a human being, there is only the anatomical-physiological analysis: it's because it warms the human brain *more than* that of animals.

The picture of the relationships between animals and human beings that appears at the end of this long analysis of Aristotle's theory of pleasure is both striking and complex. To stick with animals that are relatively close to us, that is, in terms of what interests us here, those that have five senses, we would say that all, both animals and human beings, experience pleasure through their five senses. Given Aristotle's statements ("for each sense there is a corresponding pleasure," *Nicomachean Ethics* 10.4, 1174b20), we can assume that these pleasures are proper to the senses that experience them. But two distinctions introduced by Aristotle tend to both unite and separate humans and animals: the distinction between bodily pleasures and psychic pleasures, and the distinction between per se pleasures and coincidental pleasures. We need to consider each group of senses individually.

Human beings share with animals the per se pleasures that all experience, those of touch and taste. In these domains, living beings with the power of perception can be ranked according to their level of performance, and the human being carries the day against all the others due to the finesse of his flesh. This seems to be a matter of specific superiority, that the least sensitive of human beings is superior to the most sensitive of animals, but this is not entirely certain. For the pleasures of vision and hearing, animals do not experience them per se, but coincidentally, that is, relying on tactile and gustatory pleasures. It seems that this reliance makes an appeal to memory: it's because he remembers the pleasure of having eaten beef that the lion experiences coincidentally a pleasure hearing the lowing of the ox, and also no doubt seeing the color of the ox. Self-indulgent human beings who, exactly, "enjoy like animals," can also experience coincidental pleasures, and I devised the example of pornographic images, because in them a debauched person has visual pleasures coincidentally, and one may doubtless imagine cases like that for pleasures of hearing. Thus, in contrast to some rather quick statements by Aristotle, pleasure of vision and hearing can be experienced in an intemperate way, but they are then coincidental pleasures. But—and this has not been sufficiently emphasized by the commentators—a temperate man can also experience pleasures of taste and touch. A pornographic picture cannot, in contrast, cause a coincidental pleasure in a virtuous man, because it is coincidental for a self-indulgent pleasure that he does not experience. In other words, human beings, including virtuous people, experience the entire range of pleasure experienced by animals.

Psychic pleasures, in contrast, reveal an unbridgeable gap between human and animal: the per se pleasures of vision and hearing are properly human because they depend on the possession of *logos*, and Aristotle goes so far as to say, if the *Problems* correctly reflect his positions, that they correspond to *human nature*. This is not a matter of a difference of degree. Remember that the psychic pleasures in which we are interested here are those experienced via the activity of perception, and not the pleasures that have nothing to do with the body, like the pleasure of learning or of doing geometry, that depend a fortiori on human logical capacities.

The pleasures of the sense of smell occupy an intermediate position. In fact, they belong to the psychic pleasures, and as such cannot be experienced by animals except coincidentally, while human beings can have a per se pleasure in the odor of flowers. We have abandoned the idea that animals cannot sense the odor of flowers at all: they definitely do smell them, but they don't get pleasure from it. One of the difficulties with the Aristotelian analysis of olfactory pleasures lies in the fact that, in contrast to what happens with touch, many animals have a more sensitive sense of smell than our own, which means that they sense odors that we do not, and that nevertheless they have olfactory pleasures only coincidentally. Odors that are good per se, a human privilege, perform a beneficial biological function, which is not the case with pleasant colors and harmonious sounds. In fact, they warm the human brain that needs it, because of its size and its significant coldness. Here the difference in degree slides surreptitiously, since Aristotle says that if these odors do not have such a beneficial effect on animals, it's because their brain is smaller and less cold, which registers a bit of hesitation in regard to some animals like viviparous quadrupeds, who could very well have a sort of hint (or "trace") of this per se olfactory pleasure. A symmetrical analysis seems possible for bad odors. All the odors that are good per se, which are simultaneously pleasant and bringers of health, need to be, among other things, of a certain intensity below which human beings would not perceive them and above which they would become painful to people.

Perceptibility defines the contours of an animal space in which human beings discover a location. It's impossible to arrange all sensing living beings, that is, animals in a broad sense (human beings included) on a unique scale, according to the performative plusses and minuses of their various senses, but one can according to several angles. First, a particular animal may be superior to another in regard to vision, but inferior to it in terms of hearing; next, as Aristotle says at *Parts of Animals* 3.13, for example, insects and fish cannot be compared, given the great differences in the conditions in which they exercise their function. We can return to one of the passages, already cited, that says:

> [The sense of touch] is the keenest sense in human beings. For the other senses, in fact, man cedes much to many animals, but his sense of touch is much more acute than that of other animals. That is why he is also the most intelligent (φρονιμώτατος) of animals. A sign of that is that within the human family too it is in relation to the organ of this sense that people are more or less well endowed, more than in relation to any other. In fact, those who have hard flesh are less well-endowed intellectually (τὴν διάνοιαν) and those who have soft flesh are well endowed. (*De Anima* 2.9, 421a19)

The commentators have lost themselves in guesses about this relationship between flesh, the organ of touch, and intelligence. It seems that Aristotle's reasoning is as follows: we recognize that among humans, those whose flesh is softer are more intelligent than those whose flesh is harder. Therefore, an animal with softer flesh will be more intelligent than an animal with harder flesh. But the human being is, among animals, the one with the softest flesh, and thus is the most intelligent. But the relationship between softness of flesh and intelligence is never really explained. Would it be possible, for example, to compare the sensitive flesh of quadrupeds with the insensitive flesh of crustacea? At most the absolute superiority of human beings, plus the thesis that human beings have greater tactile sensitivity, supports the idea that greater tactile sensitivity is indicative of an intellectual superiority, but this passage remains generally in the relationship of the more and the less, which is clearly indicated by the superlative φρονιμώτατος.

What happens if we add the pleasure attached to the exercise of the various senses? An animal seems to look like an incomplete human being, since a human being can experience all kinds of pleasures, both per se and coincidental, which an animal can feel, but he has in addition access to pleasures per se that animals cannot experience. In fact, this doctrine, at first glance odd, of coincidental pleasures, may be able to allow us to rethink the biological function of pleasure. The vital role that Aristotle attributes to pleasure is, in fact, fundamentally different from that which prevails in modern zoology, especially among evolutionists. According to these last, a pleasant activity will encourage an animal that engages in it to reproduce it, with sometimes important consequences, according to whether it has a positive or negative selective character. Thus, the attractiveness of sexual pleasure is a kind of ploy by nature that is particularly efficacious, in that it encourages the perpetuation of the species. This is not Aristotle's position.

In the famous chapter 2 of book 1 of the *Politics*, Aristotle describes the tendency of women and men to have intercourse, not as the result of their sexual desire, nor as a "deliberate choice" (προαίρεσις, 1252a28) that would push them to have children for various reasons (even if these two causes play a role), but "as in other animals and plants, as a consequence of the natural tendency to leave behind them a being similar to them" (1252a29). This is a general tendency involving all living things. The passage from the *De Anima* cited earlier (2.4, 415a26), which attributes to all living things a tendency to reproduce in order to participate in "divine eternity," says it even more forcefully. Similarly for Aristotle human beings do not gather in societies of various kinds in order to satisfy their needs, notably the need for security, but by virtue of a natural tendency, that which brings it about that "man is a naturally political animal." At the time anticipated by nature in her great wisdom for intercourse, a bitch will thus look for a male to couple with: whether she experiences pleasure or not does not seem to have any importance. But Aristotle claims that animals perform their two great biological functions, those related to nutrition and to reproduction,[43] with pleasure.[44] The pleasure of intercourse is for all animals a per se pleasure since, Aristotle asserts several times, Aphrodisian pleasures arise from touch,[45] while the pleasure that a carnivore experiences on seeing or smelling its prey is a coincidental pleasure. Would animals then have pleasure *in vain*?

But then what is the function of pleasure? We can get an idea from the fact that Aristotle thinks of pleasure as the success of an action to which it is related. We can see that more fully by citing two passages, both from the *Nicomachean Ethics*. The first—and we could find many similar passages—defines the relationship of pleasure to the activity that it rounds out:

> In the case of each sense, the best activity is that of the best-conditioned subject in relation to the best objects in relation to that sense. Such an activity is the most perfect and most pleasant. For to each sense there corresponds a pleasure, similarly for thought and contemplation: their most complete activity is the most pleasant and most actualized, that of the subject in the best state in relation to the best object affecting that sense; and the pleasure completes (τελειοῖ) the activity. (10.4, 1174b18)

Aristotle then explains that pleasure does not complete the activity in the way that the sensible object actualizes the capacity of sensing in the sense organ, but "as a kind of supervenient end, a bonus, like the bloom of youth is added to the power of the age" (1174b32).

The second passage is absolutely extraordinary:

> That absolutely everything, animals as well as people, pursue pleasure, is a sign (σημεῖόν τι) that pleasure is in a way the best. [A quotation from Hesiod follows.] But since neither the same nature nor the same state is best for all, it seems, not all

43. "In animals, one part of their life thus consists in activities related to reproduction, another part to those concerned with nutrition. In fact, all their life activities are concentrated in these two domains" (*History of Animals* 8.1, 589a3).

44. "What is common to all animals is an extreme desire for sexual pleasure" (*History of Animals* 6.18, 571b9).

45. Cf., e.g., *Nicomachean Ethics* 3.10, 1118a30.

> pursue the same pleasure, but all the same it's pleasure that they pursue. And maybe they pursue not the pleasure they think they are pursuing, but the same pleasure, for all have in themselves by nature something of the divine. (7.13, 1153b25)

This final statement, echoed elsewhere, like "in all natural beings there is something marvelous" (*Parts of Animals* 1.5, 645a16) and especially the no less extraordinary passage *De Anima* 2.4, 415a26, cited several times, condenses three very important theses: (i) human beings and beasts ultimately are all seeing the same pleasure, (ii) which means that they all share the same nature, (iii) witnessing that they have something divine in them, for the activity crowned by pleasure is doubtless the activity proper to divine beings. The pleasure that is the same for all, and that all seek, is to have a free and unimpeded activity. A passage a little before the *Nicomachean Ethics* passage just cited says that, and in it, Aristotle refutes in advance a Stoic position, thus showing that the question was a topic for debate well before the foundation of the Porch:

> Everyone thinks that happiness is a life accompanied by pleasure, and they closely unify pleasure and happiness, and that's reasonable, since no perfect activity is impeded, and happiness is a perfect activity. That's why someone who is happy does have, in addition to other things, a need for bodily goods, external goods, and good luck, in order not to be impeded from those directions. As for those who maintain that the person on the rack or one who is subject to extreme pains is happy provided that he is good, whether voluntarily or not they are speaking without saying anything. (7.13, 1153b14)

That pleasure is a mark of the divine in all animals (plants, of course, do not share in pleasure) does not give it a biological function: pleasure is not the main thing that gets animals going, nor what leads their actions in a positive direction,[46] but the pleasure is a *sign* that these actions have succeeded. A sign that each one, at its level, is perfect, that is, a sign that

46. Even when actions that are supposed to be pleasurable turn out to be painful, animals do them anyway: "The intercourse of horses is not painful (ἐπίπονος) like that of cattle" (*History of Animals* 6.22, 575b29). But maybe the term ἐπίπονος indicates that the coition requires hard work, not that it is not pleasurable.

Nature always achieves the best. For the dog that has a coincidental pleasure in smelling a hare, this pleasure too, as accidental as it may be, *signals* a successful activity, in this case having flushed out a prey. In fact, there is no reason to think that pleasure as success occurs only in per se pleasures.

Now we can thus see one of the major results of the Aristotelian theory of animal pleasure, especially of coincidental pleasure. Just as young people would come to the flourishing of their force, even if they never enjoyed the bloom of youth, so does and stags would have coition even if they never experienced or expected any pleasure. In Aristotle, pleasure does not have the vital function that the evolutionary biologist attributes to it, but the fact that intercourse procures per se pleasure for the stag, and that the sight of a gazelle procures coincidental pleasure for a hungry lion, is a *sign* of the perfection of Nature, and is why Aristotle sees in animal pleasure the traces of the divine. That's also why pleasure is not presented by Aristotle as a final cause (animals do not have intercourse *in order to* have pleasure, as one walks after dinner *in order to* have good health), but as a "supervenient end" (ἐπιγινόμενόν τι τέλος, *Nicomachean Ethics* 10.5, 1174b33).

Thus, the Aristotelian theory of pleasure includes all animals, both men and beasts, in a sort of common celebration of the perfection of Nature, but not to the level where the Aristotelian treatises in natural science would frequently refer to that perfection. From the biological point of view, this perfection is demonstrated by the fact that "Nature seeks the adapted" (*History of Animals* 9.12, 615a25, already cited). But pleasure is not a factor in adaptation. This theory reaffirms, on the one hand, the irreducibly animal nature of human beings (who share all pleasures with all the animals), on the other hand, the unbridgeable gap that the possession of *logos* makes between men and beasts. In any case, it cannot establish a paradigmatic function for human beings, nor serve to construct an animal *scala naturae*.

Sheep and Men

The human species is distinguished from other animal species also in that it leads an existence belonging to two different temporalities. As a species of animal, it is made up of individuals who, beyond their strong differences, all have the characteristics of never having come into existence, never being improved nor deteriorating, and never passing away. As a social species, the human race also belongs to a history full of sound and fury, something that is not true of any animal species. Here we imagine the contrast established by Claude Lévi-Strauss between societies that have for a long time been called "primitive" and he calls "frozen," and our societies called "advanced"; he compared the first to clocks and the second to steam engines. But in this contradiction between immobility and history that, in Aristotle, affects humanity, it's the immobility that is the primary perspective. We know, both by credible testimony and by allusions in the Aristotelian corpus itself, that Aristotle was committed to the idea of successive cataclysms, nothing indicating that he thought that they recurred in a regular cycle, that would decimate the human population, forcing the surviving human beings (inhabitants of mountainous places having escaped the flood) to revert to a primitive social state.[47] The history of civilization would then start over again. In my *Endangered Excellence*, I hypothesized that Aristotle relied on this periodicity to confront the dramatic events that he witnessed, and sometimes participated in, the disappearance of the *polis* into tyrannical empires. But the city represented for him the culminating condition of human society,[48] and one may hope that coming out of a cataclysm, human history would get back on track to lead again to the city. Animals too have doubtless been eliminated in great numbers by cataclysms, but they don't get a history out of that.

Obviously, Aristotle does not think that human beings are able to change in any significant way the course of the universe: he states clearly that animal and vegetable species "exist for human beings," in the sense that we sorted out in a previous chapter, but we don't see him imagining an excessive exploitation, by frenzied hunting for example, that could put the survival of a species in danger. As for minerals and metals, those are perpetually reappearing. Nevertheless, there are human interventions in the animal world that Aristotle mentions or could not ignore. Thus raising, crossing, selecting. The first interests us particularly. At *History of Animals*

47. According to a passage in Nicomachus of Gerasa preserved by Philoponus, Aristotle thought that "human beings have been destroyed in different ways: epidemics, famines, earthquakes, wars, various illnesses, and other causes, but especially by mass cataclysms, like the one that is said to have occurred in the time of Deucalion, certainly large, but not the most powerful of all. In fact, shepherds and those living in the mountains and elevated regions survived, while for the plains those who lived in them were drowned." Aristotle, *Aristotelis Fragmenta Selecta*, "De Philosophia" 8.

48. From the first lines of the *Politics*, Aristotle asserts that the city is "the most eminent of all" (ἡ πασῶν κυριωτάτη, 1.1, 1252a5) communities.

1.1, 488a26, Aristotle proposes, in his list of differences between animals, the distinction between domestic or tame animals and wild animals. He seems to hesitate between two positions. According to the first, there are animals that are always domesticated and others that are always wild, even if most of them exist in both conditions. According to the second, Aristotle says that according to "another way" (ἄλλον τρόπον, 1.1, 488a29), all domestic species also exist in a wild state. These really are two contrary positions, because the human being stands as an example of a species that is always tame according to the first position, and according to the second, can also exist in a wild state. He appears to prefer the second, since he supports it in the *Parts of Animals* (1.3, 643b4), and it is also found in the *Problems* (10.45, 895b23). And especially *Problems* 10.45 provides an explanation of the fact: there are wild animals that can never be tamed, but all the tame animals were at first wild, either in the sense that they were captured in the wild and tamed, or in the sense that they have always lived in a human milieu to which they have become adapted by training, or education, since the passage envisages that babies start off by being little savages. All that rests on the idea that the primitive nature of living things, but also of works of art,[49] is rather bad, and can be subsequently improved, even though it often gets worse. Anyway, for Aristotle, "tame animals have a better nature than wild ones" (*Politics* 1.5, 1254b10).

We must add, in support of the thesis that all tame animals also exist in a wild state, the difference between the examples that are offered in support of this thesis and those given for the opposite thesis. In the one case, the passage says, "the human being and the mule are always tame," and in the other, the examples of domestic animals that also exist in a wild state are "the horse, cattle, pigs, human beings, sheep, goats, and dogs." The examples of animals in the first list that are never wild are, one may say, biased, since human babies never live except within human families, and mules are a creation of art, at least most of them, so that these two cases concern living things that owe their birth to human intervention.

In fact, since the domestication of animals is a human activity, it should start up again, like every technical and social activity, after each cataclysm. The case is different for infants who are born wild in the very

49. The author of *Problems* 10.45 thinks that technical and artistic products, in the modern sense, were originally bad. He thinks the same thing about works of nature, which leads him to distinguish, at 896a3, an "original" nature (ἐξ ἀρχῆ) and a nature "toward which things tend" (ἐφ' ἥν), a not very Aristotelian distinction. It's the second that makes beautiful art and tame animals.

heart of human society, because it's a matter of a process that is universal and thus natural, that can even be extended to animal species. But there is a passage in the *History of Animals* that can perhaps disturb this nice equilibrium:

> The character of sheep is, in fact, simple and stupid, because of all quadrupeds it is the most foolish, they wander into lonely places for no reason and often in the depth of winter they go out from where they were and when they are surprised by the snow, if the shepherd does not move them, they don't want to go, and if they are left where they are, they will die, unless the shepherd brings the rams; then they follow along. (9.3, 610b22)

So, if the sheep cannot survive without the shepherd, they would go extinct when there is no shepherd.

There can be two ways to get out of this difficulty. The first would be to suppose that there are sheep in countries where it doesn't snow. Several objections could be made against that. In the first place, if they were animals living in warm countries, Nature would have given them their wool coats *in vain*. Secondly, the passage says that "they wander into lonely places for no reason" (610b24), showing a radical lack of adaptation of sheep to their environment, whether it snows or not. The second solution is clinched by Nicomachus of Gerasa, who tells us that shepherds are among those who survive cataclysms. So there were always sheep because there were always shepherds.

We should also think that the "domesticity" is for sheep what "abundance" is for little fishes: just as they couldn't survive if they did not have numerous offspring, similarly for the sheep if they did not have a shepherd. But that would establish a fundamental identity between the relationship of the little fish to their predators and of sheep to their shepherds. At least two other examples, both from the *History of Animals*, deal with human intervention into the course of nature. Aristotle has just said that rainy years are beneficial to shellfish, except the murex. He adds: "For other animals with shells drought is not beneficial, because they become small and less good, and that is when red scallops are born. At one time scallops were exterminated in the Pyrrhaean Strait, not only because of the dredging machine used to extract them from the sea, but also because of the drought" (8.20, 603a19). It cannot be said that Aristotle here envisages the extinction of a species, even less so an extinction due to human activity. Nevertheless, there is here a (small) hitch inflicted by human beings on natural equilibrium. But with two limitations: first, as is shown by the adverb ποτε ("at one time") and the aorist ἐξέλιπον ("were exterminated"), it's a matter of particular circumstances that are not likely to be repeated, and then one may suppose that the species of scallops recovered a stable population after this episode.

Another case, raising the same questions, is that of the does that the *History of Animals* tells us "have their fawns next to roads because of their fear of wild beasts" (6.29, 578b16). All the commentators agree that wild beasts avoid roads because they recognize a human presence, possibly by odor. But the case of the sheep and the case of the does are different. One may in fact think that despite exterminating cataclysms, there are always, in the heights, human beings following trails and leaving their odor, so that does would also find the conditions that they need, and that Nature has furnished them forever, to give birth in complete security. So, we have here an example of relationships between different animal species (human beings considered as an animal species among others); we have seen other examples in the previous chapter.

But it's not the same for the sheep, because even though dolphins are born fish-eaters, and human beings are born with a characteristic odor, human beings are not born shepherds or shepherdesses. To be sure, it is difficult to imagine a universe completely devoid of shepherds, and when Aristotle gives a list of modalities of human life in *Politics* 1.8, he seems to confer on them an air of permanence nearly like that of animal species: nomads, peasants, thieves, fishermen, hunters (1256b1), the passage having said a few lines previously that nomads live with their herds. But it's hard to see on what theoretical basis Aristotle could attribute a kind of permanence to certain human techniques, given that he says elsewhere that human societies evolve, for example from the patriarchal state of small, dispersed, groups to an urban state (cf. *Politics* 1.2, 1252b23).

If we now look at it from the point of view of the sheep, things change a bit. Sheep, it seems, cannot survive except in a domesticated state. Obviously, it's not a matter of the result of a decision or calculation by the sheep, any more than the little fish decided to be prolific. If this necessity for the sheep to be domesticated is a necessary condition for their survival, we can figure that ability of sheep to be domesticated is part of Nature's plan and thus ought to be analyzed in the same finalist way as all the traits that contribute to compensating for inferiority in an animal species: given the sheep are too stupid to survive simply on drawing from their environment, it is necessary to compensate for this detrimental characteristic.

So then, in the case of the sheep, we are at an extreme boundary between the human and animal spheres, and a different shape than the others. In the case of human beings, in fact, we have seen that their biological characteristics turn out to be useful for theoretical-practical purposes, for example for politicalness or language. In the case of the sheep, we're looking at the opposite—a human characteristic of a historical-practical sort is used to ameliorate the biological insufficiencies of an animal species.

The everlastingly unchanging character of Aristotelian nature means that human activity cannot modify natural equilibria. To be sure, for Aristotle, "art imitates nature," so nature has a role as a model, but the "by art" and the "by nature" that these expressions designate as a kind of being or causality cannot really interfere with each other: the opposition of κατὰ φύσιν and κατὰ τέχνην is one of those, among others, that frame the entire Aristotelian corpus. For philosophers who do not adhere to the general unchangingness of the universe, that is, all thinkers other than Aristotle and his "strict" adherents, a space will open up for human modification of natural equilibria. Among others, this famous passage from Lucretius is an example.

> And in the ages after monsters died,
> Perforce there perished many a stock, unable
> By propagation to forge a progeny.
> For whatsoever creatures thou beholdest
> Breathing the breath of life, the same have been
> Even from their earliest age preserved alive
> By cunning, or by valour, or at least
> By speed of foot or wing. And many a stock
> Remaineth yet, because of use to man,
> And so committed to man's guardianship.
> Valour hath saved alive fierce lion-breeds
> And many another terrorizing race,
> Cunning the foxes, flight the antlered stags.
> Light-sleeping dogs with faithful heart in breast,
> However, and every kind begot from seed
> Of beasts of draft, as, too, the woolly flocks
> And horned cattle, all, my Memmius,
> Have been committed to guardianship of men.[50]

---

50. Lucretius, *On the Nature of Things*, book 5, 855ff.

Lucretius calls attention to the animals unable to protect themselves but able to be of use to man, which will survive "until Nature reduced that stock to utter death" (v. 877). The road will be long for man to become "guardian," or even transformer, of nature, but this first step is part of the numerous breaks with Aristotelianism on which modernity has been constructed.

Conclusion

The physician René Allendy, cofounder of the Psychoanalytic Society of Paris (notably with Marie Bonaparte), also enthusiastic about homeopathy and occult sciences, wrote a work published posthumously in 1943 that made the hair of Aristotelians stand on end: *Aristotle, or the Treason Complex.*[1] This work, arguing that Aristotle's life was a series of betrayals, of Athens, Plato, Philip of Macedon, Alexander the Great, deserves better than the contempt with which it was received by historians of philosophy. Allendy's more or less psychoanalytic analysis of the love triangle he believed to have discovered between Aristotle, Hermias—tyrant of Atarneus in the Troad, Aristotle's host and coworker—and Pythias, who is sometimes presented as Hermias's pupil, whom Aristotle married and with whom he had a daughter, can seem a little quick, arbitrary, and based on slender evidence. But it deserves better than the shoulder shrugs that it has gotten. Anyway, it is neither forbidden nor useless to ask about Aristotle's bisexuality, provided that we keep in mind the paucity of our documentation on the subject. The crucial question would be, how can we make a causal connection between Aristotle's unconscious properties of the libido that we might eventually detect and the theses supported in the Aristotelian corpus?

But perhaps this "treason" of Aristotle is a dramatized image of that which I mentioned in the introduction of this book, and which, in the course of my work on Aristotelianism, seems to me more and more pregnant for one who reads Aristotle today: the Stagirite separated himself, on fundamental points, from a kind of consensus that delimited Greek

1. English title: *Treason Complex*, translated by Ruth K. Siler, published in 1949 by Social Science Publishers.

thought from the Presocratics until the end of antiquity. I tried to show that in my work on his political thought. Two examples: for Aristotle, it is a "deep structure," to speak like Noam Chomsky, that explains that human beings live in a society, reproduce themselves, and engage in slavery, and not their needs; for him it is also the case that sedition (*stasis*), which was in the eyes of all Greeks the worst of calamities that could befall cities, is in fact a normal state of affairs, sometimes even welcome, for the political body.

Here I have presented some fundamental points; I will remind you of two. First, Aristotle was the only ancient thinker who did not offer a description of the coming into being of the universe and its parts. Friedrich Solmsen is the one who has given the liveliest picture of this Aristotelian cosmology without cosmogony. Next, Aristotle, who like all ancient thinkers thought that the universe is perfect, gave this perfection an image that we might call flickering. And that is truer for living things than for other beings in the world. Aristotle often expresses his admiration for the ingenuity with which Nature has worked with various animals, but, on the one hand, there is no general plan of the animal world, because there is no discernible design behind the extraordinary blossoming of living forms, and on the other hand, the perfection of each species (because each species needs to be perfect if it is part of a perfect universe) is in a way negative: negative and unfavorable traits that endanger a species need to be sufficiently counterbalanced by positive and useful traits so that this species can fulfill the fundamental condition of every living species, which is to be able to survive forever.

So, there is no final cause of the proliferation of animal forms, no need for Nature to actualize all these gigantic potentialities, nor of a *scala naturae* exemplifying a climb toward perfection. Even more, the balance and harmony of the animal world are seen as accidental. One may seek it inside each species and at the level of species to each other. In these two domains, the traditional forms, that is, non-Aristotelian, of teleology attempt to discover the harmony behind an *apparent* disorder, and a higher purpose behind the *apparent* anarchy that seems to rule between species. The very idea of an irreducibly harmful situation or characteristic (we need to say "irreducibly" because one of the leitmotifs of this kind of thought is that evil is acceptable when it serves a higher good), or even simply useless is, from that perspective, unacceptable. Furthermore, this kind of teleology, which one finds from Plato to Leibniz, simultaneously addresses disorder in the natural world, and even more so, the realm of moral good and evil. This last point does not appear in Aristotle.

But, in Aristotelian zoology, one certainly does find analyses that could contribute to one of the schemas of one of the kinds of teleology noted earlier. For example, it's bad for the ability to walk for raptors to have curved talons, but they get a more important and even decisive good from that, since without claws they would not be able to feed themselves. There are two ways to read the example of the little fish. One might think that the presence of sharks and other gluttonous predators is purely and simply bad for the little fish, and their fecundity is in no way a superior good, but a response to an evil; but one might also figure that the presence of sharks is an evil for the sake of a superior good, namely, to allow the survival of the little fish, because if the predators were to disappear, the fishes' fecundity would result in their becoming too numerous for the insufficient resources. That's what hunters call, in excusing their carnivorous tastes, the regulation of species. If what we have tried to establish is correct, it's the first reading that is Aristotelian, because in the second, Nature would be using some species, sharks, dolphins, and others of that kind, for the safety of another group, that of the little fish. But we have seen that since the species is at the strategic level of Aristotelian zoology, that prevents thinking that one species exists for the benefit of another: even of gazelles we cannot say that they exist for the sake of the lions, or to put it another way, Nature has distributed lions in such a way that they eat gazelles, but she has not distributed gazelles in such a way that they will be eaten by lions.

I have tried to show that the anthropocentric teleology that thinks that animals, along with other beings in the world, are there to provide benefits to human beings cannot be attributed to Aristotle. For that to be the case, among the organic arrangements that characterize the horse, there would have to be traits like rideability and edibility by human beings. But as we have seen, the original nature of horses is to be wild animals, which have been converted to modes of transportation and chevaline steaks by human industry. There is no equivalent in Aristotle's animal world to the story in the Gospel of Judas that explains to us that the terrible sin of Judas served a superior good, since without him, there wouldn't have been any salvation. Or rather, a schema like that could exist, but in the domain of human practice, that is, ultimately, politics. I think that I was able to find a schema of this kind in the case of natural slavery:[2] the evil of slavery finds a superior utility in the fact that, by relieving a free and virtuous man from his material tasks and allowing him thus to engage

2. Cf. Pellegrin, *L'Excellence menacée*, 133–160; *Endangered Excellence*, 116–142.

in politics and philosophy, the slave, who is naturally unable to engage in these activities (otherwise he would deserve to be a master), participates, at the cost of the disagreeable aspects, for him, of a servile life, in the actualization of a superior good, like the Christian saint who endures the worst suffering in order to bring about God's obscure designs.

To come back to the central place that Aristotle gives to species, it has many consequences, but we first need to note that Aristotle does not justify this position. So, we need to look at it hypothetically. Because it is not incongruous to ask why it's that way, provided that we correctly locate the level at which this question is asked, and thus the level where an answer may be expected. In fact, it is hardly possible, anyway for me, to derive this Aristotelian tropism from a particular direction of profound historical tendencies in the Hellenic world, or from the fallout of an Oedipean feeling that Aristotle entertained toward Plato. In contrast, it seems to me fruitful to include Aristotle's zoology among those that one might call "biologies of diversity," where, no longer any surprise, it joins other theoretical constructions like that of Cuvier.

From that point of view, Plato and Lamarck are, as one says, on the other side of the fence, in that they see in animal diversity itself a mark of perfection, one that leads them to the claim that the animal world is *complete*, synchronically for the one, diachronically for the other. We may speak, with the expectations we have seen, of the *intention* of Nature that makes use of a material given, for example when she gives several stomachs to beasts with horns. But animal diversity does not, for Aristotle, derive from any intention of this kind: it's just a *fact*. To be sure, there is an analogical unity of functions in all animals: they all digest and reproduce, and some of them also move around, but it's impossible to deduce the existence of a tiny octopus or an axolotl by an analysis of these functions. The fact that axolotls exist is beyond explanation to the same degree as the fact that, for Cuvier, there are four branches, and not either three or five, all subject to common functions (of which the most basic is that of the nervous system), that can be included in a comparative anatomy, that is, that are analogically identical. The sign of a real biological thought is present there. Not that Lamarck and Étienne Geoffroy Saint-Hilaire weren't real zoologists. Saint-Hilaire had real Aristotelian affinities. But some of their teachings, evolutionism for Lamarck, and the unity of a general plan for animals for Saint-Hilaire, were a sign of a metaphysical and/or ideological contamination in their research; I would be prepared to argue, without doing that here, that contamination holds true, in one way or another, of *all* scientists.

I believe that I've been able to show that this winds its way through the Aristotelian zoological corpus, and it does that walking on two legs. Aristotelian teleology, which is applied primarily because more completely to animals, is, we have seen, the spinal column of Aristotelian zoology, and, as Allan Gotthelf has demonstrated, that does not attain the depths of prerational or mythic thought, which Gotthelf shows by the isomorphism that he discerns between Aristotle and Darwin. This teleology is deployed primarily in the two great zoological treatises, the *Parts* and *Generation of Animals*. But the teleological framework of a work like the *Parts of Animals*, even if it certainly rests on general theses that one should rather think of as principles than as laws (Nature always brings about the best, does nothing in vain, etc.), does not really produce its results except at the level, or mainly at the level, of the species, as we have seen. It is not without interest that it goes the same for Cuvier, for his two great laws that structure his zoology, the law of organic correlation and the law of subordination of characters, do not function completely except at the level of species. But with the *History of Animals*, the other leg of Aristotelian zoology (either second or first), we see Aristotle accounting for diversity in itself. The indispensable rehabilitation of this disrespected treatise has been proposed in this volume only as a program, one that needs now to be accomplished. It's worth remembering that Cuvier regarded the *History of Animals* as the most representative treatise of the Aristotelian science of living things.

In fact, we may say that for Aristotle, the animal world is delimited on two sides. First, by the rigidity of the system that we call physiochemical laws, which no worldly reality can possibly ignore. But, as we have just recalled, animals are also subject to what may be called teleological pressure, which brings it about that each species is perfect, in the sense of this term that we have defined. But, although the first constraint is applied universally, the teleological framing of the living world includes lacunae of several kinds. There are, in the first place, "badly made" animals, like the axolotl, which is a sort of extreme case of an animal that has disadvantageous traits. In fact, we have seen that these are not all of the same kind: there are properties that harm the animals that have them, but are ultimately more useful than harmful (thus the claws of raptors are bad for walking, but provide decisive assistance for survival); but there are some traits that are simply harmful, like the antlers of stags. We can say two things about these traits: (a) that they are often compensated for by other characteristics (thus, speed for the stags), and (b) if they are not, they don't have such a potential for harm that they put the survival of the species in danger (small octopuses survive with one rank of suckers on their tentacles). One might think that the axolotl is itself a kind of generalization of this marginal malfeasance, but we should at the same time assert that according to Aristotle the axolotl generally overcomes these huge disadvantages; that is, in very unfavorable conditions, it succeeds in exercising its vital functions so that its species is everlasting.

But there are also, as emphasized earlier, some *exceptions* to the laws regulating the animal world, and we should return to them just a little. In the first place, we need to emphasize that these exceptions can only be in relation to teleological rules, because in no case can the laws of material nature hit any snags. However, we will see that things are less simple, when we return again to the example of female bears (and female panthers too, but we'll leave those to one side).

So, female bears are more courageous than male bears. Obviously, the courage of female bears does not threaten the survival of the species, but it contradicts a general rule of the animal world that is explained as much teleologically as mechanically, and is, on its mechanical side, the application to the animal domain of a sort of general law of nature. Mechanically, a male is hotter than a female, which comes from the conditions under which he was conceived by his parents; but having, as a result, a hotter blood, he is necessarily more courageous than a female. We don't see, in Aristotle's texts, that he saw any sort of special conditions that would bring it about that such and such a female would be more intrepid than such and such a male; that's certainly the case for human beings. According to the general schema that articulates mechanism and finality, this state of affairs is useful to the species because it introduces a division of labor between the sexes: for males, relationships, often violent, with the external world, for females, the care of offspring. It's a matter of an anthropomorphic projection applying to the animal world the ultra-sexist ideology of the ancient Greeks, and runs contrary to the most superficial observation of the animal world, to the extent that one finds in it examples of bellicose females and cowardly or uninterested males. So, one cannot see how the female bears can be more courageous than the male bears without contradicting the Aristotelian theory of the sexes. We are at the point of an unraveling of the determinism that is the basis of all rationalism. Because if the male is not hotter, and the heat of the blood does not give more courage, that doesn't contradict only the intention of rational Nature, but even goes against the laws of material nature. Are the female bear embryos hotter than male bear embryos, and if so, why?

The example of the courageous female bears goes further than a badly controlled diversity: the fundamental fact that we have discovered, and which has remained largely unnoticed, is that animal diversity *overshadows* the explanatory frame on which Aristotelian zoology, and all Aristotelian natural science, is based. It's not only a matter of facts left unexplained because we still lack necessary observation—Aristotle indicates that that is the case for the reproduction of certain fish—nor is it a matter of differences that we can more or less think of as inexplicable, like different colors, sizes, looks, of various small birds: it's one of the facts that are *set aside*, because they don't fit into what Aristotle thinks of as the science of living things, and the courage of female bears is one of those facts. We can, however, assume that he could have fit a case like that of the courageous female bears into an explanatory scheme as he did, in a way that is not entirely clear to us, for the earthy matter of horns that directs itself "by necessity" upward. There too, Aristotle shows himself to be more "positive" than "metaphysical," to use the Comtean sense of these words. It would have been too uncomfortable for a truly metaphysical thought to leave behind these she-bears, so courageous, for no reason.

Similarly, we have given Aristotle positive marks for not having asked himself questions that he would have been unable to answer, even partially, like that of the origin of the universe. But we can, on the other hand, blame him for having been the only thinker of the ancient world, and in fact right to the twentieth century, to not have had any inkling of the principle of entropy at the scale of the universe, or at least for the sublunary world. On that topic, one may reread the passage from Lucretius quoted at the end of the last chapter. There are several ways to respond to this complaint. First, we should remember that one ought to not judge ancient theories according to their more or less great distance from the results of modern science, and that is why, in contrast to what Balme says, a passage that is "truer" about the brain of octopuses is not necessarily posterior to a passage that is more "false." Secondly, the way that other philosophers have evaded the problem of entropy is, one may say, magical, since they all suppose that that world, having aged and degraded, rediscovers a new youth on the occasion of an ekpyrosis or a new atomic combination occurring by chance. Finally, if contemporary cosmogony teaches us that the universe came into being about fourteen billion years ago, on the occasion of the Big Bang, that does not prevent us from asking what there was *before* the Big Bang, even if that "before" is itself an obscure notion for designating a state in which the world, and therefore time, did not exist. That's a matter of metaphysics or religion, but science, for the moment, claims that that question is outside its purview. Aristotle's approach is thus more economical, and for that reason, "more scientific" than that of his competing cosmogonists.

This Aristotelian conception of a cosmos permanently identical to itself was Christianized, especially in the West, by first-rate theologians like Augustine of Hippo and Thomas Aquinas. One may even think, when one reads for example Aquinas's *De Aeternitate Mundi*, that the creation from all eternity, as a concomitant effect of its cause of a (relatively) perfect world that is thus immutable, would be, for Thomas, more in agreement with divine omnipotence. Thus, Christian thinkers were able to break with the naïve conception of a creation occurring *after* a period in which only God existed, a conception that was still that of Philoponus, for example. Thomas Aquinas did not, however, have the audacity to think of a world without miracles, as Spinoza did. For Spinoza, the need for a miracle in the world would reveal an imperfection in this world, and thus of God.[3] Perhaps that helped bring about the domination of the Aristotelian picture of the world in the medieval Latin West, even though Aristotelianism had, before the scholastic turn of the thirteenth century, a very reduced standing in the West. It's always the world of Aristotle that Galileo attacked, and even ridiculed as "simple-minded."[4] It didn't go that way for the zoological corpus.

From the ninth century in Islamic lands, and in the thirteenth century in the Latin West, Aristotle emerged as the representative of the most advanced Greek science, apart from mathematics.[5] This made it simultaneously a corpus to critique and a stage to surpass, as the example of Galileo shows. Aristotle's zoological treatises were known to the Arabs, translated into Arabic and then into Latin (see, for example, Albertus Magnus's paraphrase in his *De Animalibus*). Why then was Aristotle's

3. Spinoza, *Tractatus Theologico-Politicus*, chap. 6.

4. In his *Dialogue Concerning the Two Chief World Systems*, the Aristotelian picture of the world is defended by Simplicio.

5. We must not forget the uninterrupted Aristotelian tradition in Byzantium.

zoological corpus enthroned in the blue like a sphinx[6] not understood for more than twenty centuries? In antiquity, after Theophrastus and his great botanical treatises, there was never a zoologist, in the sense of an author proposing an explanatory and functional approach to the animal world. I expressed earlier the perplexity Pliny the Elder's reference to Aristotle provoked, and I will say nothing about the most important Greek work on animals after Aristotle's, the treatise Περὶ Ζῴων Ἰδιότης, "That Which Is Peculiar to Animals," by Aelian, a disorganized collection of facts and anecdotes about various animals. We will also say nothing about the subtle works of Plutarch or Porphyry that refer to animals to introduce philosophical and/or moral questions, for example, comparing animal intelligence to human intelligence. In taking that road, they simply turn their backs on Aristotle.

The only area in which Aristotle's zoology survived, that is, where the texts of Aristotle's zoological corpus were studied from a theoretical point of view, was medicine. Alexandrian medicine was fundamentally Aristotelian in its anatomical and physiological explanations,[7] but the one for whom the comparison to Aristotle makes most sense, first of all, in contrast to earlier physicians, because we have many texts, is obviously Galen. In the vast Galenic corpus that has come down to us, it is the treatise *De Usu Partium* that is most useful. Without making a complete study of this work, which would be a large project, we can note some similarities and differences with Aristotle's work.

This long work (it takes up a volume and a half in the Kühn edition), which exhibits all the formal faults that readers of Galen know well, notably his verbosity and arrogance, has, I believe (but not all Galen specialists agree), a profoundly Aristotelian structure. Charles Daremberg, a preeminent historian of medicine, especially ancient, and translator of this treatise along with other works of Galen, is certainly right to translate the word χρεία in the title as "utility" rather than "usage," as some previous translators had done, thus giving the French translation the title *De l'utilité de parties du corps humain*.[8] One of the leitmotifs of the

6. Translator's note: This is an oblique reference to a line in a sonnet by Baudelaire: "As an unfathomed sphinx, enthroned by the Nile." Baudelaire, "Beauty."

7. Cf. Staden, *Herophilus*; Vegetti, "Entre le savoir et la pratique"; Pellegrin, "Ancient Medicine and Its Contribution to the Philosophical Tradition."

8. The now-standard English translation, by M. T. May, has the title *Galen on the Usefulness of the Parts of the Body*.

work is the distinction that must be made between the "utility" (χρεία) and "function" (ἐνέργεια) of the parts, that is, taking account of their "activity" (ἔργον). Galen reminds the reader several times that the best method consists of analyzing the function first, and one ought to not proceed to the "utility" before having correctly determined the function, that is, before having surmounted objections to his view. Because, for the hand, the function is obvious, it is prehension, but it's not so easy for "the functions of arteries, nerves, muscles, tendons" (I.16, III, 45, 1K). But the analysis of function presupposes the determination "of the temperament, for it is the temperament that is responsible for the characteristic essence of the parts. For the nature of the body is determined by the commingling in a certain way of the hot, the cold, the dry, and the wet" (I, 9, III, 26, 3K; May, 80). As far as the study of functions is concerned, sometimes Galen sends the reader to some of his specialized treatises, like *On the Movement of Muscles* (cf. I, 19, III, 68, 1K). This kind of research, writes Galen, "is about natural problems" (τῶν φυσικῶν ἐστι προβλημάτων, VII, 22, III, 607, 13K). That's an Aristotelian architecture, combining mechanical and final causes.

The most obvious difference between Aristotle and Galen is that the *De Usu Partium* is a treatise on "the utility of the parts of the human body," as the Greek title of the work and the first sentence of the epilogue (XVII, 1, IV, 346, 1K) tell us, hardly surprising, since Galen is a physician, writing a treatise on medicine, while Aristotle is a naturalist, constructing a comparative anatomy. We can discern in the background of Galen's text a number of thorough investigations, results of animal dissections and vivisections, and we can only mention some examples of the progress that had been achieved since Aristotle (the movements of the limbs are due to muscles, Galen has abandoned Aristotelian cardiocentrism, the nerves have been clearly identified, etc.), progress that Galen is not a little proud of . . . But this difference itself has its limits, as we can see from passages like this one: "It is not my intention to discuss the number of stomachs in ruminants or the stomach and other instruments of nutrition in each species of animal; for Aristotle has excellent treatments of them all.[9] If life were not too short for the investigation of the noblest subjects, perhaps some day I should supply what remains [to complete his observations] in this field" (IV,17, III, 328, 11K; May, 238). Elsewhere, Galen proposes, with his usual modesty, "correcting" Aristotle's zoology when he has the

9. May refers to *History of Animals* 8.1, 588b4–23 here.

time (XIV, 3, IV, 145, 4K). And, in fact, we find in Galen, especially in the *De Usu Partium*, many allusions, possibly with a sense of regret, to comparative anatomical research, or at least to comparative medical research. Thus, at III, 2, the assertion that since nonblooded animals are colder, they lack the necessary force to have strong limbs, in their case compensating for the smallness of their limbs by their number. With this remark, he casually adds, "We find in Aristotle a long and beautiful discussion of the difference presented by bloodless animals" (III, 177, 12K).

Yet nevertheless, Galen's project is far from being that of Aristotle, because, in a word, Galen does not put together a biology, that is, a study of that which one would later call "animal economy," combined with taking account of the diversity of animals. For that it was necessary to wait for Cuvier. It's all the more remarkable given that Galen obviously knew Aristotle's zoological corpus very well. From the angle of two structurally related attempts, one may see that the Aristotelian theoretical project is not adopted by Galen in at least two respects. First, in Galen, teleology mutates into a radical form. Aristotle's "flexible" teleology, which calls "the better" that which allows a species to survive forever, or in other words, that which *suffices* for attaining the perfection that permanence represents, is presented this way by Galen: "Come now, let us investigate this very important part of the man's body [the hand], examining it to determine not simply whether it is useful or whether it is suitable for an intelligent animal, but whether it is in every respect so constituted that it would not have been better had it been made differently" (I, 5, III, 9, 4; May, 72). Galen has just invoked, in the epilogue to the work, "an intelligence endowed with an admirable power [that] hovers over the earth and penetrates it everywhere" (XVII, 1, IV, 358, 11K) and speaks of a "perfect theology" (IV, 360, 13K). We are far from the admirable flourishing of Aristotle's animal world, its gaps, lacunas, and questions not addressed.

Nevertheless, we need to note in Galen some sketches, as it were, of Aristotelian procedure, consistent in analyzing how nature deals with negative traits in living things with countermeasures. Thus, after his explanation of the fact that the intestine is "so long, so narrow, so twisted" (which is because it is pierced all along by a great number of pores that permit the assimilation of food, an explanation that he offers on the authority of Plato), Galen, repeating without crediting him an analysis by Aristotle, remarks that "all animals whose intestines are not coiled but extend straight from the stomach to the fundament are greedy, gluttonous, and forever engaged in taking nourishment, like the plants" (IV, 17,

III, 328, 3K; May, 238). This imperfection of certain animals allows us to finish what we have just been talking about and, at the same time, bring up a second example of the divergence between Aristotle and Galen, the way they treat animal diversity. In the very next sentence, in fact, Galen attributes to Aristotle the *scala naturae* that we have been denying him: "But Aristotle has an excellent discussion of these matters, saying among other things that Nature, diverging gradually from her custom in creating plants, makes one animal after another, each more perfect than the preceding, until she comes to the most perfect of all, the one that is the theme of my present discourse" (IV, 17, III, 328, 7K; May, 238). From which it appears that only the human being is perfect, and that, as a consequence, the other animals have to cope with their imperfections such as having a straight intestine. Nevertheless, there is in Galen—and how couldn't there be?—the recognition of a sort of relative perfection, at least on certain points. Thus, after having dealt with, in an Aristotelian manner, the way fragile organs are protected in bipeds and quadrupeds, he writes:

> For since animals do not have hands and cannot reason as man does, they must interpose something else to protect the abdomen and thorax, and compensate for the natural weakness of the parts contained there. Hence it is better for all animals supplied with blood, other than man, to be quadrupeds and for the bloodless animals to have many feet. Conversely, it was better for man to be biped, because unlike the other animals he does not need the advantages resulting from a larger number of legs and would be handicapped in many ways if he were not a biped. (III, 2, III, 178, 14K; May, 159)

On this point too, Galen has joined the ranks, agreeing with all the other ancient philosophers: animal diversity is nothing but several ways in which animals have degenerated in comparison with man. Thus, for the question raised earlier of the twisted path of the intestines, Galen, as we say, relies on Plato, and that can only be to this passage in the *Timaeus*: "[The gods] have wound the intestines around in coils to prevent the food from passing through so quickly that the body would need more food too soon, which would make human beings insatiably voracious, and thus the human race would be unsuited for philosophy and the cultivation of the Muses, unable to obey that which is most divine in us" (73a). Like everyone else, at least until the Middle Ages (just read Albertus Magnus's *De Animalibus*), Galen considered animals to be curious beasts that need not be studied in any detail except in the framework of an apologetic argument, to praise the excellence of Man and/or God. Everyone else, except Aristotle.

Bibliography

Editions of Aristotle

Aristotle. *Aristoteles Thierkunde*. Kritisch-berichtigter Text, mit deutscher Übersetzung, sachlicher und sprachlicher Erklärung und vollsständigem Index von H. Aubert und Fr. Wimmer. Leipzig: Engelman, 1868.

———. *Aristotelis De Animalibus Historia*. Textus recognovit Leonaerdus Dittmeyer. Leipzig: Teubner, 1907.

———. *Aristotelis De Generatione Animalium*. Recognovit brevique adnotatione critica instruxit H. J. Drossaart Lulofs. Oxford: Clarendon Press, 1965.

———. *Aristotelis Fragmenta Selecta*. Recognovit brevique adnotatione instruxit W. D. Ross. Oxford: Clarendon Press, 1955.

———. *Aristotle, De Sensu and De Memoria*. Text and translation with introduction and commentary by G. R. T. Ross. Cambridge: Cambridge University Press, 1906.

———. *Aristotle, Generation of Animals*. With an English translation by A. L. Peck. Cambridge, MA: Harvard University Press, 1942.

———. *Aristotle: Historia Animalium*. In three volumes, with an English translation by A. L. Peck. Cambridge, MA: Harvard University Press (Loeb), 1965.

———. *Aristotle: Historia Animalium*. Volume I: Books I–X. Text, edited by D. M. Balme, prepared for publication by Allan Gotthelf. Cambridge: Cambridge University Press, 2002.

———. *Aristotle: History of Animals: Books VII–X*. Edited and translated by D. M. Balme. Cambridge, MA: Harvard University Press, 1991.

———. *Aristotle: On the Generation of Animals*. Translated by Arthur Platt. Oxford: Clarendon Press, 1910.

———. *Aristotle: On the Parts of Animals*. Translated with introduction and commentary by James G. Lennox. Oxford: Clarendon Press, 2001.

———. *Aristotle's De Partibus Animalium I and De Generatione Animalium*. Translated with notes by D. M. Balme. Oxford: Clarendon Press, 1972.

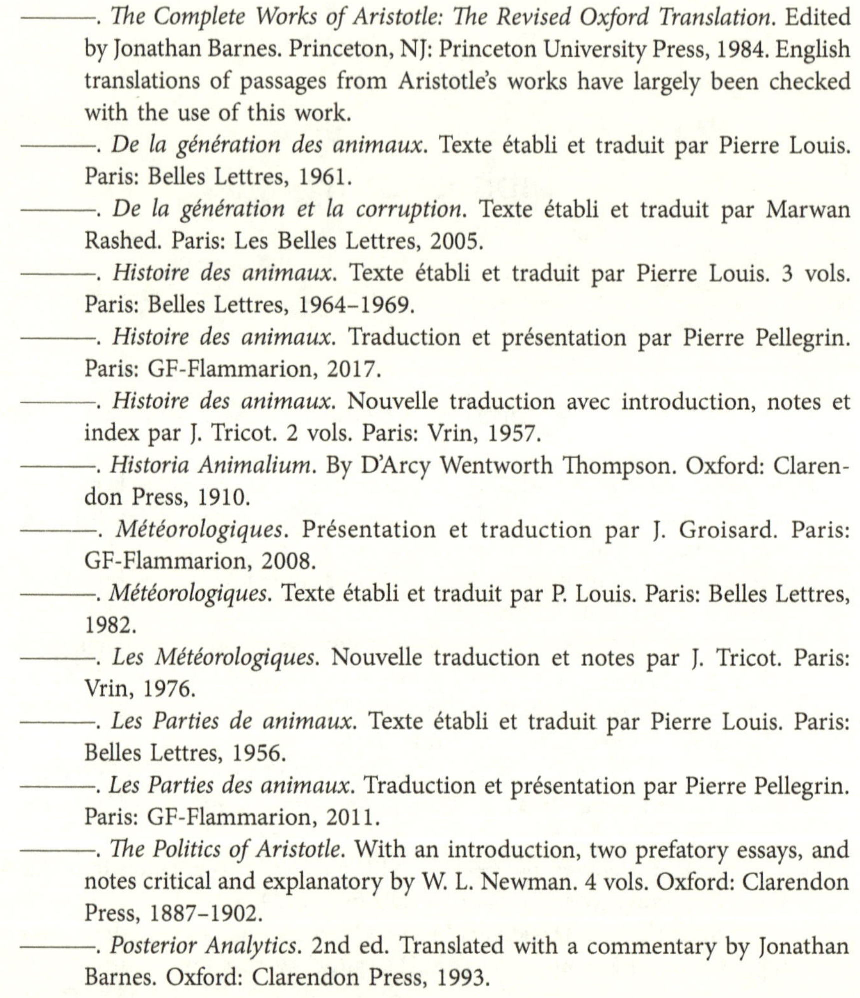

———. *The Complete Works of Aristotle: The Revised Oxford Translation.* Edited by Jonathan Barnes. Princeton, NJ: Princeton University Press, 1984. English translations of passages from Aristotle's works have largely been checked with the use of this work.

———. *De la génération des animaux.* Texte établi et traduit par Pierre Louis. Paris: Belles Lettres, 1961.

———. *De la génération et la corruption.* Texte établi et traduit par Marwan Rashed. Paris: Les Belles Lettres, 2005.

———. *Histoire des animaux.* Texte établi et traduit par Pierre Louis. 3 vols. Paris: Belles Lettres, 1964–1969.

———. *Histoire des animaux.* Traduction et présentation par Pierre Pellegrin. Paris: GF-Flammarion, 2017.

———. *Histoire des animaux.* Nouvelle traduction avec introduction, notes et index par J. Tricot. 2 vols. Paris: Vrin, 1957.

———. *Historia Animalium.* By D'Arcy Wentworth Thompson. Oxford: Clarendon Press, 1910.

———. *Météorologiques.* Présentation et traduction par J. Groisard. Paris: GF-Flammarion, 2008.

———. *Météorologiques.* Texte établi et traduit par P. Louis. Paris: Belles Lettres, 1982.

———. *Les Météorologiques.* Nouvelle traduction et notes par J. Tricot. Paris: Vrin, 1976.

———. *Les Parties de animaux.* Texte établi et traduit par Pierre Louis. Paris: Belles Lettres, 1956.

———. *Les Parties des animaux.* Traduction et présentation par Pierre Pellegrin. Paris: GF-Flammarion, 2011.

———. *The Politics of Aristotle.* With an introduction, two prefatory essays, and notes critical and explanatory by W. L. Newman. 4 vols. Oxford: Clarendon Press, 1887–1902.

———. *Posterior Analytics.* 2nd ed. Translated with a commentary by Jonathan Barnes. Oxford: Clarendon Press, 1993.

Works and Articles

Bachelard, Gaston. *La formation de l'esprit scientifique: contribution à une psychanalyse de la connaissance objective,* 1938. Translation: *The Formation of the Scientific Mind: A Contribution to a Psychoanalysis of Objective Knowledge.* Translated by Mary McAllester Jones. Manchester: Clinamen Press, 2002.

———. "Teleology and Necessity." In *Philosophical Issues in Aristotle's Biology,* edited by Allan Gotthelf and James G. Lennox, 275–285. Cambridge: Cambridge University Press, 1987.

Baudelaire, Charles. *The Flowers of Evil* (Fleurs du Mal). Translated by William Angler. Fresno, CA: Academy Library Guild, 1954.

Bélis, Annie. "Les nuances dans le *Traité d'harmonique* d'Aristoxène de Tarente." *Revue des Études grecques* 95 (1982): 54–73.

Bernardin de Saint-Pierre, Jacques-Henri. *Études de la nature*. 5 vols. Paris: Aimé Alexandre, 1825.

Berti, Enrico. "La suprématie du mouvement local selon Aristote: ses conséquences et ses apories." In *Aristoteles Werk und Wirkung, Paul Moraux gewidmet*, edited by Jürgen Wiesner, 123–150. Berlin: De Gruyter, 1987.

Bolton, Robert. "The Material Cause: Matter and Explanation in Aristotle's Natural Science." In *Aristotelische Biologie*, edited by Wolfgang Kullmann and Sabine Föllinger, 97–124. Stuttgart: Franz Steiner Verlag, 1997.

Bonitz, Hermann. *Index Aristotelicus*. Berlin: Königlichen Preussischen Akademie der Wissenschaften, 1870.

Byl, Simon. *Recherches sur les grands traités biologiques d'Aristote: sources écrites et préjugés*. Bruxelles: Palais des Académies, 1980.

Canguilhem, Georges. *La connaissance de la vie*. Paris: Vrin, 1967. Translation: *Knowledge of Life*. Translated by Stefanos Geroulanos and Daniela Ginsburg. New York: Fordham University Press, 2008.

———. "Vie." *L'Encyclopaedia Universalis*. Paris: Encyclopedia Universalis, 1973.

Carbone, Andrea L. *Aristote illustré. Représentations du corps et schématisation dans la biologie aristotélicienne*. Paris: Garnier, 2011.

Caston, Victor. "Aristote et l'unité de la psychologie: comment diviser l'âme?" In *L'Héritage d'Aristote aujourd'hui: Science, nature et société*, edited by Françoise Graziani and Pierre Pellegrin, 199–229. Alessandria: Edizioni dell'Oro, 2020.

———. "Aristotle's Psychology." In *The Blackwell Companion to Aristotle's Philosophy*, edited by M. L. Gill and Pierre Pellegrin, 316–346. Oxford: Blackwell, 2006.

Cicero. *Nature of the Gods, from the Treatises of M. T. Cicero*. Translated by Charles Duke Yonge. Bohn edition of 1878: https://topostext.org/work.php?work_id=137.

Connell, Sophia M. *Aristotle on Female Animals: A Study of the Generation of Animals*. Cambridge: Cambridge University Press, 2016.

Cooper, John. "Hypothetical Necessity and Natural Teleology." In *Philosophical Issues in Aristotle's Biology*, edited by Allan Gotthelf and James G. Lennox, 243–274. Cambridge: Cambridge University Press, 1987.

Crubellier, Michel, and Julie Journeau. "Le système de sciences aristotélicien." In *L'Héritage d'Aristote aujourd'hui: Science, nature et société*, edited by Françoise Graziani and Pierre Pellegrin, 27–43. Alessandria: Edizioni dell'Oro, 2020.

Cuvier, Georges. *Histoire des sciences naturelles depuis leur origine jusqu'à nos jours*. 5 vols. Paris: Fortin, Masson et Cie, 1841–1845.

———. *Leçons d'anatomie comparée*. Recueillies et publiée par M. Duméril. 8 vols. Paris: Crochard et Cie, 1835–1846. Translation: *Lectures on Comparative Anatomy*. Translated by William Ross, vols. 1 and 2. London: Oriental Press, 1802.

———. *Recherches sur les ossements fossiles de quadrupèdes*. Discours préliminaire, présentation, notes et chronologie par Pierre Pellegrin. Paris: GF-Flammarion, 1992.

Denniston, J. D. *The Greek Particles*, 2nd ed. Oxford: Clarendon Press, 1954.

Devereux, Daniel, and Pierre Pellegrin, eds. *Biologie, logique et métaphysique chez Aristote*. Paris: Éditions du C.N.R.S., 1990.

Dudley, John. *Aristotle's Concept of Chance: Accidents, Cause, Necessity, and Determinism*. Albany: State University of New York Press, 2012.

Duhem, Pierre. *Le Système du monde*, t. 1. Paris: Hermann, 1913.

Düring, Ingemar. *Aristotle's Chemical Treatise Meteorologica Book IV, with Introduction and Commentary*. Göteborg: Flanders Boktryckeri Aktiebplag, 1944.

———. *Aristotle's De Partibus Animalium: Critical and Literary Commentaries*. Göteborg: Flanders Boktryckeri Aktiebplag, 1943.

Eichholz, D. E. *Theophrastus, de Lapidibus*. Edited with introduction, translation, and commentary by D. E. Eichholz. Oxford: Clarendon Press, 1965.

Falcon, Andrea. *Aristotle and the Science of Nature: Unity without Uniformity*. Cambridge: Cambridge University Press, 2005.

———, ed. *Brill's Companion to the Reception of Aristotle in Antiquity*. Leiden: Brill, 2016.

Falcon, Andrea, and David Lefebvre, eds. *Aristotle's "Generation of Animals": A Critical Guide*. Cambridge: Cambridge University Press, 2018.

Fedi, Laurent. "Le prince des philosophes: Aristote vu par Auguste Comte et Pierre Laffitte." In *Aristote au XIXème siècle*, edited by Denis Thouard. Lille: Presses Universitaires du Septentrion, 2004.

Foucault, Michel. *Les Mots et les choses. Une archéologie des sciences humaines*. Paris: Gallimard, 1966.

———. *The Order of Things: An Archaeology of the Human Sciences*. New York: Routledge, 2002.

———. *Archaeology of Knowledge*. Lanham, MD: Routledge, 2002.

Frede, Michael. "Les origines de la notion de cause." Trad. Brunschwig, *Revue de Métaphysique et de Morale* 4 (1989). English original: "The Original Notion of Cause." In *Doubt and Dogmatism*, edited by Schofield, Burnyeat, Barnes, 1980. Reprinted in M. Frede, *Essays in Ancient Philosophy*, 125–150. Minneapolis: University of Minnesota Press 1987.

Freud, Sigmund. *Introductory Lectures on Psychoanalysis*. Edited by James Strachey in collaboration with Anna Freud. London: Hogarth Press, 1963.

Galen. *Galen on the Usefulness of the Parts of the Body*. Translated from the Greek with an introduction and commentary by Margaret Tallmadge May. Ithaca, NY: Cornell University Press, 1968.

Gill, Mary Louise, and Pierre Pellegrin, eds. *A Companion to Ancient Philosophy*. Malden, MA: Blackwell, 2006.

Gotthelf, Allan, ed. *Aristotle on Nature and Living Things*. Pittsburg, Bristol: Mathesis, Bristol Classical Press, 1985.

———. "Darwin on Aristotle." *Journal of the History of Biology* 32 (1999): 3–30.

Gotthelf, Allan, and James G. Lennox, eds. *Philosophical Issues in Aristotle's Biology*. Cambridge: Cambridge University Press, 1987.

Graziani, Françoise, and Pierre Pellegrin, eds. *L'Héritage d'Aristote aujourd'hui: Science, nature et société*. Alessandria: Edizioni dell'Oro, 2020.

Grmek, Mirko D., dir. *Histoire de la pensée médicale en Occident*. Paris: Seuil, 1995.

Groisard, Jocelyn. "Hybridity and Sterility in Aristotle's *Generation of Animals*." In *Aristotle's "Generation of Animals": A Critical Guide*, edited by Andrea Falcon and David Lefebvre, 153–170. Cambridge: Cambridge University Press, 2018.

Guthrie, W. K. C. *In the Beginning: Some Greek Views on the Origin of Life and the Early State of Man*. London: Methuen, 1957.

Hatzimichali, Myrto. "Andronicos of Rhodes and the Construction of the Aristotelian Corpus." In *Brill's Companion to the Reception of Aristotle in Antiquity*, edited by A. Falcon, 81–100. Leiden: Brill, 2016.

Hintikka, Jaakko. *Time and Necessity: Studies in Aristotle's Theory of Modalities*. London: Oxford University Press, 1973.

Humbert, Jean. *Syntaxe grecque*. 3rd ed. Paris: Klincksieck, 1972.

Jaeger, Werner. *Aristotle: Fundamentals of the History of His Development*. Translation by R. Robinson. London: Oxford University Press, 1948.

Johnson, Monte Ransome. "Aristotelian Mechanistic Explanation." In *Teleology in the Ancient World: Philosophical and Medical Approaches*, edited by Julius Rocca, 123–150. Cambridge: Cambridge University Press, 2017.

———. *Aristotle on Teleology*. Oxford: Clarendon Press, 2005.

Joly, Robert. "La Biologie d'Aristote." *Revue Philosophique* (1968): 219–253.

Keaney, John J. "Two Notes on the Tradition of Aristotle's Writings." *American Journal of Philology* 84 (1963): 53–54.

Kuhn, Thomas S. *The Structure of Scientific Revolutions*. Chicago: University of Chicago Press, 1962.

Kucharski, Paul. "Sur la théorie des couleurs et des saveurs dans le 'De sensu' aristotélicien." *Revue de Études Grecques* 67 (1954): 355–390.

Kullmann, Wolfgang. "Different Concepts of the Final Cause in Aristotle." In *Aristotle on Nature and Living Things*, edited by Allan Gotthelf, 169–175. Pittsburg, Bristol: Mathesis, Bristol Classical Press, 1985.

Labarrière, Jean-Louis. "De la phronèsis animale." In *Biologie, logique et métaphysique chez Aristote*, edited by Daniel Devereux and Pierre Pellegrin, 406–428. Paris: Éditions du C.N.R.S., 1990.

———. *La Condition animale. Études sur Aristote et les Stoïciens*. Louvain-la-Neuve: Peeters, 2005.

Lamarck, Jean-Baptiste. *Système des animaux sans vertèbres*. Paris: Deterville, 1801.

Lameere, William. "Au temps où Franz Cumont s'interrogeait sur Aristote." *L'Antiquité Classique* 18 (1949): 279–324.

Lee, H. D. P. "Place Names and the Date of Aristotle's Biological Works." *Classical Quarterly* 481 (1948): 61–67.

———. "The Fishes of Lesbos Again." In *Aristotle on Nature and Living Things*, edited by Allan Gotthelf, 3–8. Pittsburg, Bristol: Mathesis, Bristol Classical Press, 1985.

Lefebvre, David. "Aristote sur le sommeil de l'embryon et du nouveau-né, GA V,1,7778b20–779a26." In *L'Héritage d'Aristote aujourd'hui: Science, nature et société*, edited by Françoise Graziani and Pierre Pellegrin, 167–197. Alessandria: Edizioni dell'Oro, 2020.

———. *Dynamis: Sens et genèse de la notion aristotélicienne de puissance*. Paris: Vrin, 2018.

Lennox, James G. "Aristotle's Biological Development: The Balme Hypothesis." In *Aristotle's Philosophical Development: Problems and Prospects*, edited by William Wians, 229–248. Lanham, MD: Rowman & Littlefield, 1996.

———. *Aristotle's Philosophy of Biology*. Cambridge: Cambridge University Press, 2001.

Lovejoy, Arthur Oncken. *The Great Chain of Being: A Study of the History of an Idea*. Cambridge: Harvard University Press, 1936.

Ménétrier, Pierre-Eugène. "Comment Aristote et les anciens médecins hippocratiques ont-ils pu prendre connaissance de l'anatomie humaine?" *Bulletin de la Société française d'histoire de la Médecine* (1930): 254–262.

Milhaud, Gaston. "Le Hasard chez Aristote et chez Cournot." *Revue de Métaphysique et de Morale* 10, no. 6 (1902): 667–681.

Natali, Carlo, ed. *Aristotle: Metaphysics and Practical Philosophy. Essays in Honour of Enrico Berti*. Louvain-la-Neuve: Peeters, 2011.

Olympiodorus. *Olympiodori in Aristotelis Meteora Commentaria*, vol. 12, part 2. Edited by Wilhelm Stüve. Berlin: Reimer, 1902.

Pellegrin, Pierre. "Ancient Medicine and Its Contribution to the Philosophical Tradition." In *A Companion to Ancient Philosophy*, edited by Mary Louise Gill and Pierre Pellegrin, 664–685. Malden, MA: Blackwell, 2006.

———. *La Classification des animaux chez Aristote: Statut de la biologie et unité de l'aristotélisme*. Paris: Belles Lettres, 1982. Translation: *Aristotle's Classification of Animals: Biology and the Conceptual Unity of the Aristotelian Corpus*. Translated by Anthony Preus. Berkeley: University of California Press, 1986.

———. "De la tradition aristotélicienne." In *L'Héritage d'Aristote aujourd'hui: Science, nature et société*, edited by Françoise Graziani and Pierre Pellegrin, 19–26. Alessandria: Edizioni dell'Oro, 2020.

———. "De l'explication causale dans la biologie d'Aristote." *Revue de Métaphysique et de Morale* 2 (1990): 197–219.

———. *L'Excellence menacée. Sur la philosophie politique d'Aristote*. Paris: Garnier, 2017. Revised English edition: *Endangered Excellence: On the Political Philosophy of Aristotle*. Translated by Anthony Preus. Albany: State University of New York Press, 2020.

———. "Parties de la cité, parties de la constitution." In *Aristotle: Metaphysics and Practical Philosophy: Essays in Honour of Enrico Berti*, edited by Carlo Natali, 177–200. Louvain-la-Neuve: Peeters, 2011.

———. "Le plaisir animal selon Aristote." *Chôra: Revue d'études anciennes et médiévales* 17 (2019): 145–162.

———. "What Is Aristotle's *Generation of Animals* About?" In *Aristotle's "Generation of Animals": A Critical Guide*, edited by Andrea Falcon and David Lefebvre, 77–88. Cambridge: Cambridge University Press, 2018.

Philoponus, John and Miachel of Ephesus. *In Libros De Generatione Animalium Commentaria*, Commentaria in Aristotelem Graeca v. 14 pars 3. Berlin: G. Reimeri, 1903.

Plato. *Timaeus*. Translated by Donald J. Zeyl. Indianapolis: Hackett, 2000.

Pliny the Elder. *The Natural History*. Translated by John Bostock et al. London: G. Bell, 1890.

Preus, Anthony. *Science and Philosophy in Aristotle's Biological Works*. Hildesheim: Olms, 1975.

Rashed, Marwan. "A Latent Difficulty in Aristotle's Theory of Semen: The Homogeneous Nature of Semen and the Role of the Frothy Bubble." In *Aristotle's "Generation of Animals": A Critical Guide*, edited by Andrea Falcon and David Lefebvre, 108–129. Cambridge: Cambridge University Press, 2018.

Rocca, Julius, ed. *Teleology in the Ancient World: Philosophical and Medical Approaches*. Cambridge: Cambridge University Press, 2017.

Sedley, David. "Is Aristotle's Teleology Anthropocentric?" *Phronesis* 36, no. 2 (1991): 179–196.

Smith, Jean Chandler. *Georges Cuvier: An Annotated Bibliography of His Published Works*. Washington, DC: Smithsonian Institution Press, 1993.

Solmsen, Friedrich. *Aristotle's System of the Physical World: A Comparison with His Predecessors*. Ithaca, NY: Cornell University Press, 1960.

———. "The Fishes of Lesbos and Their Alleged Significance for the Development of Aristotle." *Hermes* 106, no. 3 (1978): 467–484.

Sorabji, Richard. *Necessity, Cause, and Blame*. London: Duckworth, 1980.

Spinoza, Benedict de. *Tractatus Theologico-Politicus*. London: Trübner, 1862.

Staden, Heinrich von. *Herophilus: The Art of Medicine in Early Alexandria*. Edition, translation and essays. Cambridge: Cambridge University Press, 1989.

Thompson, D'Arcy Wentworth. *A Glossary of Greek Fishes*. Oxford: Oxford University Press, 1947.

Vegetti, Mario. "Entre le savoir et la pratique: la médecine hellénistique." In *Histoire de la pensée médicale en Occident*, edited by Mirko D. Grmek, 67–94. Paris: Seuil, 1995. Paris: Seuil, 1995.

Wieland, Wolfgang. *Die aristotelische Physik. Untersuchungen über die Grundlegung der Naturwissenschaft und die sprachlichen Bedingungen der Prinzipienforschung bei Aristoteles.* Göttingen: Vandenhoeck & Ruprecht, 1962.

Wiesner, Jürgen, ed. *Aristoteles Werk und Wirkung, Paul Moraux gewidmet*. Berlin: De Gruyter, 1987.

Zucker, Arnaud. "Sur un prétendu anthropomorphisme aristotélicien en zoologie. Le 'modèle' humain en anatomie comparée." *Revue des Études Grecques* 130, no. 1 (2017): 43–71.

Index

Adaptation, 9–10, 107, 174, 176, 226, 283, 285, 286
Alexander of Aphrodisias, 123, 275n
Allendy, René, 291
Anaxagoras, 56, 61, 245
Andronicus of Rhodes, 47
Antlers, 91, 100, 102, 106, 109
See also Horn
Athenaeus, 47
Axolotl, 108, 225, 295, 296

Bachelard, Gaston, 8, 20ff
Balme, David, 34f, 48, 75 n.23, 78, 98 n.39, 99, 132, 140
Barnes, Jonathan, 61
Barthélemy-Saint-Hilaire, Jules, 46, 49
Bees, 22, 30, 105f
Bernardin de Saint-Pierre, Jacques-Henri, 45, 58, 108, 233
Berti, Enrico, 190 n.13
Bile, 35, 80, 86
Bios, life, 112–113
Birds, coloring, 44, 109
Bipedalism, 82
Bison, 101, 107
Bladder, 40–41
Blood vessels, 239ff
Boar, 38 and n. 33
See also Pig
Bolton, Robert, 174f
Bones, 36f, 58, 90, 93, 95, 116, 170
Brain, 34, 36, 99, 249, 268–270, 275, 277, 278
Buffon, Georges-Louis Leclerc, de, 38, 110
Byl, Simon, 22

Callippus of Cyzicus, 227
Camel, 37, 103f, 107, 226
Canguilhem, Georges, 6, 10, 20, 44, 113, 141f, 173, 177
Carbone, Andrea, 242, 251 n.17
Caston, Victor, 183 n.8
Causality, 5, 25–29, 39,43, 45, 51–59, 68–73, 74, 76, 80, 93, 95, 126, 143, 146, 158–165, 167, 169, 172, 174–178, 179, 213, 220, 223, 229, 236, 283, 288, 300
Chance, 57, 58, 66, 67, 122, 155
Chronology of Aristotle's writings, 28, 32ff
Cicero, *On the Nature of the Gods* 2.42, 220
Concoction, 77, 85, 86, 128, 147, 148, 152, 166
Connell, Sophia, 22, 96 n.36, 125, 127, 133, 143, 146
Cooper, John, 74, 78ff, 84, 86f, 176
Cosmology, Cosmogony, 49, 60–64, 111, 216–217, 223, 237, 292, 298

Cosmos, 60–63, 113, 200, 228, 237, 241, 242, 298
Courage of female animals, 41, 44, 125, 139, 234, 296–297
Crocodile, 36, 42, 180
Crombie, Alistair, 20
Cuckoo, 255f
Cuvier, Georges, 4ff, 9, 11ff, 20ff, 39, 42ff, 49f, 98, 182, 196, 200, 211, 249f, 294

Darwin, Charles, 9ff., 57, 233
Deer, 100, 103, 234, 251
See also Antlers
Democritus, 21, 51, 60, 67, 93f, 140
Dilthey, Wilhelm, 3
Diversity, animal, 39
Drossaart Lulofs, Hendrik Joan, 138 n.23
Dudley, John, 143
Duhem, Pierre, 228 n.44
Düring, Ingemar, 47, 164
Dynameis (δυνάμεις), faculties, powers, 81–83, 98, 117–119, 183–199, 254–256

Eagle, 210–211
Ears, 92
Eggs, 128
Eichholz, D. E., 165f
Elephant, 38, 260
Empedocles, 57f, 61, 70, 113, 221
Entropy, 20, 61–62, 297
Erect stature, 241ff
Eternal, everlasting, 61–67, 71, 78, 184, 247, 280
Ether, 217–225
Eudoxus of Knidos, 227
Eyelid, 88, 90, 106
See also Hair
Eyes, 81f, 88, 106

Falcon, Andrea, 229
Females, 22–24, 41, 44, 139f
Female material, 51, 68–70, 90–96, 119, 125–148, 153–154, 160, 163, 169, 236
See also Courage of female animals
Feminine species names, 37 n.32
First Mover: see Unmoved Mover
Fish, 222
Fish predation, 42, 230
Fish reproduction, 131
Foam bubble, 148, 149, 155, 219
Form, 71, 181, 201, 208, 211
See also Species
Foucault, Michel, 12ff, 20, 23, 37, 42, 44, 124
Frede, Michael, 68
Freud, Sigmund, 237

Galen, 49, 299ff
De Semine, 121 n.7
De Usu Partium, 49, 299–301
Genesis (coming to be), 49, 69, 94, 124, 140, 169, 176
Genos, family, group, 71, 208, 229
Gotthelf, Allan, 9f, 110, 295
Groisard, Jocelyn, 164, 166f
Guthrie, W. K. C., 111

Hair, 38f, 44 n.40
Hands, 245
Harvey, William, 27
Hatzimichali, Myrto, 47
Heart, 17, 40, 70, 82–85, 95, 99, 115–116, 128, 166, 170, 190, 192, 195, 265 n. 33
Herophilus of Chalcedon, 49
Hintikka, Jaakko, 67 n.15
Hippocratic Corpus, 29, 240
On the Nature of Bone, 240
The Nature of Man, 240

Homoiomeries and anhomoiomeries, 83, 92ff, 105, 114ff, 127, 164ff
Hoof, 17, 37–38, 43, 45, 98, 102, 182, 196
Horns, 18, 38, 40, 41, 91, 92, 94–98, 100–114, 132–135, 222–223, 243, 256, 270, 294, 297
See also Antlers
Hybrids, 102, 125, 152
Hylomorphic, 96
Hylozoism, 131–132, 152
Hypothetical necessity, 74ff, 103
See also Necessity

Insects, 18, 28, 180, 210

Jaeger, Werner, 32f
Joly, Robert, 22

Kant, Immanuel, 60, 114
Keaney, John, 47f
Kucharski, Paul, 272
Kuhn, Thomas, 21ff
Kullmann, Wolfgang, 71

Labarrière, Jean-Louis, 255
Lamarck, Jean-Baptiste, 10, 16, 23, 171 n.58
Lameere, William, 217f
Language, 258ff
Lee, H. D. P., 32f
Lefebvre, David, 212
Lennox, James, 27 n. 20, 28 n. 21, 75 n.23, 84, 87, 95 n. 35, 135, 143
Lesbos, 33
Lévi-Strauss, Claude, 284
Lobster, 208, 243f
Logos, 74, 96, 100, 116, 119, 121, 163, 246, 256, 260, 266, 271, 283
Louis, Pierre, 44, 164, 197 n.15, 214 n.32, 224
Lovejoy, Arthur, 66 n.15, 211, 215
Lucretius, *On the Nature of Things* 5, 855ff, 288

Material Causality, 26, 39, 45, 51–52, 59, 73–74, 116, 174–177, 220–226, 229, 246
See also Causality
Mayr, Ernst, 10
Mechanistic explanation, 51, 54ff, 66, 100, 161, 175, 191
Metals, 165ff
Michael of Ephesus, 86 n.34, 218
Milieu, 141f
Monsters (*terata*), 90, 138, 160
Moon, 213, 225, 235
Animals in the, 216f
Moriology, 18

Nature as cause, 52, 66, 96ff, 109, 227, 230, 233, 246f, 256, 258
Naxos (Sheep), 35
Necessity, 59, 71ff, 134, 191
Hypothetical, conditional, 73ff, 201
ἐξ ὑποθέσεως, 77
Nicomachus of Gerasa, 284 n.47, 286
Nomos ("law," principle of explanation), 19, 79
Nutritive Soul, 117ff, 190, 199, 209

Octopus, 109
Odors, 269ff
Ogle, William, 11, 206, 211
Olympiodorus, 166
Organism, 13, 15, 21, 43, 76–79, 107, 119, 130, 142, 148 n. 37, 170, 181, 182
Oryx, 102
Ousia ("substance," entity, essence), 24, 69, 82, 97, 109

Parmenides, 54, 61, 71
Peck, A. L., 65
Perception, 25, 81–84, 99, 105, 112, 144, 186, 187, 191, 194, 209, 229, 239, 250, 260, 261, 265, 272, 276–279
Phainomena, Φαινόμενα (observables), 25
Phallocratism, 238
Philoponus, 284 n.47
Physis (nature), 55
  See also Nature
Pig, wild, 38 n.33, 232
  See also Boar
Pittendrigh, Colin S., 10
Plato, 54, 72, 224, 257
  Phaedo 96aff, 55ff, 174
  Protagoras, 64
  Republic, 229
  4, 238 n.2
  Timaeus, 56, 67, 72, 100, 185
  29a, 67
  30c4, 225
  36d, 227
  39e, 211
  69b, 67
Pleasure, 263ff
Plenitude, Principle of, 66,67 n. 15
  See also *Scala naturae*
Pliny, *Natural History,* 47
Pneuma, 68, 149ff, 162, 169
Principle, 7, 26, 52, 69, 70, 82, 83, 89, 99, 122, 123, 126, 129, 132–133, 145, 147, 154, 186, 191, 219, 235, 236, 239
  teleological principle, 39, 91, 98, 101, 103, 104, 259, 295
Pyrrha, Strait, 154
Pythagoras, Pythagorean, 228, 272

Raptor talons, 9, 109, 293, 295
Rashed, Marwan, 116
Resemblance of offspring to parents, 135ff
Residue (περίττωμα), 125, 149ff
Respiration, 76
Rhinoceros, 98, 102
Ross, Sir David, 160

Saint-Hilaire, Étienne Geoffrey, 14, 46, 49, 294
Salamander, 215f
Scala Naturae, 18, 117, 200f, 215, 242, 247, 283, 292
Sea water, 150, 152, 156, 213–214
Sedley, David, 53, 238
Sensitive faculty, 187, 195ff
Serpents, 42, 232
Sharks, 293
Sheep, 154 n.46, 283, 286ff
Shellfish, 153, 162f, 172, 209, 215, 248
Simplicius commentary on *De Caelo,* 63
Sleep, 88, 93
Smell (sense of) 38, 107–108, 189, 265–270, 273–283
Solmsen, Friedrich, 33, 62, 292
Sorabji, Richard, 84 n.30
Species, 9–10, 45, 49, 50, 57, 61, 65–68, 71, 93, 102–119, 140, 143–162, 179, 200–215, 221–241, 247, 253–257, 283–288, 292–296, 301
  See also *Genos*
Sperma (seed), 133
Spleen, 85f
Spontaneous Generation (τὸ αὐτόματον), 102, 104, 111ff, 143ff, 159, 261
Stars, 219f
Stoic, 68, 229
Stomach, 41–43, 102–109, 196, 206–211, 294, 300–301

Strabo, 47
Substance, 70
  See also *Ousia*
Sun, 219, 235
Symmetry, 85 n.33

Taxonomy, 12–16
Teeth, 37, 40f, 43f, 57f, 95, 101ff, 106, 134, 234, 239, 258
Teleology, 51ff, 67, 71, 108, 114
Teleonomy, 10
Thales, 55
Themistius, 122 n.8, 160
Theophrastus, 48, 165, 299
  Causes of Plants 5.18.1, 137
  History of Plants 3.1.4, 161 n.49
Thompson, D'Arcy Wentworth, 29, 32, 131 n.17
Tongue, 104, 135, 258, 260
Tricot, Jules, 145, 164, 166 n.55, 205

Unmoved Mover, 63f, 112, 124

Van Leeuwenhoek, Antoni, 127
Vascular system, 240
Vitalism, 16, 19–20, 89, 113, 177–178
Von Baer, Karl Ernst, 127

Wieland, Wolfgang, 52, 113
Wind eggs, 96, 126ff, 156, 197

Zucker, Arnaud, 253 n.18

References to Aristotle's Works

Aristotle, *Prior Analytics* 2.27, 70a14, 107

Aristotle, *Posterior Analytics,* 7
- 1.3, 72b13, 77
- 1.13, 78a25ff, 44 n.38
- 2.17, 99b5, 80

Aristotle, *Physics*
- 2.1, 192b9, b14ff, 52, 122
- 2.2, 194b14, 218 n.39, 235 n.50
- 2.4–6, 158
- 2.4, 196a1, 159
- 2.4, 196a29, 161
- 2.5, 197a33, 159
- 2,6, 197b30, 161
- 2.6, 197b32, 159, 161
- 2,6, 197b36, 161
- 2.7, 198a16, 175
- 2.7, 198a22, 73, 162
- 2.7, 198a33ff, 69, 151 n.44
- 2.8, 198b10, 59
- 2,8, 198b12, 54
- 2.8, 198b16ff, 56f, 175
- 2,8, 198b29, 57
- 2.8, 198b35, 102 n.44
- 2.8, 199b28, 66
- 2.8, 199b32, 52
- 2.9, 199b34, 74
- 2.9, 200a14, 88
- 6.1, 231a22, 118
- 6.7, 238b6, 248 n.14

Aristotle, *De Caelo*
- 1.1, 268a14, 19 n. 14
- 1.1, 268b7, 227
- 1.1, 268b10, 227
- 1.4, 271a33, 107
- 1.10, 279b12, 62
- 2.2, 242 n.6
- 2.9, 290b21, 228

Aristotle, *Generation and Corruption,* 115, 271
- 2.10, 336b18, 235

Aristotle, *Meteorologica*
- 1.2, 339a15, 216
- 1.3, 340b6, 8, 217
- 1.3, 341a12, 165 n.51
- 1.4, 341b7ff, 165
- 3.6, 165
- 3.6, 378a15, 167
- 3.6, 378a 20,21, 166
- 3.6, 378a28ff, 166
- 3.6, 378b1, 168
- 4.1, 379a16, 126 n.16
- 4.4, 382a7, 223f
- 4.6, 383a30, 168
- 4.8, 384b30 33, 167
- 4.10, 388a10, 164

Aristotle, *Meteorologica* (continued)
4.10, 388a13, 114f, 164ff
4.8, 388a20, 21, 164
4.10, 389a7, 167
Aristotle, *De Anima,* 25
1.1, 403a29, 195
2.1, 412a13, 113
2.2, 413a22, 112
2.2, 413a33, 118 n.5
2.2, 413b11, 183
2.2, 413b22–25, 184, 264
2.3, 183
2.3, 414a26ff, 246
2.3, 414a31, 184, 188
2.3, 414b19, 127, 184f, 193
2.3, 414b28, 185
2.3 415a1, 189
2.3, 415a3, 186, 189
2.3, 415a9, 190
2.4, 415a23, 117f
2.4, 415a26, 152, 280, 282
2.4, 416a18, 119
2.4, 416b2, 119
2.9 421a7, 276
2.9, 421a10, 107, 266, 276f
2.9 421a11, 277
2.9 421a19, 266
2.9, 421a20, 109, 279
2.11, 422a23, 83
3.1, 81
3.1, 425a19, 83
3.2, 426a27ff, 265, 272
3.2–4, 186
3.3, 427a21, 186
3.3, 428a1, 187
3.3, 428a5–16, 188
3.3, 429a1, 187
3.4, 429a13–29, 186f
3.7, 431a16, 188
3.9. 432a31, 188
3.9–11, 189
3.9, 432b7, 192
3.9, 432b14, 191f
3.9, 432a31, 188
3.9 432b26, 192
3.10, 433a21, 192
3.10, 433b1, 118 n.5
3.11, 433b31, 189, 193
2.11, 434a4, 193
Aristotle, *Parva Naturalia,* 25
Aristotle, *Sense and Sensible Objects*
3, 439b22–440b18, 271
4, 441a2, 108 n.49
4. 441a3, 266
4, 441b25, 275 n. 41
5, 443a29, 165
5; 443b1, 267 n. 36
5, 443b28, 268
5, 444a14ff, 268, 274f
5, 444a28ff, 267,269
5, 445a2, 268
Aristotle, *On Memory* 1, 449b30, 188 n.10
Aristotle, *On Sleep*
1, 454a11, 198f
2, 455b25, 88
Aristotle, *On Dreams*
1, 453b23, 218 n.36
1, 459a17, 187
1, 459a15–22, 187f
Aristotle, *Divination in Sleep*
2.463b17, 218 n.36
Aristotle, *Length and Shortness of Life*
5, 466b14, 150 n.42
Aristotle, *On Respiration*
19, 477a21, 213, 219ff, 241
19 477a25, 275 n.42
Aristotle, *History of Animals,* 11, 24ff, 35
1.1, 486a16ff, 181f
1.1, 486b5, 182 n. 5
1.1, 486b9, 181 n. 3
1.1, 486b17, 17
1.1, 487a10, 31, 179f

1.1, 487b6–33, 180
1.1, 488a8, 255
1.1. 488a10, 254f
1.1, 488a26ff, 285
1.1, 488a29, 285
1.1, 488b11, 180
1.2, 489a3, 40
1.5, 490a17, 28
1.6, 490b27, 40
1.6, 491a6ff, 26
1.6, 491a19,20, 31, 238
1.6, 491b2, 36
1.15, 494a26, 243
1.15, 494b16, 108 n.49, 239
1.16, 484b19, 239
1.16, 484b28, 99
1.16, 495b24, 207
1.17, 496b25ff, 35
2.1, 37f
2.1, 498b11, 34
2.1, 500b16, 34
2.1, 501a4, 251
2.1, 501a19, 40
2.2, 499a9, 38
2.3, 501b22, 41
2.17, 507b16, 207 n.26
2.17, 508a3, 207
2.17, 508a9, 208
3.1, 511a27ff, 41
3.2, 511b11ff, 239
3.2, 511b31, 240 n.4
3.3, 513a19, 241
3.3, 513b34, 241
3.7, 36
3.7, 516a24, 36
3.7, 516b5, 170
3.12, 519a16, 154 n46
3.17, 520a5, 181 n. 3
4.1, 523b21, 247
4.1, 524b4, 99
4.5, 530a32, 28
4.8, 533a32, 266
4.8, 533b5, 266
4.8, 535a2, 266
4.9, 535a31ff, 258
5.1, 539a1,2, 144
5.1, 539a8ff, 145
5.1, 539a27ff, 157
5.1, 539a31, 127
5.1, 539b2ff, 146
5.1, 539b7, 104
5.1, 539b15, 145
5.15, 547b19, 153, 158
5.19, 552b15, 216
5.31, 556b25, 147
6.18, 571b6, 150 n 42
6.18, 571b9, 281 n. 44
6.22, 575b29, 282 n.46
6.29, 578b16, 287
8.1, 588a18ff, 252, 256, 262
8.1, 588b4, 117
8.1, 588b11, 117
8.1, 589a3, 271 n.43
8.2, 589b28, 109
8.2, 591a17, 42
8.3, 593b25, 42
8.3, 593b25, 42
8.3, 593b30, 222 n.42
8.4, 594a10, 42
8.20, 603a19, 286
8.28, 606b22, 102 n.43, 152 n.45
8.29, 607a19, 242
9.1., 608a21, 253
9.1., 608a33, 41
9.1., 608a34, 139
9.1, 608b4ff, 253f
9.1, 608b26, 212
9.1, 608b30, 233
9.1, 609b28, 232
9.1, 610a12, 212f
9.2, 610b12, 232
9.3, 610b22, 286
9.12, 615a25, 9, 283
9.29, 618a25, 255

Aristotle, *History of Animals* (continued)
9.32, 210
9.39, 623a8, 256
9.44, 629a8, 233 n.48
9.45, 630a20, 101
9.50, 632a4, 18
10, 35
Aristotle, *Parts of Animals,* 11, 24ff
1.1, 639b21ff, 75, 78f
1.1., 640a4, 79
1.1, 640a10ff, 69f
1.1, 640a14,15, 25, 27f
1.1, 640a18ff, 20
1.1, 640a33, 84, 158
1.1, 640b28,29, 97
1.1, 641A&, 97
1.1, 641a25, 97
1.1, 641b5, 189 n.11
1.1, 642a1,2, 59, 73ff
1.1, 642a6, 78
1.1, 642a13,15, 72
1.1, 642a16, 177
1.1, 642a18, 64 n.12, 177
1.1, 642a21, 161
1.1, 642a29, 54f
1.1, 642a32ff, 76
1.3, 643b4, 285
1.5, 645a16, 282
1.5, 645a22, 108
1.5, 645b14, 170
2.1, 646a8, 25
2.1, 646a12, 115
2.1, 646b10,11ff, 92, 115, 170
2.1, 646b24ff, 93, 115
2.1, 646b30, 115
2.1, 647a5, 83
2.1, 647a9, 83
2.1, 647a24, 82
2.1, 647a31, 83
2.1, 647b8, 115
2.2, 647b29ff, 105f
2.2, 648a12, 44, 108
2.2, 648a15, 106
2.2, 648a16, 108
2.7, 652a26,31, 99
2.7, 652b16, 99
2.7, 653b13, 119
2.8, 654a10, 18
2.9, 36
2.12, 92
2.13, 657a25, 88
2.14, 34, 39
2.14, 658a22, 38
2.14, 658a23, 28, 32, 35
2.14, 658b10, 39
2.15, 38
2.16, 659a16ff, 259
2.16, 659b32, 239, 259
2.16, 660a13, 108
3.1, 661b23, 40
3.1, 661b26ff, 91
3.2, 663a8ff, 101
3.2, 663a10, 100
3.2, 663b23, 75, 95, 97f, 100, 103
3.2, 663b24, 103 n. 45
3.2, 663b31ff, 133ff
3.2, 664a1, 97
3.2, 664a6, 102
3.3, 664b32, 100 n.41
3.2, 665a8, 100 n. 41
3.4, 665b13, 100 n.41
3.5, 668a6, 118
3.6, 669a35ff, 244f
3.6, 669b5, 245
3.7, 669b17, 85 n.33
3.7, 669b26ff, 85f
3.7, 670a13, 84f
3.7, 670b25, 84
3.8, 670b32ff, 40f
3.12, 673b29, 109
3.13, 279
3.14, 674a29ff, 103f
4.2, 676b25, 80, 84

4.2, 677a1ff, 35
4.2, 677a11ff, 80, 175
4.2, 677a35, 80
4.5, 678b7, 209
4.5, 681a12, 117
4.5, 681b13, 209
4.5, 682a9, 210
4.5, 682a30, 145 n.32
4.6, 683a20, 259
4.8, 684a26ff, 243f
4.8, 684b1, 208
4.9, 684b6ff, 209f
4.9, 684b17ff, 248
4.9, 685b14, 109
4.10, 686a25ff, 245f
4.10, 686b1ff, 245, 251
4.10, 687a6, 245 n.9
4.10, 687a17, 227
4.10, 689a33, 34
4.12, 693b5, 82
4.13, 696b26,27, 230
Aristotle, *Movement of Animals,* 25, 190
4, 699b19, 216f
6, 700b19, 188, 191f
9, 702b20, 191
Aristotle, *Progression of Animals,* 25, 190
4, 705b18ff, 243
4, 706a19ff, 241
12, 711a19, 66
Aristotle, *Generation of Animals,* 22, 24ff
1.1, 715a11, 27f
1.1, 715a25ff, 150
1.2, 716a7, 133
1.16, 721a7, 105
1.17, 721b29, 144
1.18, 722b30, 171
1.19, 726b35, 127
1.21, 729b22, 150 n.43
1.21, 730a30, 130
1.23, 730b33, 150 n.42
1.23, 731a1, 149 n. 41
1.23, 731a24-b9, 246 n.10
2.1, 731b20, 73, 173f
2.1, 731b28, 19, 123
2.1, 731b31ff, 65, 110
2.1, 732a1ff, 126
2.1, 732a6, 154
2.1, 732a20, 150 n.42
2.1, 732a29, 53 n.4
2.1, 733a33, 212
2.1, 734b31ff, 89, 95, 116, 125, 132
2.1, 735a17, 118, 246 n.10
2.1, 735a27, 124 n.11
2.3, 736b8, 197f
2.3, 736b29ff, 218f
2.3, 736b37, 225
2.3, 737a4, 218
2.3, 737a27, 139
2.4, 737b25, 254
2.4, 738b26, 124
2.4, 740b24ff, 120
2.4, 740b32, 120
2.5, 741a6ff, 129f
2.5, 741a19, 126f, 132
2.5, 741a21, 132
2.5, 741a23, 127
2.5, 741a32ff, 130f
2.6, 744b16, 86
2.6, 745a31, 100 n.41
2.7, 745b30, 41
2.7, 746a31, 102 n.42
3.1, 749b1, 2, 126 n. 15, 128
3.1, 749b35, 157
3.1, 750b3ff, 128
3.9, 758b20, 53 n.4
3.10, 760a35, 246 n.10
3.11, 761a13, 218
3.11, 761b9ff, 156, 213ff, 223f
3.11, 761b14–15,18, 215
3.11, 761b20ff, 216f
3.11, 762a8–762b21, 146f, 150ff, 169

Aristotle, *Generation of Animals* (continued)
3.11, 762a32, 143
3.11, 762b26,28, 112, 151
4.1, 766a15, 136
4.2, 766b32, 35, 136
4.2, 767a34, 136
4.3, 767b5, 139
4.3, 767b10, 140
4.3, 767b13, 119
4.3, 767b18, 120
4.3, 768a14ff, 136ff
4.3, 768b5ff, 137
4.3, 770a6, 138
4.4, 772b6, 236
4.10, 777b16ff, 235
4.10, 778a2, 235f
5.1, 778b10ff, 80ff, 87
5.1, 778b21, 199
5.1, 779a2, 199 n. 19
5.1, 779a7, 198
5.2, 781b22, 92
5.4, 784b8, 126 n.16
[Aristotle], *Problems*
10.45, 895b23, 285
10.45, 896a3, 285 n. 49
19.38, 920b30, 273
Aristotle, *Metaphysics*
Alpha.1, 980b26, 260 n.28
Epsilon.1, 1025b24, 257 n.24
Zeta.9, 1034b4, 151
Eta.4, 1044b12, 52
Lambda.3, 1070a6, 158
Lambda.5, 1071a15, 236
Lambda.7, 1072a19, 63
Aristotle, *Nicomachean Ethics*
3.9, 1117b1, 264
3.10, 1117b28ff, 263f, 270
3.10, 1118a1–12, 270, 273f
3.10, 1118a16ff, 266
3.10, 1118a30, 281 n. 45
3.10, 1118a33, 266
5.1, 1129a26, 112 n.2
7.1, 1145a20, 274 n.40
7.11, 1152b1, 263
7.13, 1153b14ff, 282
7.13, 1153b25, 247, 281f
10.4, 264
10.4, 1174b18, 281
10.4, 1174b20, 264, 277, 281
10.5, 1174b33, 283
10.5, 1175a25ff, 263f
Aristotle, *Politics*
1.1, 1252a5, 284 n.48
1.2, 1252a28, 29, 280
1.2, 1252b1, 259 n.27
1.2, 1252b23, 287
1.2, 1253a15, 260
1.5, 1254b10, 285
1.8, 1256b1, 287
1.8, 1256b10–22, 53
1.13, 1260a20, 123 n. 9
1.11, 30
2.6, 1264b17, 257
3.9, 1280a30, 255 n.20
4.3, 1289b27, 204
4.4, 200, 226, 250
4.4, 1290b25, 195f, 201, 204
4.4, 1290b33, 209
4.4, 1290b34–37, 206
4.4, 1290b36, 20
4.4, 1290b37–1291a7, 204
4.14ff, 203
4.14, 1297b35, 203
4.15, 1300a22, 202 n.21
6.7, 206
8.7, 1342a18ff, 273
Aristotle, *Poetics* 4, 1448b20, 273

www.ingramcontent.com/pod-product-compliance
Lightning Source LLC
LaVergne TN
LVHW050148080826
844660LV00002B/117

* 9 7 8 1 4 3 8 4 9 1 4 7 9 *